AF815455

UNDERSTANDING EARTHQUAKE DISASTERS

UNDERSTANDING EARTHQUAKE DISASTERS

Amita Sinvhal
Professor
Department of Earthquake Engineering,
Indian Institute of Technology, Roorkee
Uttarakhand, India

Tata McGraw Hill Education Private Limited
NEW DELHI

McGraw-Hill Offices

New Delhi New York St Louis San Francisco Auckland Bogotá Caracas
Kuala Lumpur Lisbon London Madrid Mexico City Milan Montreal
San Juan Santiago Singapore Sydney Tokyo Toronto

Tata McGraw Hill

Published by Tata McGraw Hill Education Private Limited,
7 West Patel Nagar, New Delhi 110 008.

ISBN (13): 978-0-07-070709-2
ISBN (10): 0-07-070709-X

Managing Director: *Ajay Shukla*
Head—Professional and Healthcare: *Roystan La'Porte*
Executive Publisher—Professional: *R Chandra Sekhar*
Production Executive: *Rita Sarkar*
Manager—Sales and Marketing: *S Girish*
Deputy Marketing Manager—Science, Technology and Computing: *Rekha Dhyani*
General Manager—Production: *Rajender P Ghansela*
Asst. General Manager—Production: *B L Dogra*

Typeset at Tej Composers, WZ-391, Madipur, New Delhi 110063, and printed at Gopsons, A-2 & 3, Sector-64, Noida, U.P. 201 301.

RXAYCRDZDCXAR

Cover Design: Kapil Gupta, New Delhi

The McGraw·Hill Companies

To

Swapnil, Yash, Manya

Preface

Earthquakes take a heavy toll on human life and property as witnessed after the recent earthquakes, like the Kashmir earthquake of 2005, the Sumatra earthquake of 26 December 2004, the Kutch earthquake of 26 January 2001, and the Latur earthquake of 1993. This book is a consequence of the many lectures that followed these earthquakes. These lectures were for audience from diverse backgrounds—who were interested in and wanted to understand more about earthquakes and the disasters caused by them—such as planners, policy makers, decision makers, administrators, builders, teachers, technical institutions, researchers, and students of civil engineering, earthquake engineering, seismology, architecture, planning, management, geology and geophysics, and many others. Because of the diverse and complex nature of the subject, it was necessary to strike a fine balance between damage surveys, seismology and engineering aspects of earthquakes.

Some of the larger cities in India, with an increasingly large stock of brick masonry and multistory buildings, are at risk due to ground tremors from distant and near earthquakes. A distant and small earthquake is 'usually' less hazardous compared to a near and large one. The Kutch earthquake of 2001 was an eye-opener in this sense. Rural houses of random rubble stone masonry near the epicenter were as vulnerable as new multistory buildings at large epicentral distances in an urban landscape. What happened in a prosperous city like Ahmedabad, which is at an epicentral distance of more than 250 km, and Surat, which is even further away, serves as a lesson for making earthquake-resistant multistory buildings in future at other similar seismically vulnerable sites. The study of the damage inflicted by every earthquake, therefore, imparts important lessons, which if learnt, can go a long way in minimizing future human misery.

This book deals with elementary concepts of earthquakes, such as global seismicity, causes of earthquakes, plate tectonics, seismic waves, magnitude, intensity, faults in the earth, seismo-tectonics, seismicity, seismic zoning, how earthquakes are monitored and how data is used as a tool to understand the earth's interior. Further, it deals with several damaging effects of earthquakes, such as ground failure, landslides, tsunamis and the human habitat. Photographs accompanying the text tell their own stories. It is my sincere hope that this book will help the reader demystify earthquakes and the disasters they cause, which may lead to an urge to know more about how to minimize the disastrous effects of future earthquakes.

The subject has been given a simple treatment. Most terms used are defined in a glossary at the end of the book and mathematical treatment has been kept to a bare minimum. A reader interested in more rigorous treatment of any particular topic will find references and several online resources provided at the end of this book useful.

I acknowledge with deep gratitude, the facilities and continued support I received from the Head, Department of Earthquake Engineering and my colleagues at this Department. With some I had the opportunity to carry out fieldwork on various earthquakes, in difficult post-earthquake conditions, inhospitable and rugged terrain, inclement weather, and the threat from terrorists along the Line of Control (LoC). Learning from these experiences was adventurous, exciting and, at times, very disturbing and frightening.

For a stimulating interaction in course of my research, gratitude is due to many, especially to my colleagues at the Department of Earthquake Engineering, to the late Professors Jai Krishna and L.S. Srivastava; to Professors H.R. Wason, Ashwani Kumar, A.D. Pandey, A.K. Mathur, R.N. Dubey, Daya Shanker, D.K. Paul, M.L. Sharma, and G.I. Prajapati. I cherish my long association with Professors Pratima Rani Bose, Vipul Prakash, B.C. Mathur, V.H. Joshi, Sachin Pore and Sudarshan Singh, who were former colleagues at this Department; to late Professor K.N. Khattri and to Professors A.K. Awasthi, V.N. Singh, A.K. Jain, A. Joshi and S. Singh, from the Department of Earth Sciences; and to Professor Amit Bose, who was formerly at the Department of Architecture and Planning at IIT Roorkee. I am grateful to all individuals, establishments, and district administrations who helped me in field expeditions, gave their time freely to share their experiences, and provided relevant information. A special gratitude is also due to the Indian Army for their logistic support in several field investigations.

A special thanks to Mr. Subodh K. Saini who typed most drafts and made illustrations. Ms Swapnil Sinvhal, Mr. Manish Jain, Dr. Ila Gupta and Mr. Abhishek Singh helped in drawings in earlier drafts of the manuscript. Many thanks too to Professor M.P. Jain and Ms Revathi Bhaskar for their continued support and encouragement. Except for quoted matter, I take full responsibility for any errors and omissions in this manuscript.

Finally, Professor Harsha Sinvhal, my family, friends, and publishers, who believed in me and did not let me give up, deserve my thanks.

AMITA SINVHAL

email: amitafeq@iitr.ernet.in

amitafeq@gmail.com

Contents

List of Symbols

a	Ground amplitude in microns
g	Acceleration due to gravity
h	Focal depth
m_b	Body wave magnitude
p	Initial ascent of P-wave to the surface
s	Initial ascent of S-wave to the surface
A	Maximum trace amplitude of an event to be measured
A_0	Maximum trace amplitude of a zero-magnitude earthquake
c	Wave reflected at boundary of outer core
E	Epicenter
E	East
E	Energy in an earthquake
E_0	Energy released in earthquake of magnitude, zero, $= 2.5 \times 10^{11}$ ergs
E_M	Energy released in earthquake of magnitude M
FF	Fault (as shown in figure)
I	Intensity
I	Reflection of P-wave from boundary of inner core
J	Shear wave that travels through the inner core
k	Bulk modulus of elasticity of a medium
K	Compressional wave that comes from the outer core
L	Length of fault
M	Magnitude
M, M_L	Richter magnitude
M_0	Seismic moment
M_S	Surface wave magnitude
M_W	Moment magnitude

N	North
O	Centre of the earth
P	Longitudinal waves or compressional waves
P, S, L	These symbols are seismological shorthand for three successive group of waves recorded on seismograms of normal earthquakes
R	Radius of the earth
R	Hypocentral distance
S	Point of observation, station
S	Transverse waves
S	South
$(S–P)$	Time difference between arrival of S- and P-waves
T	Period of wave in seconds
T_P	Travel time of P-wave
T_s	Travel time of S-wave
V	Volume
V_P	Velocity of P-waves
V_S	Velocity of S-waves
W	West
W	Work done
Greek Characters	
α	Constant = 1.656, used in magnitude computation
β	Constant = (1.818 + C); C is station constant
Δ	Epicentral distance
μ	Rigidity
ρ	Density g/cc, kg/m^3
σ	Stress
λ	Wavelength
π	Pi
Conversion Factors	
1 micron, (μm)	0.001 mm, 10^{-4} cm, 10^{-6} m
1 megaton	1 million tons
1°	111 km
km	Kilometer
m	Meter

List of Acronyms

BIS	Bureau of Indian Standards
EMS	European Macro Seismic Scale
EQRD	Earthquake Resistant Design
FFT	Frontal Foothill Thrust
GMT	Greenwich Mean Time
GPS	Global Positioning System
GSI	Geological Survey of India
IMD	India Meteorological Department
ISI	Indian Standards Institution
IST	Indian Standard Time
ITSZ	Indus Tsangpo Suture Zone
MYA	Million years ago
MBT	Main Boundary Thrust
MCT	Main Central Thrust
MI	Mercalli Intensity Scale
MMI	Modified Mercalli Intensity Scale
MSK	Medvedev Sponhouer Karnik Intensity Scale
NEIC	National Earthquake Information Center
ONGC	Oil and Natural Gas Commission
RBC	Reinforced Brick and Concrete
RCC	Reinforced Cement and Concrete
RF	Rossi Forel Scale of Intensity
UCT	Universal Coordinated Time
USGS	United States Geological Survey
RRSM	Random Rubble Stone Masonry

1 CHAPTER

Global Seismicity

INTRODUCTION

Earthquakes are one of the most devastating natural phenomena. Every year thousands of people are rendered homeless, displaced, injured, or even killed all over the world due to earthquakes. Growing population and global urbanization is increasing the threat of earthquakes. Man from time immemorial has experienced earthquakes. It was generally believed that like all other natural phenomena, large animals like Sheshnag in Indian mythology or the catfish in its Japanese counterpart caused these. However, the common theme in all these explanations was that an earthquake occurred when the earth shook violently.

A very large number of earthquakes occur throughout the world every year; in fact earthquakes occur more often than one might tend to believe. However, spatial distribution of earthquakes shows that some regions have more earthquakes than other regions, while large areas are almost free of seismicity. Seismicity is the distribution of earthquakes in time and space. Any region, which has frequent earthquakes, is considered seismically active. Seismicity is concentrated along certain narrow, semicontinuous geographical regions called seismic belts. These are shown in Figure 1.1. Seismic belts are of particular interest as frequent earthquakes occur in these regions, induce large-scale damage repeatedly, and make large populations vulnerable. Two prominent seismic belts can be identified on the globe. These are the Circum Pacific Belt and the Alpine-Himalayan Belt, as discussed in the following sections.

THE CIRCUM PACIFIC BELT

The Circum Pacific belt, also known as the ring of fire, is long and narrow. It exists along the Pacific coast of North and South America and continues into

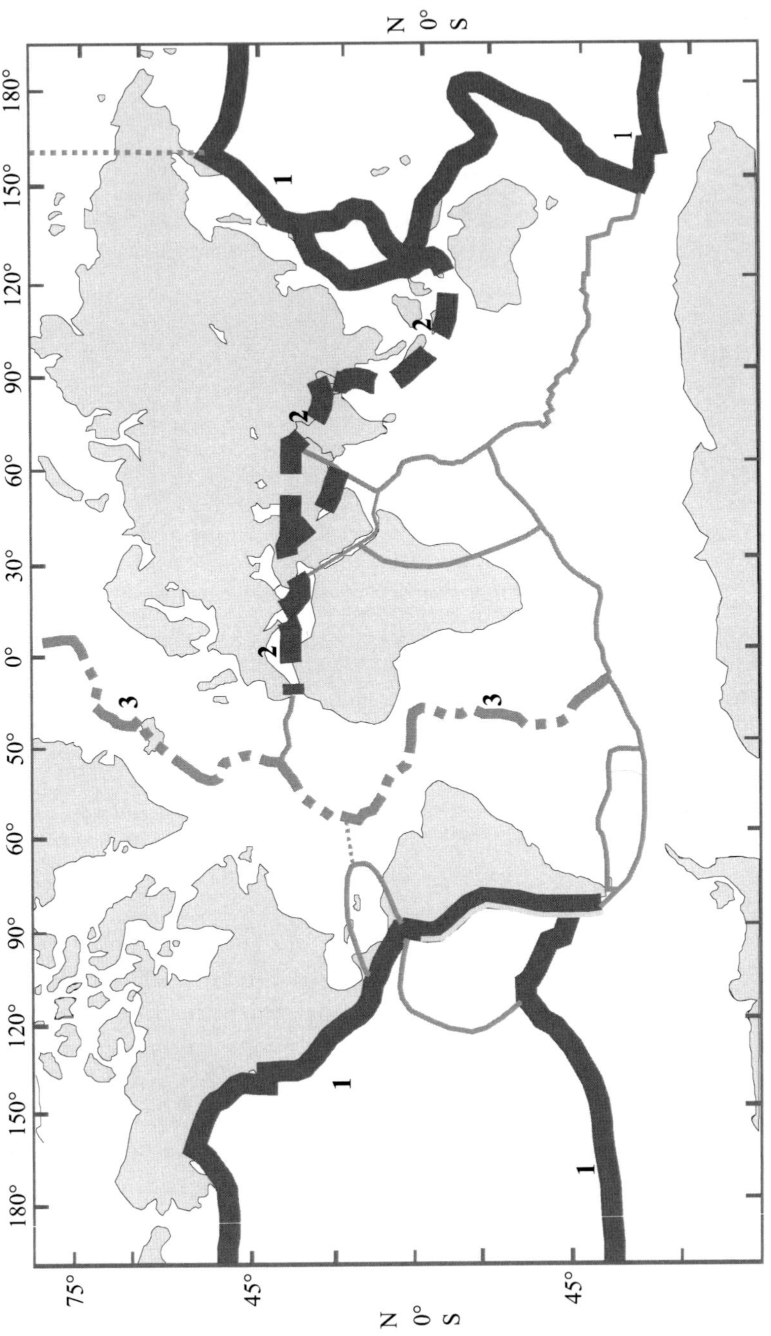

Fig. 1.1 The two main seismic belts are: (1) the Circum Pacific belt, and (2) the Alpine-Himalayan belt. (3) The Mid-Atlantic Ridge, forms a third, less active belt. (See color figure also.)

the Pacific coast of Asia. It is the most active of all seismic belts and has the largest concentration of devastating earthquakes. It contributed more than three quarters of world seismicity; in fact between 1904 and 1952, it gave off 75.6% of global seismic energy (Gutenberg and Richter, 1954).

This belt comprises of, starting from the 12 o'clock position assumed to be at the Bering Strait and going anticlockwise, the Aleutian Islands, Alaska (Good Friday earthquake of 1964, M = 8.6, 131 casualties); Canada; U.S.A., including the states of Washington and California (San Francisco earthquakes of April 18, 1906, M = 8.3, 700 dead; and February 1971, M = 6.6, 65 dead; Loma Prieta earthquake of 1989, M = 7.1, 63 dead; North Ridge earthquake of 1994, M = 6.7, 61 dead); Mexico (September 1985, M = 8.1, 9500+ dead); Central America; Columbia (January 25, 1999, M = 6.0, 1171 dead); Ecuador (January 31, 1906, M = 8.9); Nicaragua El Salvador (2001, M = 7.7, 700 dead); Guatemala and countries within the Andes Mountains of South America, e.g., Peru and Chile (January 24, 1939, M = 8.3, 128000 dead; May 22 1960, M = 8.5). Then on the east coast of Pacific Ocean are New Zealand, Kermadec, Tonga and Fiji islands (Samoa earthquake of June 26, 1917, M = 8.7); East Indies, Papua New Guinea, and Philippines; Japan (Kwanto earthquake of September 1, 1923, M = 8.3, 143,000 dead; Sanriku earthquake of March 2 1933, M = 8.9; Kobe earthquake of January 17, 1995, M = 7.2, 5000+ dead); Taiwan (September 1999, M = 7.6, 2400 dead); the Kamachatka peninsula (November 10, 1938, M = 8.7); and many other places in between.

The Circum Pacific belt is very complex and includes special topographic features such as island arcs, oceanic trenches, and mountain ranges. It has intermediate and deep focus earthquakes, together with shallow focus earthquakes.

THE ALPINE-HIMALAYAN BELT

The Alpine-Himalayan belt is the next most active belt. It contributed 22.1% of seismic energy given off on the globe between 1904 and 1954. This seismic belt is more diffused than the Circum Pacific belt. Topographic features associated with this belt are mountain ranges on continents and island arcs and deep trenches in oceans. It includes the mountainous regions of Alps in Europe, Zagros in Iran, Sulaiman and Kirthar ranges in Pakistan, Hindu Kush and Pamir regions, the Himalayas in Asia, and extends toward the East Indies, via the Arakan Yoma mountain ranges and continues eastward into Indonesia and Philippines. It includes the mountain ranges that radiate from the Pamir knot, such as Karakoram, Kunlun, Altyn Tagh, and those that stretch into Tibet, China, and Mongolia.

Going from west to east, it covers the countries of south Europe around the Alps and the Mediterranean Sea, such as Portugal (Lisbon earthquake of 1755, M = 8.6, 70,000 dead), Spain, Italy (Messina earthquake in South Italy, 1908, 200,000 dead; L'Aquila in Central Italy, 6 April 2009, M = 6.3, 290 dead), Greece, Yugoslavia, Rumania and Bulgaria, Armenia (Spitak earthquake of December 7, 1988, M = 7.0, 25,000 dead), Russia (1995, M = 7.6, 2000+ dead). Countries in North Africa afflicted by earthquakes within this belt are Algeria (October 10, 1980, M = 7.7, 3500 dead; North Algeria, May 21, 2003, M = 6.8, 2300 dead), Morocco (SW Atlantic coast, February 29, 1960, M = 5.7, 12,000 dead) Libya, and Egypt (Cairo, 1992, M = 5.9, 550 dead).

Some countries afflicted by earthquakes within this belt in Asia are Turkey (Erzincan, December 26, 1939, M = 7.9, 33,000 dead; 1992, M = 6.8, 570 dead; Izmit, August 17, 1999, M = 7.4, 17,000 dead), Iran (South Iran, April 04, 1972, M = 7.1, 5054 dead; NE Iran, September 16, 1978, M = 7.7, 25,000 dead; Manjil, June 21, 1990, M = 7.3, 40,000 + dead; 1997, M = 5.5, 554 dead; 1997, M = 7.3, 2400+ dead; S W Iran, December 26, 2003, M = 6.8, 30,000 dead), Afghanistan (N Afghanistan–Tajikistan region, February 4, 1998, M = 6.1, 5000+ dead; N. Afghanistan, March 25, 2002, M = 5.8, 1000 dead), Pakistan (Quetta, 31 May, 1935, M = 7.6, 50,000 dead), Nepal, China (1556, Shanxi Province, M = 8.0, casualties 1,000,000; Kansu, July 23, 1905, M = 8.7; Tien Shan, January 3, 1911, M = 8.7; Yunnan Province, 1970, M = 7.7, 15,621 dead; Tangshan, 1976, M = 8.0, 242,000 dead; Lijiang, 1996, M = 6.5, 304 dead; Sichuan, May 12, 2008, M = 7.8, 70,000 dead), Bangladesh, Myanmar, Indonesia (Sumatra earthquake of December 26, 2004, Ms = 9.3, more than 2,30,000 dead), and Philippines.

In India, this belt covers the entire Himalayan range, from Kashmir to Arunachal Pradesh (Kashmir earthquake of October 08, 2005, M = 7.6, more than 86,000 dead; Kangra earthquake of 1905, M = 8.6; Bihar–Nepal earthquake of 1934, M = 8.4; Assam earthquakes of 1897 and 1950, M = 8.7) and then turns sharply southward (Calcutta earthquake of October 11, 1737, 300,000 dead), straddling the Andaman and Nicobar Islands in the Bay of Bengal (North Andaman earthquake of June 26, 1941, M = 8.7).

The Circum Pacific belt and the Alpine-Himalayan belt intersect in the region comprising of the Philippines island arc and trench system.

OTHER REGIONS OF SEISMICITY

Besides the Circum Pacific Belt and the Alpine-Himalayan Belt, other regions of reduced seismicity also exist on the globe. These comprise of mid-oceanic ridges, continental rifts, marginal areas, regions of old seismicity, and stable masses. Regions of old seismicity refer to pre-Cambrian shields of Africa, India, Siberia, Fenoscandia, Australia, Canada, and Brazil.

TOPOGRAPHY

Seismicity and seismic belts are concentrated along large-scale regional features with high topographic relief such as young mountain ranges on continents; and ridges, trenches, and island arcs in oceans. It is, therefore, necessary to dwell for a while on these topographic features. The following physical and topographic features may be encountered while moving from the highest region on a continent toward the deepest part of an ocean: mountains, plains, continental margins, and abyssal plains, Mid-oceanic ridges, trenches, and island arcs. These are shown in Figure 1.2.

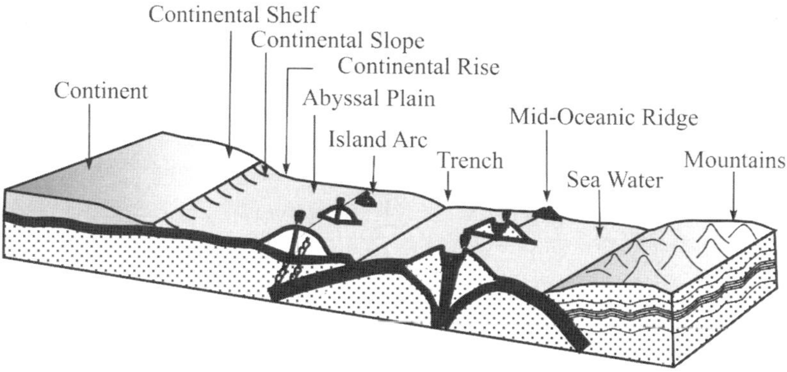

Fig. 1.2 Topographic relief showing generalized cross-section through the oceanic and continental crust, including mountains, continental shelf, continental slope, abyssal region, island arc, trench, and mid-oceanic ridge.

A continental margin is covered with water and extends from the shoreline to the deep ocean. It is divided into three regions—shelf, slope, and rise. A continental shelf is regarded as a portion of continental crust that is submerged in seawater. It is that portion of the sea floor that adjoins a continent and over which maximum depth of seawater is 200 m. It may be about 1000 km wide. Most offshore oil and gas is pumped from here. Its outer margin is the continental slope, which dips very steeply, may have as much as 1200 m of water above it, may be about 20 km wide and extends to the abyssal region. A continental rise is a gently sloping area that begins at the end of the slope and extends to the deep ocean. An abyssal plain is the deep and flat area of an ocean floor and may have a water column of 5000 m above it.

A major linear elevated landform, which resembles a mountain range and is submerged in the sea, is known as a mid-oceanic ridge. It is a long, continuous mountain chain, where the length may vary from 200 to 20,000 km. It may consist of many small, slightly offset segments. The crest of a ridge may rise 2–4 km above the abyssal plain. If it is high enough to be exposed above the water level, it may become an island. Near the axis, the ridge slopes away almost symmetrically on both sides of the crest. A mid-oceanic ridge is characterized by a rift valley, as shown in Figure 1.3. A rift valley is a fault trough formed in a divergence zone or in an area of tension. Sometimes these give off lava. Mid-oceanic ridges exist in all oceans. In the Indian Ocean, these exist as the South West Indian Ocean ridge, the South East Indian Ocean ridge, the Central Indian Ocean ridge, and the Carlsberg ridge. The Mid-Atlantic Ridge is submerged below the Atlantic Ocean except in places where it appears as islands, such as at Iceland and the Azores. It continues northward as the Reykjanes ridge. Some prominent mid-oceanic ridges in the South Pacific are the Macquarie ridge, Pacific Antarctic ridge, East Pacific rise, and the Chile rise, shown in Figure 1.4. Ridges also exist in the Arctic Sea and the Red Sea. Chapter 2 on plate tectonics explains how and why the ridges were formed.

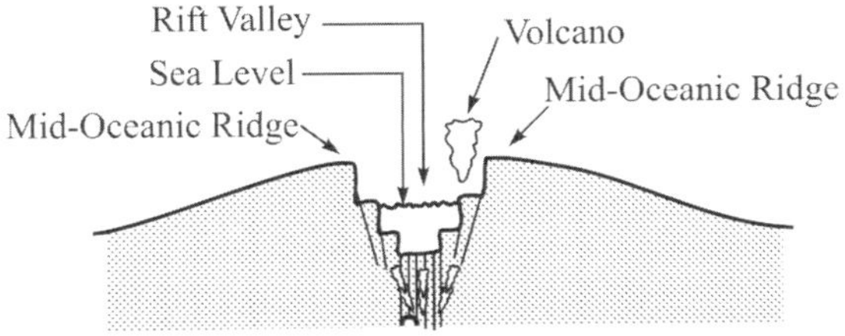

Fig. 1.3 Cross-section through a mid-oceanic ridge showing a rift valley. Volcanoes form on rift edges and the rift floor sinks below sea level.

A trench is a long, deep, narrow, and arcuate depression in the ocean floor. It may be several thousand kilometers long and 8–10 km wide. Some well-known trenches in the Pacific Ocean are the Aleutian trench, Japan trench (also known as Ryukyu trench), the Mariana trench, Tonga trench, Kermadec trench, New Hebrides trench, Middle America trench, also known as Mexico trench, and Peru Chile trench. The Andaman–Sumatra–Java–Sunda trenches are in the Indian Ocean, while the Caribbean trench is in the Atlantic Ocean. These are shown in Figure 1.4. The Mariana trench at 11.04 km is the deepest trench and is situated off the coast of Philippines. It may be interesting to note that even if Mount Everest (height = 8.85 km) were submerged in the Mariana trench, there would still be a column of nearly 2.1 km of water above it. The Challenger Deep, at the southern end of the Mariana trench, plunges almost 11 km deep into the earth's interior. See Chapter 2 for how and why the trenches were formed.

An Island arc is an arcuate chain of volcanic islands close to a trench. Andaman and Nicobar Islands in the Bay of Bengal, Japan, Aleutian Islands, and the Caribbean Islands provide examples of island arcs.

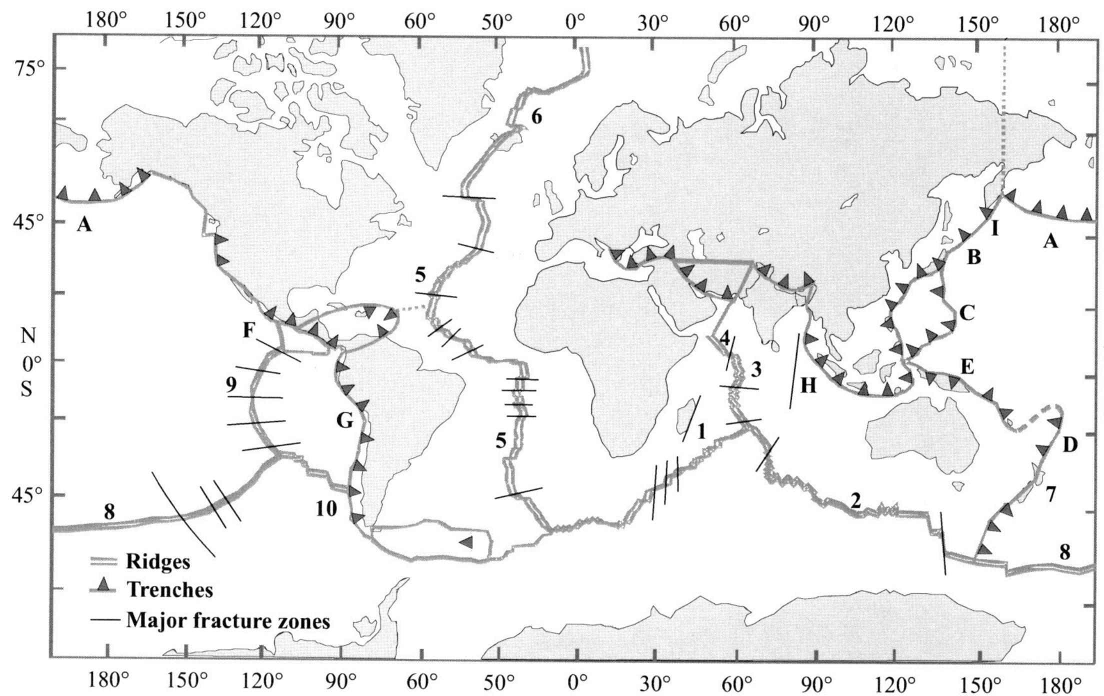

Fig. 1.4 Map showing the position of mid-oceanic ridges, trenches and major fracture zones. Parallel thick lines indicate crest of the mid-oceanic ridge system. The ridges shown on this map are: (1) South West Indian Ocean Ridge, (2) South East Indian Ocean Ridge, (3) Central Indian Ocean Ridge, (4) Carlsberg Ridge, (5) Mid Atlantic Ridge, (6) Reykjanes Ridge, (7) Macquarie Ridge, (8) Pacific-Antarctic Ridge, (9) East Pacific Rise, and (10) Chile Rise. Thick lines with teeth indicate deep-sea trenches. These are: (A) Aleutian Trench, (B) Japan Trench, (C) Mariana Trench, (D) Kermadec-Tonga Trench, (E) New Hebrides Trench, (F) Middle America Trench, also known as Mexico Trench, (G) Peru-Chile Trench, (H) Andaman–Sumatra–Java–Sunda Trench, and (I) Kurile Trench. Thin solid lines indicate major fracture zones or transform faults. (See color figure also.)

EARTHQUAKE CATALOGS

Several earthquake catalogs give good comprehensive data on earthquake parameters, casualties, major effects, etc. Gutenberg and Richter (1954), Richter (1958), the United States Geological Survey, (USGS), Oldham (1870, 1928), Tandon and Srivastava (1974), Bapat (1982), India Meteorological Department, IMD, and Rao & Rao (1984) give good useable catalogus.

CONCLUSION

The two major seismic belts, the Circum Pacific belt and the Alpine-Himalayan belt, are of particular interest as repeated destructive earthquakes in these regions make large populations and the built environment vulnerable. The next chapter will show how these belts are related to plate margins.

REFERENCES

Bapat, A., R. C. Kulkarni and S. K. Guha, 1983, *Catalogue of Earthquakes in India and Neighborhood*, Indian Society of Earthquake Technology, Roorkee.

Gutenberg, B. and C. F. Richter, 1954, *Seismicity of the Earth and Associated Phenomena*, Princeton University Press, New Jersey.

Oldham, R. D., 1870, *A Catalogue of Indian Earthquakes: from the Earliest Times to the End of 1869 A D*, Memoirs of Geological Survey of India, 63 p.

Oldham, R. D., 1928, *The Cutch (Kachh) Earthquake of 16th June 1819 with Revision of the Great Earthquake of 12th June 1897*, Memoirs Geological Survey of India, Volume 46, p 71–147.

Rao, B. R. and P. S. Rao, 1984, Historical seismicity of Peninsular India, *BSSA*, **74**(6), p 2519–2533.

Richter, C. F., 1958, *Elementary Seismology*, W. H. Freeman and Co., San Francisco, 768 p.

Tandon, A. N. and H. N. Srivastava, 1974, Earthquake occurrence in India, in *Earthquake Engineering*, Jai Krishna Sixtieth Birth Anniversary Commemoration Volume, p 1–49, Sarita Prakashan, Meerut.

USGS: United States Geological Survey.

2 CHAPTER

Plate Tectonics

INTRODUCTION

After understanding global seismicity, it is relevant to know what causes an earthquake. An earthquake can be caused due to several reasons. Some of these are landslides, volcanic eruptions, and collapse of subsurface cavities. Earthquakes can also be caused due to man-made reasons like mining and nuclear explosions, etc. However, such earthquakes are usually small and few in numbers. More than 99% of all earthquakes are tectonic in origin. Tectonic means large-scale deformation of the earth's crust resulting from forces deep inside the earth. These forces include folding and faulting of rocks and their metamorphosis. Tectonic earthquakes are those that result from sudden release of energy stored within the earth due to major deformations in the earth's crust.

PLATES

Plate tectonics gives a geological model of the surface of the earth, i.e., the lithosphere, which is divided into several rigid segments called plates. These models deal with different aspects of plates like creation and destruction of plate, and movement and interaction between plates. Plate tectonics unifies several global phenomena like global seismicity, volcanic activity, continental drift, and sea floor spreading. It also explains the origin of several large topographic features of the earth such as young mountain belts and rift valleys on continents, and ridges, rift valleys, trenches, and island arcs in oceans.

The surface of the earth, i.e., the lithosphere, is divided into several plates. The crust is like the cracked shell of a hard-boiled egg and consists of several large and small pieces called plates. These are in constant motion with respect to each other. Most earthquakes occur at boundaries of these plates, and are confined along narrow geographical regions called seismic belts. The theory of plate tectonics explains where, how, and why most earthquakes occur on the globe. The six large plates are African, American, Antarctica, Eurasian, Indian, and the Pacific plates. Among the many smaller plates some prominent ones are Arabia, Caribbean, Cocos, Nazca, Philippines, Scotia, Iran, and Somalia. Several smaller plates also exist, e.g., Juan de Fuca on the Pacific coast of North America and the Andaman plate in the Bay of Bengal. Figure 2.1 shows several large and small plates.

A plate is a thin rigid body with a large horizontal dimension. At some depth, usually between 40 and 150 km, plates are decoupled from the underlying material. Each plate has a different horizontal dimension; it may be as broad as 10,000 km, e.g., the Pacific plate, or as small as a few hundred kilometers, e.g., the Andaman plate. A plate may be made entirely of either continental crust or oceanic crust or a combination of both. The Pacific plate consists entirely of oceanic crust, whereas the African plate comprises of the entire continent of Africa and part of the Indian and Atlantic oceans around it, as well as part of the Mediterranean Sea. Chapter 3 on seismic waves deals with internal structure of the earth, as revealed by seismic waves, and also different kinds of crust, mainly continental and oceanic crust.

Plate Margins

A plate margin is the marginal part of a particular plate. Margins of two plates meet at a common boundary. Plate boundary is the surface trace of the zone of motion between two plates. These are regions where damaging earthquakes occur repeatedly and claim a heavy death toll. Global seismic belts, i.e., the Circum Pacific belt, the Alpine-Himalayan belt, and the Mid Atlantic Ridge, define margins of most plates. Most seismic and tectonic activity is localized at plate margins.

The three types of plate margins are constructive, destructive, and conservative, and these are shown in Figure 2.2. Plates move away from each other at constructive margins, move toward each other at destructive margins, and slip past each other at conservative boundaries.

Constructive Plate Margin

This kind of margin is also referred to as a creative plate margin or a source zone, as new crust is created here. It is also known as a divergence zone as the two plates move away from each other. Mid-oceanic ridges characterize these margins. Description of mid-oceanic ridges is given in Chapter 1.

Formation of Mid-Oceanic Ridges

The, source material that makes a mid-oceanic ridge comes from the upper mantle. Heat from beneath the lithosphere initiates thermal expansion and

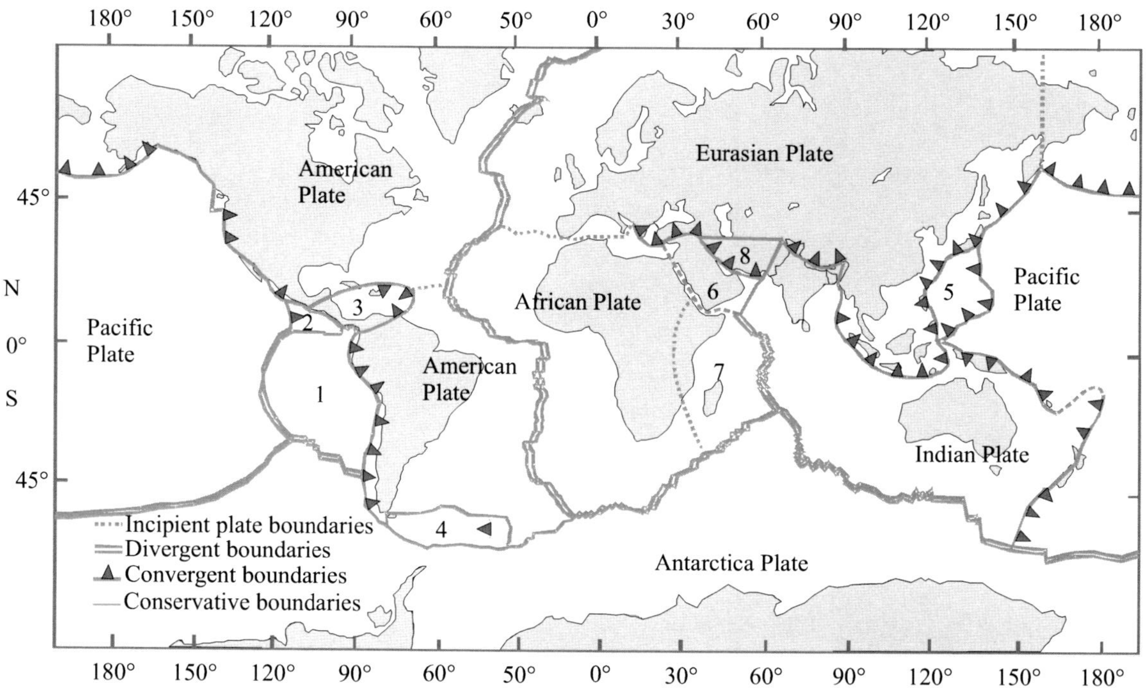

Fig. 2.1 Six major plates—African, American, Antarctica, Eurasian, Indian and Pacific—are marked on this map. The minor plates are: (1) Nazca plate, (2) Cocos plate, (3) Caribbean plate, (4) Scotia plate, (5) Philippines plate, (6) Arabian plate, (7) Somalia Plate, (8) Iran Plate.

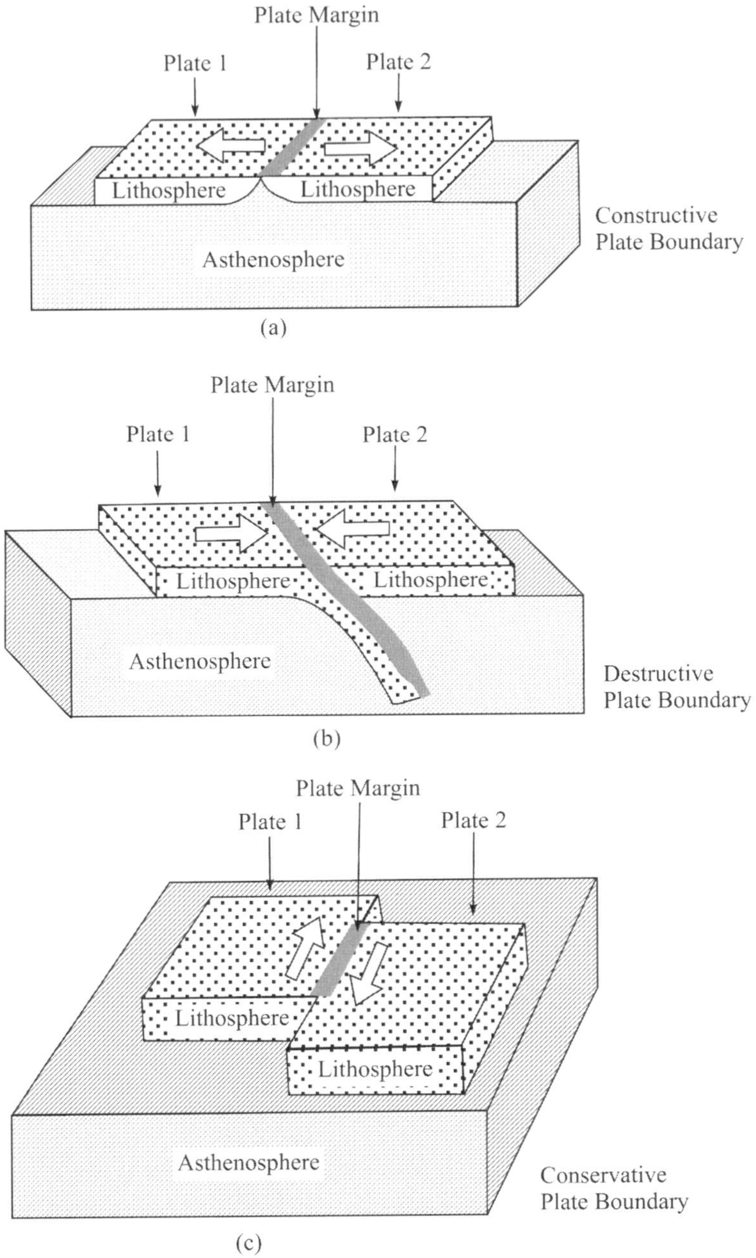

Fig. 2.2 Bold lines highlight the three kinds of plate boundaries: constructive, destructive and conservative. Large arrows show direction of motion of each plate. Plates (a) move apart and grow at constructive boundaries (b) compress and are destroyed at destructive boundaries and (c) slide past one another at conservative boundaries, neither creating nor destroying plate material.

domes the surface. Eventually, hot and molten magma comes out at the surface through the dome. The surface responds by normal faulting and later by formation of a rift within the dome. Magma is deposited symmetrically on both sides of a center that spreads to form a ridge. Horizontal extent of a ridge may be several hundred km in length. Volcanoes may form on edges of a rift. Magma gradually cools and solidifies along slopes of ridges to form the new crust. This is shown in Figure 2.3.

The new material is continuously added to that edge of the existing plate that is nearest to the ridge axis. Therefore, the youngest crust is nearest to a ridge axis, and the age of the crust increases as this distance increases. This gives a variable age to the oceanic crust. In this process, the oceans spread and this is the concept of sea floor spreading. Sea floor spreading is the process by which adjacent plates move apart to make room for new oceanic crust. Thus, the spreading process at the ridge produces crust that is new, thin, and layered. Moreover, it is oceanic in character, i.e., it is basaltic and contains the minerals olivine and pyroxene in abundance. This process continues intermittently, at rates that vary from 0.5 to 10 cm per year, through many geological periods.

Since the earth's magnetic field acts like a magnetic dipole, magnetic material within the up welling magma tries to align itself along the direction of the earth's magnetic field while it cools down. This process gives rise to magnetic anomalies in the crust, as shown in Figure 2.4. Since magnetic anomalies are parallel to the ridge axis, these are linear in form. Linear magnetic anomalies were observed in several oceans; their spacing is different in all oceans because spreading rates vary, but each ocean shows almost the same sequence. Correlation of anomalies in different oceans was instrumental in formulating the theory of plate tectonics. Rates of sea floor spreading were established by magnetic anomalies. These indicate that magnetic poles have reversed their position 171 times in the past 76 million years.

Mid-oceanic ridges exist in all oceans and some prominent ones are shown in Figure 1.4. The Mid Atlantic ridge spreads at the rate of approximately 2.5 cm per year. Sea floor spreading over the past 100–200 million years caused a small inlet of water to grow gradually into the vast Atlantic Ocean between Europe, Africa, and the Americas. Why are mid-oceanic ridges created? To understand this, refer to convection currents in the mantle as given later in this chapter.

When divergent boundaries exist in continental regions, these are known as rift zones. In the East African rift zone, the spreading process has already separated Saudi Arabia from the rest of the African continent, forming the Red Sea. A diffuse band of earthquakes in East Africa contains active volcanoes and long narrow lakes. A new spreading center may be developing in the N–S direction along the River Nile. Some other notable rift zones are

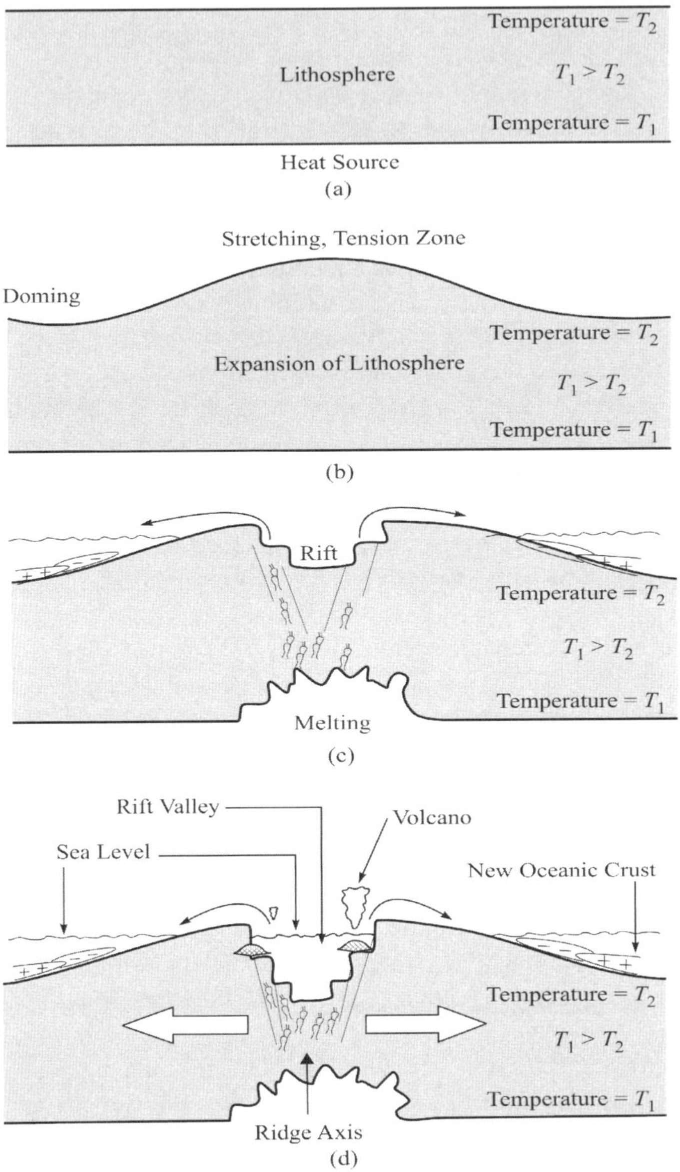

Fig. 2.3 Mechanism of formation of a mid-oceanic ridge and a rift valley. T_1 and T_2 are temperatures and T_1 is greater than T_2. Temperature increases with depth in the lithosphere and the rocks below expand due to excessive heat. This results in stretching and doming of lithosphere and ultimately in the formation of a mid-oceanic ridge and a rift valley. Different stages of this process are shown in this figure. (a) Break up of lithosphere is initiated by heating from beneath. (b) This causes thermal expansion that domes and stretches the surface. + and – indicate magnetic anomalies. (c) The surface of the lithosphere responds by normal faulting and formation of a rift valley. (d) Volcanoes form on rift edges. The rift floor may

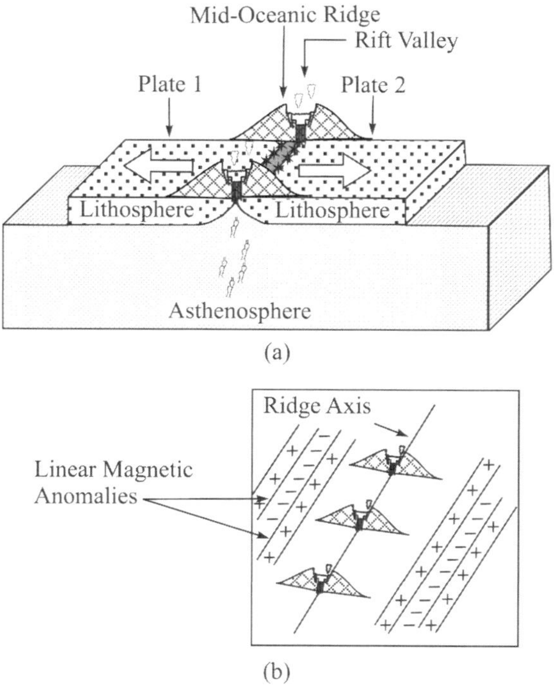

Fig. 2.4 (a) A mid-oceanic ridge with a rift valley is formed at a constructive plate boundary. Large arrows show direction of motion of plate. Star shows earthquake foci. (b) A schematic diagram of the process by which linear magnetic anomalies are formed parallel to the ridge axis.

along the Rhine valley, and the Baikal rift zone of Europe, and the Narmada and Tapti rift zone in India.

Oceanic ridges and rift zones give rise to shallow focus earthquakes, where depth is usually between 2 and 8 km. Magnitude is usually moderate; magnitude 6 or more is rare. This is because the lithosphere at these boundaries is very thin and weak, so sufficient strain cannot accumulate to cause large-sized earthquakes. Normal faults exist in this region, implying extension away from the ridge axis. Volcanic activity exists along ridge axis. Seismically active ridges are characterized by high heat-flow values. With increasing distance from the ridge crest, the heat flow falls until it reaches the average level for oceans. Ridges are close to isostatic equilibrium.

Destructive Plate Margin

At these margins, crust is destroyed or consumed by the mantle. These regions are known as convergence zones as plates move toward each other, and also as sinks, as the lithosphere sinks or subducts into the mantle. Island arcs and deep trenches in the ocean characterize destructive plate boundaries. Their description is given in Chapter 1, and the well-known trenches are

Tectonic forces cannot destroy continental crust, but oceanic crust is disposable. Convection currents in the mantle play an important role. In convection currents shown in Figure 2.11 cooler parts of the convection current join together and descend from the surface of the earth downward into the mantle. These drag along with them old oceanic crust and in this process, creates trenches on the surface of the earth. These places are associated with the downward motion of the lithosphere. The rate of destruction varies between 5 and 15 cm per year. This gives rise to friction between the subducting plate and the asthenosphere, which melts part of the subducted plate and also the asthenosphere above it. This hot molten material rises to the surface and manifests as volcanoes and volcanic islands. These islands are parallel to the trench axis and situated on the overriding plate, as shown in Figure 2.5. Unlike ridges, sinks are not symmetrical features.

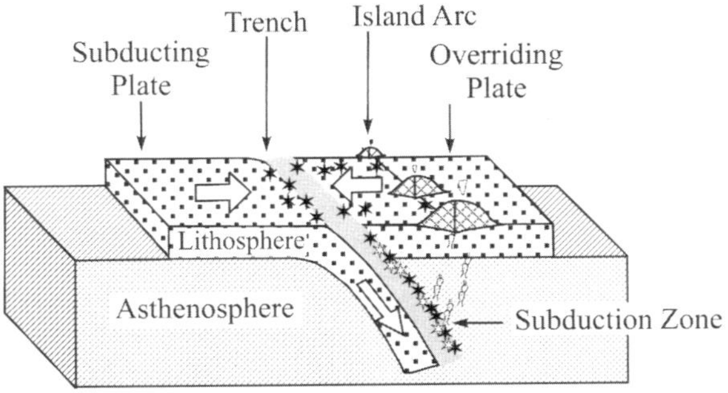

Fig. 2.5 At a destructive plate boundary, the subducting plate 1 sinks below the overriding plate, intersection between the margins of these two plates manifests as a trench. Island arcs form on the overriding plate. ★ shows large magnitude earthquakes.

Intermediate and deep focus earthquakes characterize these margins, i.e., seismicity is recognized down to a depth of 300–400 km and may even extend beyond that, to 700 km, the maximum depth at which seismicity has been recognized. Shallow focus earthquakes are also common. Frictional resistance develops between the surface of the descending plate and the asthenosphere, leading to accumulation of strain and its eventual release as an earthquake. The foci are restricted within a narrow zone, 80–100 km wide, which curves both along its length and down dip. This zone meets the surface of the earth close to the line of the ocean trench and dips away beneath the island arc. These inclined zones of seismicity characterize all active island arc systems and are known as subduction zone or Benioff zone. A subduction zone comprises of a narrow (tens of kilometers thick) dipping margin of ocean descending into the earth away from a trench. The angle of subduction (inclination) varies between 30° and 85° but is commonly close to 45°.

The ten largest earthquakes in the last century occurred along subduction zones. The Sumatra earthquake of December 2004 also originated in the subduction zone defined by the Andaman–Sumatra–Java–Sunda trench system. This earthquake, of submarine origin, claimed almost 2,30,000 human lives in coastal regions of the Indian Ocean due to the tsunami generated after the earthquake. Chapter 10 deals with tsunamis and the destruction caused by it.

In the Benioff zone, principal stresses are aligned parallel to the direction of dip. This suggests that the descending plate is under compression parallel to its length, and that earthquakes take place within it. Besides intermediate and deep focus earthquakes, there may be some shallow focus earthquakes also. In these regions, shallow earthquakes show either normal faulting or thrust faulting. The former occur parallel and just outside the trench and probably indicate an extension of the upper surface of the lithosphere as it descends into the mantle. Deep focus earthquakes with thrust faulting occur on the island arc side of the trench. These are probably caused by slip between the oceanic plate and the rocks above it. Oceanic trenches have an abnormally low heat flow, but a short distance away in the adjacent island arc, the heat flow is high. Trenches are filled with soft sediments and show the largest negative gravity anomaly on earth.

Convergence between two plates can occur in three ways: (i) between two oceanic plates, (ii) between an oceanic and a continental plate, and (iii) between two continental plates.

Convergence between two oceanic plates

This is the simplest kind of a convergent boundary. Since both the plates have a similar density and thickness, therefore either plate can sink below the other, and tectonic forces in the region will decide which plate subducts. At the Mariana trench, situated off the coast of Philippines, the faster moving Pacific plate converges into the slower moving Philippine plate, as shown in Figure 2.6.

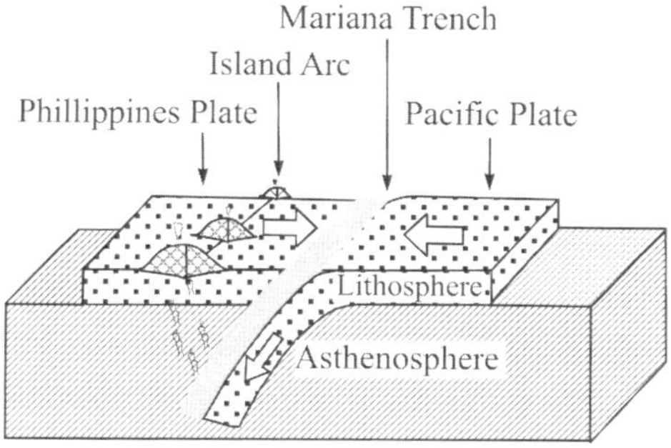

Fig. 2.6 Two oceanic plates, the Philippines plate and the Pacific plate, converge at a destructive boundary. The Mariana Trench is a surface manifestation of

Convergence between an oceanic and a continental plate

When the continental part of a plate arrives at a sink, its buoyancy (lighter, density 2.85 g/cm^3) with respect to the mantle (3.55 g/cm^3) disturbs downward motion of the subducting plate. This leads to some changes in the pattern of inter-plate motion. Continuation of the motion crumples up the continental crust on the surface and gives rise to rapid uplift of mountains, volcanic activity, and large earthquakes. The heavier oceanic plate sinks into the mantle. Destructive earthquakes that occur in the Andes Mountains in South America provide a good example of this condition. The Chile earthquake of 1960 is one such example. The Pacific plate, which is oceanic in character, is subducting below the South American part of the continental plate as shown in Figure 2.7. Surface manifestation of this process is the Peru-Chile trench, parallel to the Pacific edge of South America.

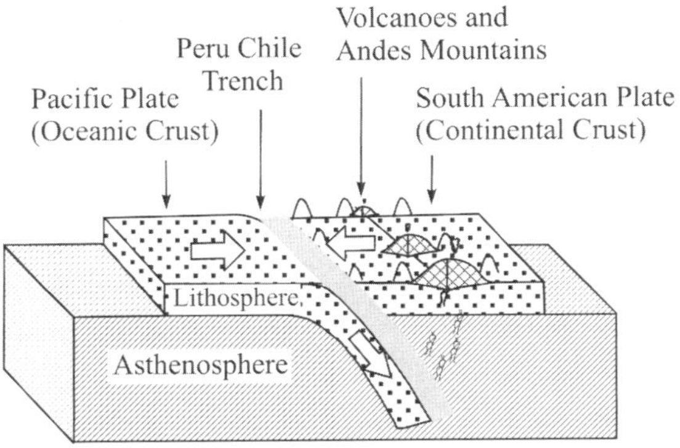

Fig. 2.7 Convergence between an oceanic (Pacific) and a continental (American) plate has caused the formation of the Peru–Chile trench and the Andes mountains in South America. As the oceanic plate is heavier therefore it sinks below the lighter continental plate.

Convergence between two continental plates

When two continental plates approach a sink neither plate is subducted, because each is light and has almost the same density and, like two colliding icebergs, resist downward motion. Continental crust readjusts on the colliding edge of both the plates. This gives rise to "continental collisions." A classic example is provided by the collision of the Indian plate with the Eurasian plate. It caused the oceanic part of the Indian plate to subduct below the Eurasian plate in the geological past. Currently, the continental crust of the Indian plate is juxtaposed against the continental crust of the Eurasian plate. Slow continuous convergence of these two continental plates shortened and crumpled the intervening crust and gave rise to several mountain ranges including Himalayas and the Tibetan plateau, parts of which are composed of oceanic crust, as shown in Figure 2.8. Tectonic evolution of the Indian plate is

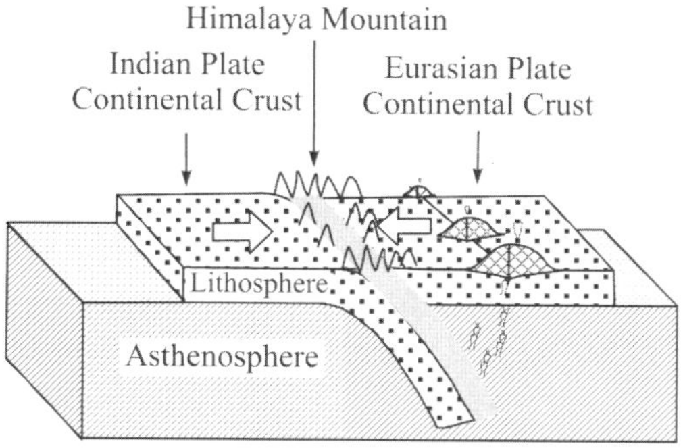

Fig. 2.8 Convergence between two continental plates, the Indian plate and the Eurasian plate, has caused the formation of the Himalaya Mountains.

given in Chapter 5. Large destructive earthquakes are common in such a situation. Four great earthquakes have occurred in the Himalaya Mountains within a span of 53 years in the last 110 years alone. The Kashmir earthquake of October 8, 2005 (magnitude 7.6) was also a result of this process. Seismicity of India is discussed in Chapter 6.

Conservative Plate Margin

This is also known as a transform plate boundary and it is characterized by a large system of transform faults. These large faults, also known as fracture zones, connect two constructive plate boundaries or, less commonly, destructive plate boundaries. These conservative plate boundaries help large rigid plates to move large distances without any significant internal deformation. At these plate margins, two adjacent plates slip or slide past each other, without creating new or destroying old plate material, as shown in Figures 2.9 and 2.10. The slip manifests itself as horizontal displacement and is observed in linear magnetic anomalies in the oceanic crust.

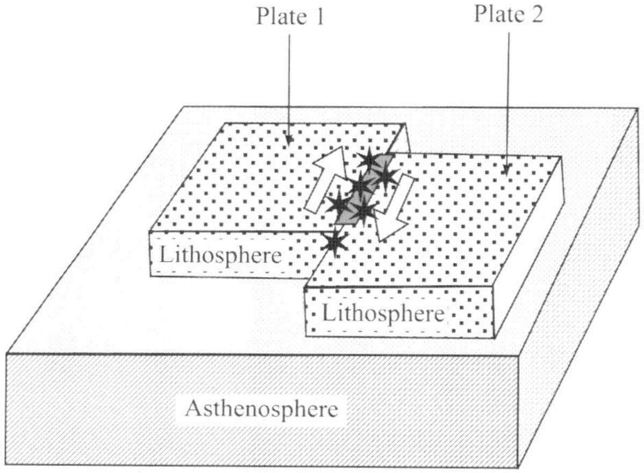

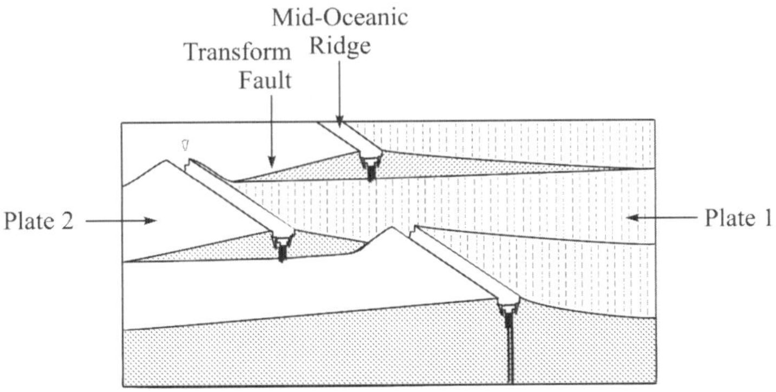

Fig. 2.10 A Conservative plate boundary connecting two segments of a mid-oceanic ridge that are displaced with respect to each other. A ridge-ridge transform fault appears between two segments of a ridge that are displaced from each other.

A transform fault, or a fracture zone, is a strike slip fault connecting the ends of an offset in a mid-oceanic ridge. Most transform faults are found in oceans. However, a few occur on land, for example the San Andreas Fault system in California. Along this fault, the Pacific plate is slipping past the North American plate at an average rate of about 5 cm/year. The Pacific plate is moving in the northwest direction and the American plate is moving westwards. Earthquakes at these boundaries have a shallow depth of focus, and there is no volcanic activity. A large amount of friction is generated at these plate margins at shallow depth, which causes large strains to accumulate at several places along the fault, which, in turn, is relieved by several large earthquakes. The 1906 San Francisco earthquake occurred in this way along the northern edge of the San Andreas Fault.

SOME ASSUMPTIONS IN THE THEORY OF PLATE TECTONICS

The theory of plate tectonics, like any other theory, is based on several assumptions. There are no constraints on number, shape, and size of plates, and these factors keep changing with geological time. Distance that separates sinks and sources is highly variable. In some cases, new crust may travel only a few hundred kilometers before it is consumed. In other cases, the sink and source may be 2000 or 3000 km apart. Only those parts of a plate, which are capped by oceanic crust, can participate in construction or destruction of a plate. A plate may be surrounded by any combination of boundaries and margins–constructive, destructive, and conservative. A triple junction is a region where three plates meet. As the three margins involved can be constructive, destructive, or conservative, many combinations are possible.

Three spreading ridges form the simplest triple junction. An example is provided in the Indian Ocean.

The earth is considered as a closed surface with a constant surface area. This implies that at any given time, the total area of plates generated at a creative plate boundary is equal to the total area of plates destroyed at destructive plate boundaries, taking the entire earth as a system. However, individual plates may increase or decrease in area, or some plates may get totally destroyed or new ones may get created.

Plates move on the surface of a spherical earth over a deep interior. They can move large distances without undergoing significant internal deformation. Motion of all plates is interdependent. Therefore, any change in velocity and direction of motion of one plate affects the motion of other plates. Differential motion may exist between adjacent plates. Plates are continuously in motion with respect to each other, and with respect to the earth's axis of rotation. They travel at the slow and variable rate of about 1–15 cm per year, which is equivalent to the growth rate of a fingernail.

Plate movements follow Euler's geometrical theorem, which implies that every displacement of a plate from one position to another on the surface of a sphere can be regarded as a simple rotation of the plate about a suitably chosen axis of rotation that passes through the center of the sphere. This axis is called the axis of rotation. Points where this axis cuts the earth's surface are called poles of rotation.

CAUSES OF PLATE MOTION

What is the mechanism that causes plates to move? A definite answer is not available because information on temperature, pressure, distribution of radioactivity, and physical properties of the interior of the earth are uncertain. Two schools of thought provide the answer to this. Gravitational forces, such as subduction of the cold dense lithosphere drive the plates to move, or may be the outpouring of lava generates enough momentum to push plates away from the ridge crest.

The existence of mantle-wide thermal gradients associated with high heat flow at ridges and low ones at trenches suggests that convection currents exist within the solid earth, as shown in Figure 2.11. The surface of the earth has normal temperature and pressure. A large thermal gradient exists between the surface of the earth and the hot outer core, and because of this reason and due to high pressure in the core, molten material in the outer core tries to escape toward the surface of the earth via the mantle. The heat is picked from the outer core, transported via the mantle, and lost at the surface of the crust. Due to the heat from the core, fluid moves across the hot lower surface of the mantle. This mechanism gives rise to convection currents in the mantle.

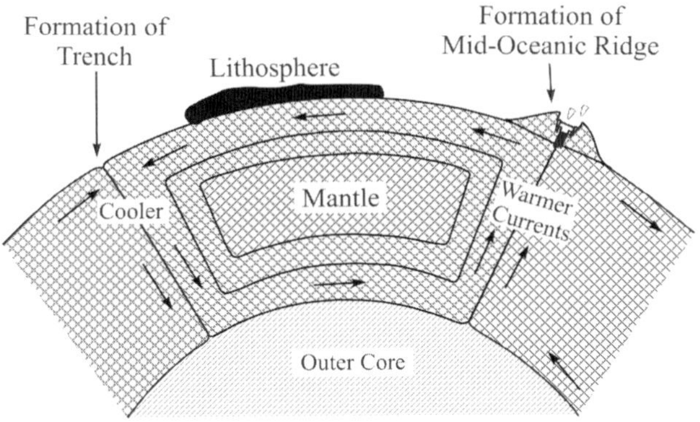

Fig. 2.11 Convection currents in the mantle give rise to mid-oceanic ridges at constructive plate boundaries and trenches at destructive plate boundaries. Arrows show general pattern of flow of convection currents.

Boundaries of two warm currents rise from the liquid core, join together in the mantle, rise to the surface of the earth, and split the lithosphere and form mid-oceanic ridges. Similarly, and elsewhere on the surface of the earth, two cold currents join together, are pulled into the mantle, and drag the lithosphere into the mantle and give rise to trenches. These set up convection currents in the mantle and may also be the main cause of earthquakes. Plates ride on a softer substratum, the asthenosphere, drifting laterally a few cm per year. At these slow rates, the asthenosphere is ductile. Mid-oceanic ridges and trenches are distributed irregularly on the globe, indicating that the pattern of convection cells is not simple. Moreover, these keep migrating in space and in geological time.

INTERPLATE AND INTRAPLATE EARTHQUAKES

Seismicity results from failure within the lithosphere, or slip along margins of adjacent plates. Where the lithosphere is shallow, only shallow seismicity occurs, where it descends into the deeper mantle, intermediate and deep seismicity is found. Current seismic activity lies along plate boundaries as at these margins, accumulated strains are released as earthquakes. This constitutes interplate seismic activity. More than 99% of global seismicity is an interplate activity. About 1% of global seismicity is due to intraplate earthquakes, which occur far from plate margins. The Latur earthquake of 1993 and the Jabalpur earthquake of 1997 are two such examples.

CONCLUSION

This chapter discussed some salient features of the theory of plate tectonics, mainly the tectonic model of the surface of the earth and origin of tectonic earthquakes. Since earthquakes originate inside the earth, there is a need to understand what lies in the interior of the earth. This aspect is revealed by seismic waves, as discussed in the next chapter.

REFERENCES

Please see the Bibliography

3 CHAPTER

Seismic Waves

INTRODUCTION

The previous chapter explained some salient features relevant to the theory of plate tectonics. Since most tectonic earthquakes originate on plate margins and have a bearing on what lies inside the earth, there is a need to understand what lies within the interior of the earth, which is revealed by seismic waves. On reaching the surface of the earth, seismic waves not only shake the ground but also the built environment supported on it. Sometimes these become disastrous; this aspect of seismic waves is discussed in this chapter.

SEISMIC WAVES

Most earthquakes occur when strains accumulated in rocks exceed their elastic limit and rocks rupture. This releases a tremendous amount of energy at the fault rupture in a very short span of time, i.e., within a few seconds. Energy spreads in all directions, away from the source, in the form of seismic waves. The medium through which seismic waves travel is assumed to be infinite in size, homogeneous, isotropic, endowed with elastic properties, and where displacements and strains are infinitesimal. Velocity with which seismic waves travel in a medium, such as rock, depends on several factors; some of the most important ones are density of the medium ρ, rigidity μ, and bulk modulus of elasticity k of the medium.

Body Waves

The earth transmits seismic waves in two ways: body waves and surface waves. Body waves travel through the body of the medium and are further classified as primary and secondary waves. In contrast to body waves, surface waves travel along the free surface of the earth and are further classified as Rayleigh and Love waves.

Primary waves are known as longitudinal, compressional, irrotational, and also as push or *P*-waves. The symbol P stands for primary. Particle motion associated with these waves is similar to sound waves and consists of alternating compressions and rarefactions during which adjacent particles of the solid, i.e., transmitting particles are closer together and farther apart during successive half cycles. The motion of particles is always in the direction of wave propagation.

If a pressure is suddenly applied at a point inside a homogeneous elastic medium of infinite size, as by an impact, the region of compression will move outward from the disturbance as an expanding spherical shell, the increase of radius having the compressional wave velocity, V_p. Behind this, another expanding shell may develop representing rarefaction and later, at an approximately equal distance, the second compressional pulse may develop, as shown in Figure 3.1. The *P*-wave velocity, V_p, depends on density of the medium ρ, bulk modulus of elasticity k, and rigidity μ and is given by Equation (3.1):

$$V_p = \{(k + 4/3\ \mu)/\rho\}^{1/2} \tag{3.1}$$

As *P*-waves are the fastest of all seismic waves, they are the first ones to reach any point on the surface of the earth. Both solid and liquid materials in the earth's interior can transmit these.

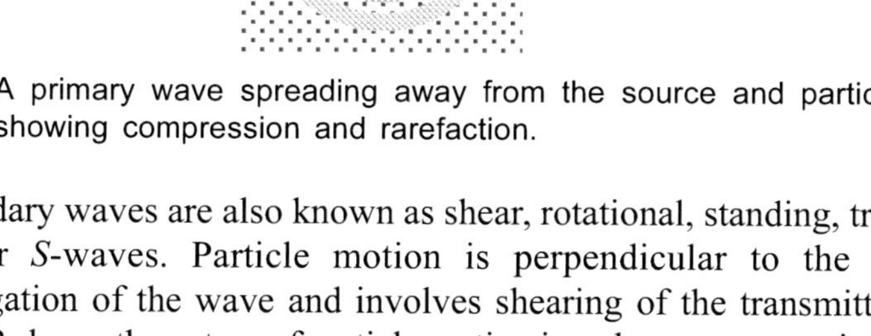

Fig. 3.1 A primary wave spreading away from the source and particle motion showing compression and rarefaction.

Secondary waves are also known as shear, rotational, standing, transverse, shake, or *S*-waves. Particle motion is perpendicular to the direction of propagation of the wave and involves shearing of the transmitting rock. Figure 3.2 shows the nature of particle motion in a shear wave passing through an elastic medium. Velocity, V_s, of shear waves is given by

$$V_s = (\mu/\rho)^{1/2} \tag{3.2}$$

S-waves are slower than *P*-waves; therefore at any place, these always arrive after *P*-waves. The ratio of compressional to shear wave velocity is given by $V_p/V_s = (k/\mu) + 4/3^{1/2}$. This expression shows that the compressional wave velocity is always greater than shear wave velocity in any medium. The

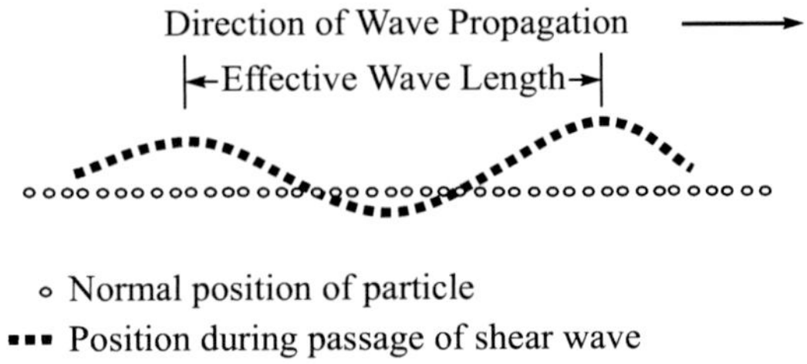

Fig. 3.2 Diagrammatic representation of particle motion in shear waves. The actual movement in the material is perpendicular to the direction of wave propagation.

radical must be greater than 1 because k and μ are always positive. For most consolidated rock materials, V_p/V_s is between 1.5 and 2.0. Shear waves travel only through solid material within the earth. As shear deformation cannot be sustained in a liquid (as $\mu = 0$ for a perfect liquid), shear waves will not propagate in liquid materials. Transverse waves can oscillate in any plane and exhibit the property of polarization. Polarization is the process by which oscillations occur in one plane only. *S*-waves polarized in the horizontal plane are classified as *SH*-waves. If polarized in the vertical plane, they are classified as *SV*-waves.

Surface Waves

In contrast to body waves, surface waves travel along the free surface of the earth. Surface waves are further classified into *Rayleigh waves* and *Love waves*. These arrive at a place after the *P*- and *S*- waves have passed through it. Love and Rayleigh waves disperse into long wave trains while traveling, and at a substantial distance from the source, these cause maximum shaking felt during earthquakes.

For Rayleigh waves, the particle motion, always in a vertical plane, is elliptical and retrograde with respect to the direction of propagation. This is shown in Figure 3.3(a). The amplitude of motion decreases exponentially with depth below the surface. Velocity of Rayleigh waves is less than that of body waves, being about $9/10^{th}$ that of shear waves in the same medium. Stoneley waves are surface waves of Rayleigh type for the case of a finite layer overlying an infinite substratum. Those surface waves, which are observed only when a low-speed layer overlies a higher-speed substratum, are called Love waves. Their particle motion is horizontal and transverse to the direction of propagation (Figure 3.3b). These waves propagate by multiple reflections between the top and bottom surface of the low-speed layer.

PROPAGATION OF SEISMIC WAVES

As a seismic wave spreads away from its source, its energy reduces due to several factors. Heterogeneity within the earth is the main reason for this, and

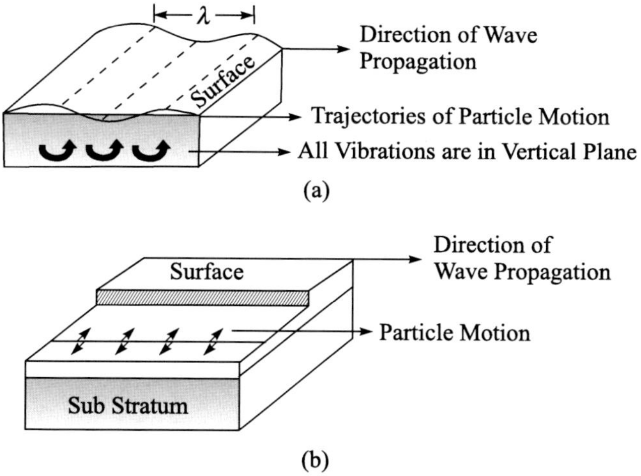

Fig. 3.3 Particle motion of: (a) Rayleigh waves, and (b) Love waves, traveling along the surface of a solid.

the traveling seismic wave is modified accordingly. When a seismic wave reaches a boundary, it is modified considerably depending on the nature of the boundary. In the simplest case, the boundary is plane and horizontal, and the two media on either side of it are in welded contact so that stresses and displacements are continuous across the boundary. It can be reflected or refracted, depending on the angle of incidence of the wave. Laws of reflection and refraction of seismic waves are analogous to those in geometrical optics. When an *S*- or *P*-wave strikes an interface at an angle other than 90°, a phenomenon known as mode conversion occurs. If a *P*-wave strikes an interface, four propagation modes may result: reflected and refracted *P*-wave and reflected and refracted *S*-wave. This is shown in Figure 3.4. Similarly, if a shear wave strikes an interface, the same four modes occur in different proportions. Besides reflection and refraction, a propagating seismic wave looses energy due to other means also such as by dispersion, absorption, attenuation, friction, conversion into heat, etc.

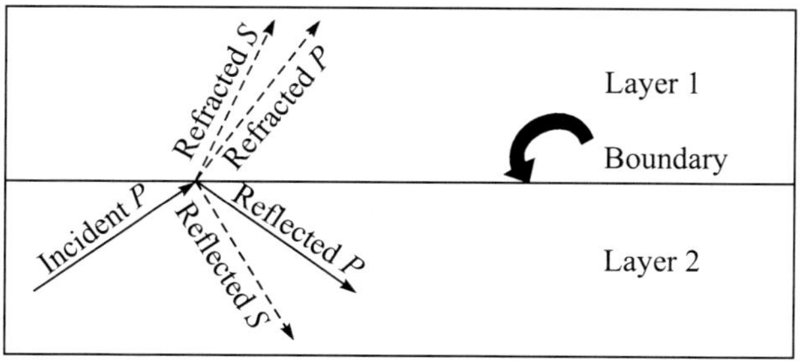

Fig. 3.4 An incident *P*-wave at a boundary is reflected and refracted as a *P*-wave. In addition, transformation to *S*-wave also occurs at the boundary, and these are reflected and refracted as *S*

As understanding about earthquakes increases, more realistic versions, other than homogeneous, of the medium in which the earthquake originates and through which seismic waves propagate are considered. These can be a stack of horizontal layers, as shown in Figure 3.5, inclined layers, gradual change in properties of different layers, presence of subsurface structures like anticlines, synclines, faults, domes, etc. Characteristics of the propagating wave are considerably modified due to these and other complexities in the subsurface.

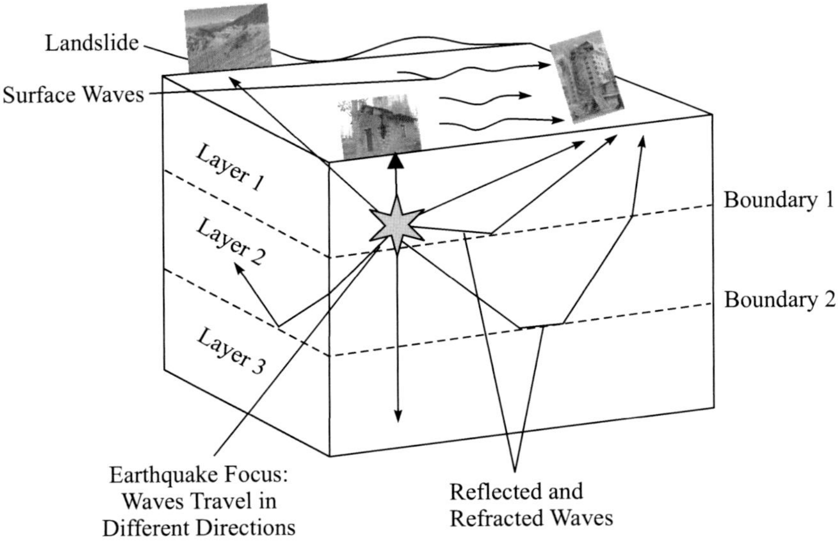

Fig. 3.5 Seismic waves originate from the focus of an earthquake, shown by a star, and travel in different directions. When these reach the surface of the earth either directly, after reflection and refraction from boundaries, or as surface waves, they shake and damage the ground and the built environment in many different ways.

Since waves of different kinds, i.e., body and surface waves, travel at different velocities, and there may be refractions, reflections, and multiple reflections at different boundaries, a disturbance that was nearly instantaneous at the source results in a train of seismic waves arriving at the point of observation for a considerable length of time. In general, the larger the distance between the source and the receiver the larger is the duration of the train of waves. Seismic waves are received on a sensitive instrument called a seismometer and recorded on a seismograph. The recorded data are called a seismogram. A seismogram records the particle motion at the recording station and shows the amplitude of body and surface waves as a function of time this record is a composite of what is happening at the source, the transmission path between the source and the receiver, and characteristics of the receiving station and the receiver. Thus, a seismogram shows complex oscillations, which include reflection, refraction, dispersion, and attenuation of the traveling seismic wave.

Different phases of seismic waves, i.e., *P*-, *S*-, Rayleigh, and Love waves, their time of arrival, and difference between the times of arrival of different type of waves, on the seismogram yield very useful information about the properties of the media through which the waves travel, and in revealing the interior of the earth. Figure 3.6 shows the *P*-, *S*- and surface waves on a seismogram.

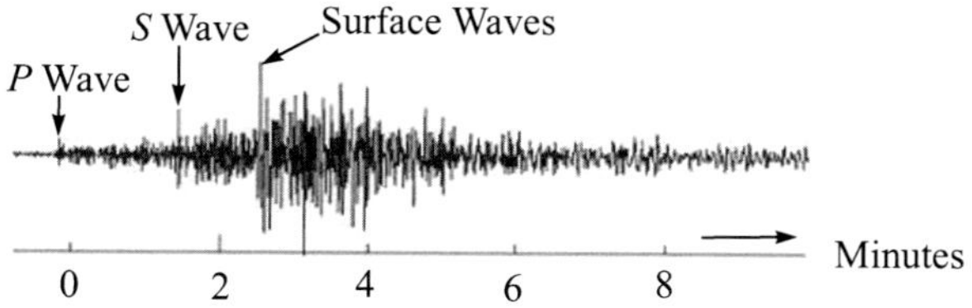

Fig. 3.6 Typical *P*, *S*, and surface waves are shown on a seismogram. Time markings are shown on the *X*-axis.

INTERNAL STRUCTURE OF THE EARTH

Once enough is known about damage and destruction caused by an earthquake, questions arise about what causes an earthquake. A proper answer to this can be sought only after it is known what constitutes the, internal structure of the earth.

Factual evidence about the composition of the earth is restricted to its surface and to samples taken from mines and bore-holes or wells, none of

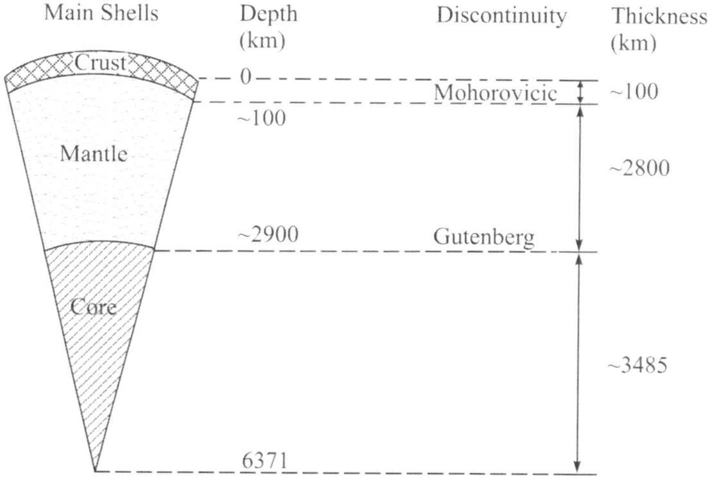

Fig. 3.7 A schematic section through the earth showing the three main shells: the crust, mantle, and core. The Mohorovicic discontinuity separates the crust from the mantle. The mantle is separated from the core at a depth of about 2900 km by the Gutenberg discontinuity.

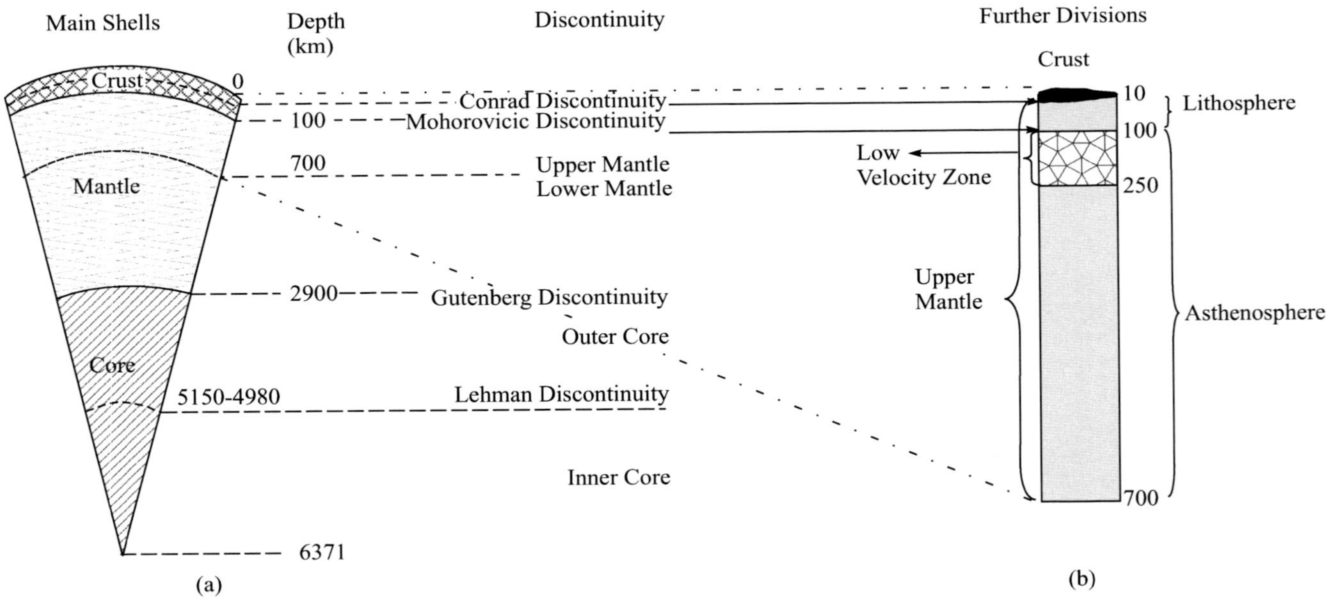

Fig. 3.8 (a) Subdivisions of the main shells—depth at which these occur and the Conrad and Lehman discontinuities. (b) Expanded section shows simplified relationship between the lithosphere, the asthenosphere, and the upper mantle.

which penetrate more than 10 km into the earth's interior. Geological processes on the surface of the earth can expose rocks that come from a depth not more than 20–25 km, and volcanoes throw up pieces of rock that may once have been part of the earth's upper mantle. Apart from these scanty data, there is no direct evidence concerning composition of the earth's interior.

Estimates of depth, velocity, density, rigidity, compressibility, pressure, temperature, and mineral content within the earth can be derived by indirect evidence only, from the study of seismic waves. These reveal models of the earth's interior, which helped in locating several shells inside the earth and in estimating their physical properties. In the simplest model, the spherical earth consists of three concentric shells: the crust, the mantle, and the core, the crust being the outermost and the core being the innermost. These shells are separated by distinct boundaries or discontinuities. As technology progressed, and recording, computing, and analysis techniques improved remarkably, more shells and minor discontinuities were identified within each shell. These advancements led to frequent revisions and refinements of density and velocity of different shells and depth of these discontinuities. These are given in Figures 3.7 and 3.8 and Table 3.1.

Table 3.1 The earth's internal layering, showing depth, density, and pressure.

Name of Layer	*Depth (km)*	*Density (10^3 kg/m^3)*	*Pressure (kilo bars)**
	0	2.8	
Crust		3.0	
	33	3.3	9
Upper mantle		↓	
	700	4.3	260
Lower mantle		↓	
		5.5	
	2890	10.0	1350
Outer core		↓	
	5150	12.3	
		13.3	3340
Inner core		↓	
	6371	13.6	3700

The Crust

The outermost shell, the crust, is a thin shell of variable thickness. It is further subdivided into two types of crust, continental crust and oceanic crust. Continental crust is lighter (2.85 g/cm^3), thicker, older, and geologically more complex than the oceanic crust.

When compared to continental crust, the oceanic crust is denser (3.55 g/cm^3); thinner, almost 5–10 km thick below the oceans; younger, the oldest ocean floor is only 200 million years old; and geologically simpler than the continental crust. The procedure by which the new oceanic crust is formed is given in the section on constructive plate boundaries in the chapter 2 on plate tectonics.

In continental regions, the crust is about 30–40 km thick, and it gets thicker in mountainous regions, almost 100 km below the Himalayas. The oldest continental regions are nearly 3 billion years old (compared to the age of the earth, which is about 4.6 billion years). Examples are the Precambrian shields of Africa, India, Siberia, Australia, Canada, and Brazil, the upper continental crust of which are dominated by igneous rocks such as granite or by metamorphic rocks such as gneiss and granodiorite.

Geological complexity is indicated by seismic data, which reveal that in continental regions the lower 15–20 km of crust has higher seismic velocities and densities compared to the upper crust. These are separated by the Conrad discontinuity. The upper crust manifests as rocks exposed on the continental land surface, which show regional variations in geological and chemical composition. For example, younger margins of continents consist largely of sediments derived from continued erosion of the continental surface and transported to the coast where most of it is deposited in shallow water on the continental shelf. Such sediments may be accumulations many kilometers thick.

The Mantle

The mantle is a solid shell that lies between the crust and the core. It extends down to a depth of about 2900 km. Although this is less than half the earth's radius (6371 km), the mantle forms 83% of the earth by volume and about 68% by mass. Velocity and density increase gradually with depth. Despite the fact that the mantle is physically inaccessible, an understanding of its nature is extremely important because mantle is the source region responsible for several global phenomena like major earthquakes, sea floor spreading, continental drift, and orogeny. On the basis of seismic velocities, mantle can be further divided into two shells: the upper mantle and the lower mantle. The upper mantle exists between the crust mantle boundary and 700 km, and the lower mantle exists between 700 km and the boundary to the core. The upper mantle is again divided into two shells, from Moho down to 200 km, and again from 200 to 700 km. A low-velocity zone exists at a depth of about 100–250 km below the surface. This is shown in Figure 3.8(a).

The lithosphere, meaning rock layer, is the outermost rigid shell of the earth and consists of the entire crust and adjacent part of the upper mantle. It extends from the surface of the earth to a depth of about 100–200 km. It lies over the asthenosphere, which is solid and part of it is molten. Relative to the material above and below, the upper part of the asthenosphere (from about

100–250 km depth) is a soft plastic solid and corresponds roughly with a low-velocity zone. This is shown in Figure 3.8(b). Low seismic wave velocities and strong seismic attenuation characterize it. It may be the site of convection and magma may be generated here (Monroe and Wicander, 2001). The lower part of the asthenosphere gradually becomes harder at a depth of about 700 km.

The, lithosphere is slightly lighter than the asthenosphere. Therefore, mountains sink deeply into the asthenosphere, like an ice cube extends far deeper into the water than it shows above. Since continental crust is the lightest part of the lithosphere, the crust below the mountains in continental regions is the thickest. Thus, the crust below the Himalayas and the Tibetan plateau extends downward to more than 70 km. The lithosphere is deep below old continental areas (craton), where it can exceed depths of 200 km, and thinnest in areas of recent tectonic activity and young ocean floors where it may be only a few km thick.

The Core

The Earth's core is a sphere that extends inward from the core mantle boundary at a depth of about 2900 km to the center of the earth. The core contains two distinct shells, the inner core and the outer core. The transitional layer between the two is about 150 km thick and is known as the Lehman discontinuity. It is marked by a rapid increase of *P*-wave velocity. The outer core is more homogeneous than all other shells. It is molten, and behaves like

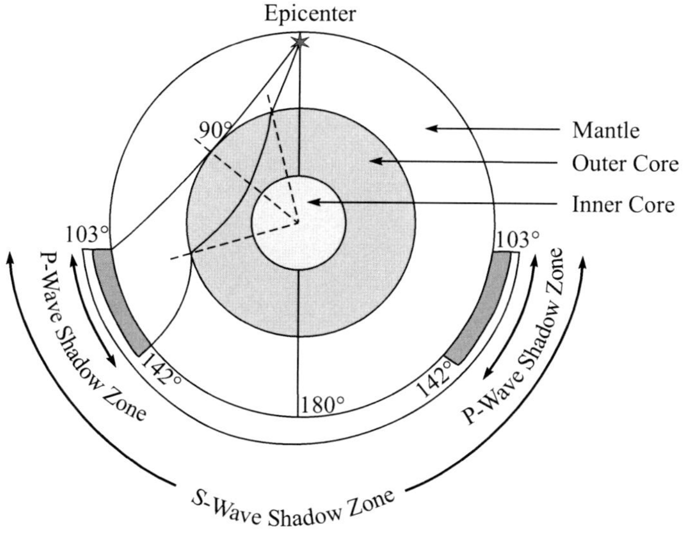

Fig. 3.9 The *P*- and *S*-wave shadow zone. As no *S*-waves pass through the core, the core is apparently liquid in nature. This makes a shadow zone for the *S*-wave. The star indicates the earthquake focus. *P*-wave shadow zone occurs between 142° and 103°.

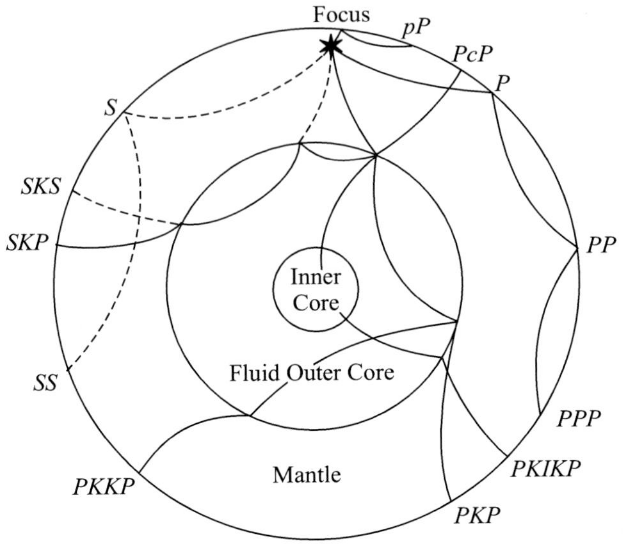

Fig. 3.10 Nomenclature of different seismic wave paths as they come to the surface after traveling through the mantle, outer core, and inner core.

a viscous fluid, even though its density is approximately that of lead. Because of its liquid nature, it does not transmit shear waves emanating from earthquakes. The inner core, from a depth of about 5150 km to the center of the earth, is solid. It is about the size of the moon and is fairly isolated from the rest of the earth. It is more than twice as dense as the mantle, and although it is only 16% of the Earth by volume, it has about 32% of its mass.

There is a region on the surface of the earth where *S*-waves are absent after an earthquake. This is the *S*- wave shadow zone and its size is the primary evidence of a liquid core. Similarly, there is a *P*-wave shadow zone. The two zones overlap partially, and neither *P*- nor *S*-waves are received in this region.

Discontinuities

Seismic waves from earthquakes reveal that physical properties change at boundaries of all these shells. The boundary between the crust and the mantle exists at a depth of about 100 km and is called the Mohorovicic discontinuity, often abbreviated to "Moho" or the "M-discontinuity". At the base of the crust velocity of seismic waves increases abruptly, to more than 8 km/sec. The core mantle boundary, at about 2900 km, is known as the Gutenberg discontinuity or Wiechert-Gutenberg discontinuity. At this discontinuity, there is an abrupt and sharp change in velocity of seismic waves, the *P*-wave velocity reduces considerably, and the *S*-waves disappear. Density of material on either side of the discontinuity is also very different; in the mantle it is about 5.5×10^3 kgm^3, whereas in the core it increases tremendously to about 10^4 kgm^3.

DIFFERENT PHASES OF SEISMIC WAVES

Seismic waves that originate from an earthquake reflect and refract at seismic boundaries. Prominent boundaries are the surface of the earth, the Mohorovicic, and Gutenberg discontinuities, and minor discontinuities are the Conrad and Lehman discontinuities. These are shown in Figure 3.8(a).

When a *P*-wave that leaves the focus in a direction away from the surface of the earth is reflected once at the surface and remains within the mantle, it is denoted as a *PP* phase. A further reflection from the surface gives rise to the *PPP* phase. If the *PP* phase reflects and then transforms into an *S* type of wave, then this is the *PPS* phase. In addition, there are phases such as *p*, *pP*, *pPS*, *pPP*, *sPP*, *sPS*, etc., the symbol p refers to an initial ascent of the *P*-wave to the surface of the earth, and s refers to its *S*-wave counterpart. A few phases are illustrated in figure 3.10. When these waves are instrumentally recorded and recognized on a seismogram they are identified as different phases of seismic waves.

Waves that are reflected and refracted from the core mantle boundary, i.e., the Gutenberg discontinuity, give rise to important phases on seismograms. The symbol c is used for denoting an upward reflection from this discontinuity. Thus, if a *P*-wave is incident on such a discontinuity, the upward reflection is denoted by the phase *PcP*. If the *P*-wave penetrates the core, it is denoted by the symbol K. Thus, the phase *PKP* corresponds to a wave that starts as the *P*-wave, is refracted into the core as the *P*-wave and is refracted back into the mantle as the *P*-wave, and emerges on the surface as such. Thus, the *PKKP* phase corresponds to a wave that suffered an internal reflection at the Gutenberg discontinuity. Some phases corresponding to different phases of waves are indicated in Figure 3.10.

When the *P*-wave reflects upward at the Lehman boundary (i.e., at the boundary between the outer and the innercore), the symbol used is *i*, and *I* corresponds to reflections of the wave path that has penetrated the innercore. Thus, *PKIKP* refers to one that has penetrated and is reflected into the innercore. By combining the symbols *P*, *S*, *p*, *s*, *c*, *K*, *i*, and *I* in various ways, notation for main phases associated with body waves can be set. Some of these are shown in Fig. 3.10. Travel times for different phases of seismic waves for an earthquake that originates at the surface are given in Jeffrey Bullen tables (1940, 1958).

EARTHQUAKE DAMAGE AND SEISMIC WAVES

In most earthquakes, it has been observed that the worst affected area is at or close to the epicenter, and damage decreases as epicentral distance increases. But sometimes earthquakes cause disasters even at large epicentral distances. When seismic waves arrive at the free surface of the earth, they vibrate the ground and any structures supported on it. These vibrations depend on several factors; some of the better-understood factors are frequency content of seismic waves and natural frequency of the structure, together with local geology and soil conditions. When a fault ruptures, seismic waves are propagated in all directions, causing the ground to vibrate at frequencies ranging from about 0.1 to 30 Hz.

Body Waves

The first waves to arrive at any place after an earthquake are *P*-waves; these are followed by *S*-waves. Body waves are high-frequency waves. Like all other high-frequency waves, their amplitude attenuates very fast as distance increases. Therefore, their amplitudes are pronounced at a small epicentral distance. Moreover, any structure in the epicentral region, which has a natural frequency of vibration in the same range, is liable to be set into vibration, sometimes in near resonance mode. If the structure cannot withstand these vibrations, it may deform, damage, or even collapse. Since low-height structures are short-period structures, they fall in this category. Therefore, in the epicentral region, body waves inflict maximum damage to low-height structures. Therefore, brick masonry houses whether single, double, triple, or

(a) (b) (c) (d)

Fig. 3.11 Damaging effects of Kutch earthquake of January 26, 2001 on low-height structures at different epicentral distances. (a) Bhachau, (b) Ratnal, (c) Bhuj, and (d) Mandvi.

four storey high, stone masonry houses, and other similar structures, which are devoid of any earthquake-resistant measures, collapse even in moderate-sized earthquakes, and claim a heavy death toll in the epicentral region. Moreover, damage to such low-height structures decreases as epicentral distance increases. Seismic performance of houses made of random rubble stone masonry is more dismal than that of brick masonry.

This has been brought out repeatedly in several recent earthquakes, like the Uttarkashi earthquake of October 20, 1991, Latur earthquake of September 30, 1993, Kutch earthquake of January 29, 2001, and Kashmir earthquake of October 8, 2005. Short-period effects at close epicentral distances for different kinds of low-height structures are shown in Figure 3.11(a) for Bhachau, which was the epicenter of the Kutch earthquake of 2001, and witnessed total devastation of random rubble stone masonry. A four storey building in the same figure shows that the entire structure settled to the ground after columns in the soft storey collapsed. Figure 3.11(b) A similar situation prevailed at Ratnal, at an epicentral distance of almost 35 km. Figure 3.11(c) At Bhuj, at an epicentral distance of almost 70 km, a three-story house on stilts overturned, and destruction of random rubble stone masonry houses was widespread. Figure 3.11(d) At Mandvi, at an epicentral distance of 100 km, gable walls were damaged in several stone masonry houses (Bose et. al., 2001).

Surface Waves

Compared to body waves, surface waves are long-period waves; therefore these travel a larger distance and with large amplitudes. Moreover, Love and Rayleigh waves disperse into long wave trains, and cause maximum shaking felt during earthquakes. Therefore, a structure that is located even at a large epicentral distance and has a natural frequency of vibration in the range of surface waves is liable to vibrate, sometimes in the resonance mode. If the structure cannot withstand these high amplitude vibrations, it may be prone to damage: may deform, damage, or even collapse partially or totally. Tall and long structures are long-period structures, and are liable to be adversely affected by long-period waves at large epicentral distances if adequate earthquake-resistant measures are not provided in the structure. Therefore, tall buildings, tall chimneys, elevated water tanks, flyovers and long span bridges are liable to damage even at large epicentral distances by surface waves.

In addition if such structures are founded on soft alluvium, unconsolidated sediments, or on filled or reclaimed ground, amplitude of surface waves can amplify considerably. Strong shaking caused by this makes long-period structures more susceptible to damage and to local high intensity. This was one of the main contributory factors for partial collapse of several multistory

Fig. 3.12 Damaging effect of Kutch earthquake of January 26, 2001, on tall buildings located at large epicentral distances. One interconnected tower has fallen off in (a) Ahmedabad and (b) Surat.

buildings at a large epicentral distance of 250 km in Ahmedabad and 350 km in Surat due to the 6.9 magnitude Kutch earthquake of 2001, as shown in Figure 3.12. This was mainly because of the long-period effects of surface wave, (together with several other inherent defects, some of which are discussed in the chapter on multistory buildings). The destructive effect of surface waves on long-period structures has been brought out repeatedly in several earthquakes, and for multistory buildings is shown in Chapter 12.

CONCLUSION

This chapter discussed how seismic waves not only reveal what lies inside the earth but also help in understanding how these propagate, shake the surface of the earth, and the built environment supported on it. Depending on their frequency content and the natural frequency of the structure through which these waves pass, these can sometimes become disastrous not only at small but also at large epicentral distances. In the next chapter, we will see the relation between the origin of an earthquake at a plate margin, more precisely at a fault, and disastrous aspects of a fault.

REFERENCES

Bose, P. R., A. Sinvhal and A. Bose, 2001, Traditional construction and its behavior in Kutch earthquake, in *Proceedings of the Workshop on Recent Earthquakes of Chamoli and Bhuj*, May 24–26, 2001, Roorkee, p 151–158.

Jeffreys, H. and K. E. Bullen, 1940, 1958, *Seismological Tables*, British Association, Gray-Milne Trust, 50 p.

Monroe, J. S. and R. Wicander, 2001, *The Changing Earth Exploring Geology and Evolution* (Third Edition), Thomson Learning Academic Resource Center, USA, 733 p.

4 CHAPTER

Earthquakes and Faults

INTRODUCTION

Most tectonic earthquakes originate either on preexisting faults or create new faults at the time of the earthquake, i.e., earthquakes and faults are deeply interrelated. When elastic energy, which is stored in rocks due to accumulation of strain, is released at the time of tectonic earthquakes, rocks break, are displaced, thus causing faults. Most active faults are located in interplate environments. An earthquake may affect nearby faults and may subject rocks on both sides of the fault to deform. On the surface of the earth, these can sometimes cause topographic changes, surface distortions, regional warping of ground, uplift, submergence of coastlines, and many other associated phenomena. The built environment supported on this kind of damaged ground is adversely affected. For this reason, it is very important to know where faults exist and their potential of getting seismically activated in the near future.

WHAT IS A FAULT?

A fault is a fracture along which observable displacement of blocks in the crust occurs parallel to the plane of break (Hills, 1959). The fracture may be a plane or a gently curved surface across which there is relative displacement of rock material. An example is shown in Figure 4.1. A plane that best approximates the fracture surface of a fault is called a fault plane.

The angle between true north and the horizontal line contained in this fault plane is called the strike of the fault. The angle that the fault plane makes with the horizontal is called the dip of fault plane. These are shown in Figure 4.2. This angle is measured in a plane perpendicular to the strike of fault. The angle between the fault plane and the vertical plane is called hade. Hade is complement of dip of fault plane. Slip is relative displacement of formerly

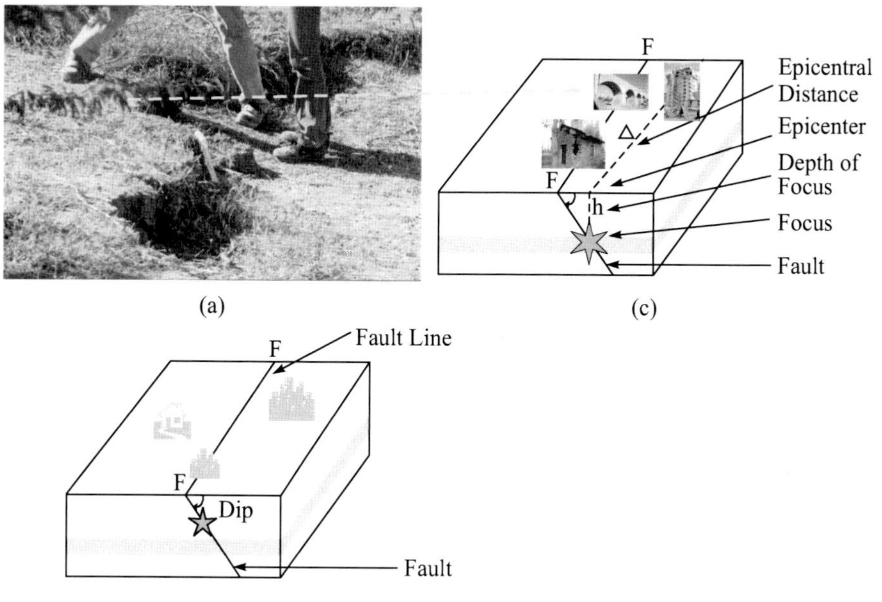

Fig. 4.1 (a) A surface fault, (b) Concept of origin of an earthquake at a fault is shown here. Star depicts the earthquake on the fault, surface manifestation of which is shown as a fault line, FF, (c) Elementary earthquake terminology such as focus, epicenter, depth of focus, (h), and epicentral distance,(Δ), are shown here.

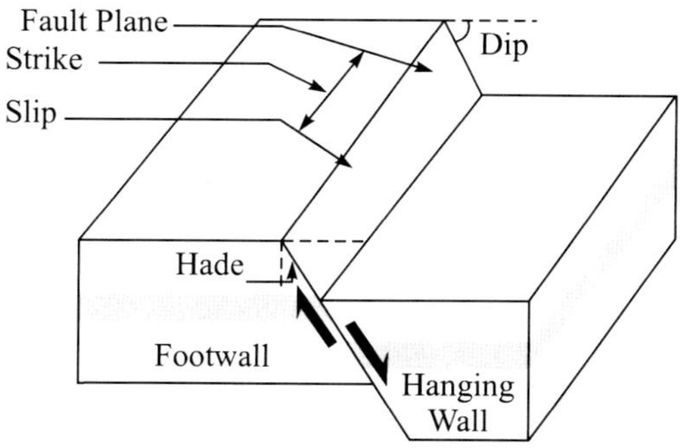

Fig. 4.2 Illustration of various terms used in description of a fault.

adjacent points, measured along the fault plane. Net slip is the resultant of strike slip and dip slip.

Strike slip is the slip component parallel to the strike of the fault, and dip slip is the slip component parallel to the dip of fault. That face of the rock, which lies below the fault plane, is called footwall. That face of the rock that lies above the fault plane is called hanging wall. Throw and heave are apparent displacements as seen in a cross-section normal to the fault plane. Throw is the vertical distance separating the faulted parts of a bed, and heave is the horizontal distance. In a strike slip fault, relative displacement is purely horizontal, i.e., predominantly parallel to strike of the fault. A strike slip fault connecting the ends of an offset in a mid-oceanic ridge is referred to as a transform fault, a trans-current fault, or a fracture zone.

In a dip slip fault, movement is parallel to dip of the fault. This kind of a fault is further classified into a normal fault and a reverse fault. A dip slip fault in which the block above the fault moves downward relative to the block below is called a normal fault. A dip slip fault in which the upper block, above the fault plane, moves up and over the lower block so that older strata are placed over younger ones is called a reverse fault. A reverse fault may also be called a thrust fault if the slip makes a low angle with the horizontal. An oblique slip-fault has both dip-slip and strike-slip components, of almost equal amplitude. Different kinds of faults are shown in Figure 4.3.

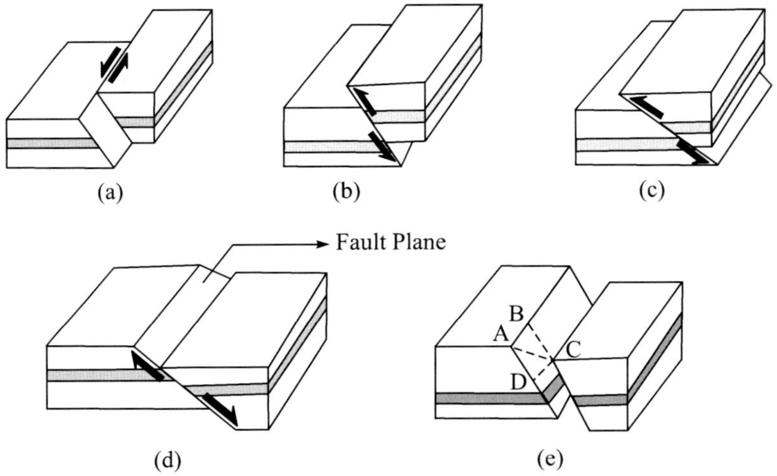

Fig. 4.3 Large arrows show movement in different kinds of faults. (a) In a strike–slip fault it is parallel to strike of fault plane, (b) Reverse fault, (c) Thrust fault, i.e., a low angle reverse fault, (d) Normal fault, and (e) Oblique normal fault. AC—net slip; AB—strike–slip; AD—is dip–slip.

DIFFERENT ASPECTS OF FAULTS

Faults may be buried deep inside the earth, and there may be no evidence of their existence on the surface. On the other hand, sometimes the fault may be close to the surface of the earth and may even be exposed on the surface. The latter is then known as a surface fault. This is shown in Figure 4.1(a). However, surface faults may sometimes be covered by thick vegetation, alluvium, snow, lakes, or sea, or may be hidden in some other way and its surface evidence may not always be obvious. Faults that have not shown any perceptible seismicity for a long geological time are dormant faults.

Geologically young fracturing may occur below and near the surface of the earth. The surface trace of a fault is usually represented as a single line on a map. However, in actual practice, a fault line is not necessarily confined to a single linear plane, but this is usually the best way of expressing a diffused zone of several linear and minor fault traces very close to each other. Sometimes a fault may exist as several broken sections or as discontinuous segments. The zone of disturbed rocks between fault blocks is the fault zone. The displacement of a surface fault is confined within a narrow zone and may sometimes be as large as a few hundred meters. Their damage potential increases as size of displacement increases.

The great Assam earthquake of 1897 gave rise to several spectacular surface faults such as the Chedrang fault and Samin fault. The 20-km long Chedrang fault, trending NWN–SES, was the most spectacular of all faults, with a vertical displacement of more than 12 m on the surface at several places. This is one of the largest known displacements for a single earthquake. The Samin Fault was 15-km long and showed displacements of 3 m. These faults followed the trend of the Chedrang River and other meandering streams, suggesting reactivation of an old line of weakness in crystalline rock. Numerous lakes, waterfalls, and pools were formed along these faults (Oldham, 1899). More details of the Assam earthquake of 1897 are given in Chapter 6.

Faults can vary in linear dimensions, from several thousands of kilometers in length, in which case they are referred to as mega faults, to a few kilometers only, in which case they are minor faults. Between these two kinds of faults, there may be faults that are hundreds of kilometers in length, in which case they may be referred to as major faults. Subsidiary faults may occur in the vicinity of large faults. It is not necessary that an entire fault ruptures in an earthquake, only a portion of it may rupture, and at times the rupture may be more than 300 km for a single large earthquake. Empirical relations between linear dimension of a fault and magnitude of the earthquake it can support are given in Chapter 7 on earthquake magnitude.

When a fault ruptures, seismic waves are propagated in all directions, causing the ground to vibrate at frequencies ranging from about 0.1–30 Hz. When a fault moves, rocks on both sides of it are subject to deformation and displacement. These can cause topographic changes, surface distortions, regional warping of ground, and uplift and submergence of coastlines. Many strong earthquakes have produced regional distortions, often with displacement on several small faults.

Three mega faults in the Himalayas extend from Kashmir in the west to Arunachal Pradesh in the east. These are the Main Central Thrust (MCT), the Main Boundary Thrust (MBT), and the Frontal Foothill Thrust (FFT) and are shown in Figure 4.4. These are associated with numerous subsidiary faults.

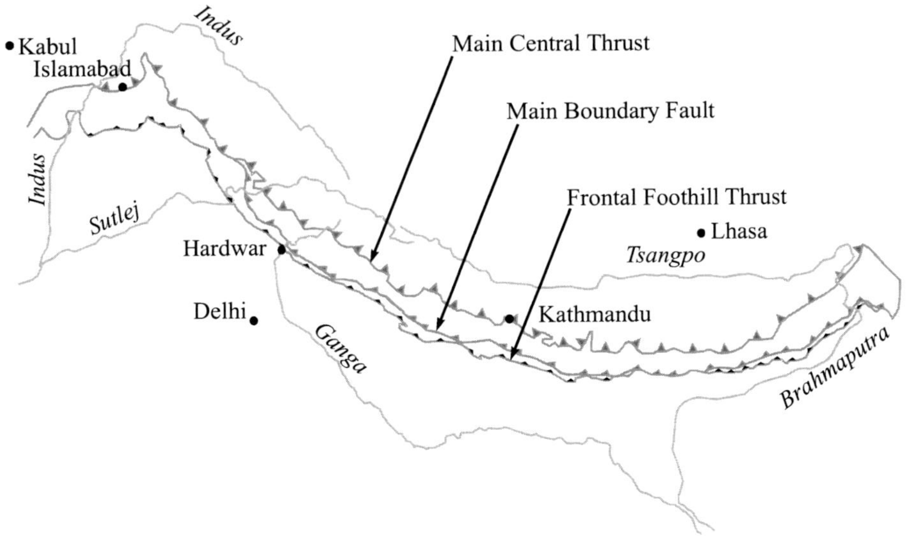

Fig. 4.4 Three mega thrusts along the Himalayas extend from Kashmir in the west to Arunachal Pradesh in the east. These are the Main Central Thrust, Main Boundary Thrust, and Frontal Foothill Thrust. (See color figure also.)

These faults are in the vicinity of the margin of the Indian plate, and are associated with current seismic activity, neotectonics, surface deformation, and a tremendous amount of earthquake-induced damage. The Uttarkashi earthquake of 1991 and the Chamoli earthquake of 1997 originated on the MCT and Kashmir earthquake of 2005 originated on the MBT.

It may sometimes be possible to locate the hypocenter of an earthquake on a fault; it is then referred to as the causative fault. In most cases, causative faults may not have surface manifestations and may be hidden in the subsurface. The causative fault for the Kutch earthquake of June 16, 1819 had surface manifestations as the Allah bund fault. The latter is the earliest well-documented example of surface faulting during an earthquake. This normal fault had an east–west strike. This surface fault was in the form of a low ridge, about 80-km long, 25-km wide, and with a maximum vertical offset of about 3 m. It blocked the flow of the Indus for several days; the dam was later cut by the Indus River and revealed marine shells, indicating transgression of sea. It was formed about 8 km north of Sindri; sea waves inundated this town by a column of 4 m of water. The locals later called this fault 'the Allah bund' or the Mound of God. Several spectacular ground effects, such as liquefaction, fissures, and earthquake fountains, were reported in this earthquake (Oldham, 1928). In the Rann of Kutch region 2–2.5-m-high fountains of sand and water spouted from ground fissures.

Casualty figures in this sparsely populated barren area was as high as 10,500, of which more than 2000 people were killed in Bhuj alone.

The causative fault for the Kutch earthquake of 2001 was the region between Adhoi Fault and Kutch Mainland Fault. Buried under the thick alluvium of Samakhiali and Lakadia plains are several more faults and their interlocking could have increased existing stresses, which were released during this earthquake. These are shown in Figure 7.8. The most spectacular surface fault was observed north of Mandvi. Vertical displacement of about 20 cm in soft alluvium was traced for about 3 km parallel to the Rukmavati River, and is shown in Figure 4.1(a). Numerous northwest–southeast trending ground fissures were observed in the vicinity of this fault. Cross-fissures developed at the confluence of the river (Sinvhal et al., 2003).

The causative fault for the Uttarkashi earthquake of 1991 was part of the MCT, shown in Figure 4.5. Genesis of the shallow focus, depth 6–8 km, Latur earthquake was associated with the formation of a new fault on river Terna, a tributary of the Godavari River. The causative fault for this earthquake had surface manifestations as a heave in the Talni region (Pande et al., 1995). The Jabalpur earthquake of 1997 originated on a fault associated with the Narmada River.

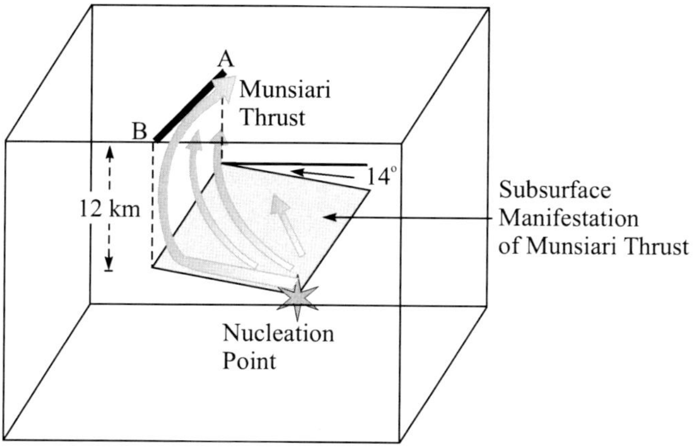

Fig. 4.5 Model of rupture propagation for Uttarkashi earthquake of 1991. Star indicates the hypocenter on the fault plane. AB indicates surface manifestation of Munsiari thrust.

EVIDENCE OF ACTIVE FAULTS

Presence of faults can be estimated by several methods. Geological field surveys sometimes reveal evidence of surface faults. Stratigraphic evidence such as marker beds, contacts, or unique structures, which match across the fault zone and appear to have once been continuous, are offset relatively on two sides of the fault. High relief is usually an indicator of active faults and seismicity. Features of small scale indicate geologically recent activity of a fault. Vertical or dip–slip displacement gives rise to fault scarps, which are considered as a good visual evidence of faulting.

Strike slip fault is often evident by offset streams and other watercourses (Stoffer, 2006). An individual feature of this kind might be due to the fault zone acting as a channel, plus possible local tilting. Emergence of water as fountains and springs through the crushed rock of the fault zone is common during an earthquake. This was indicated along the Kutch Mainland Fault and in many other regions in the Kutch earthquake of 2001 and in several other earthquakes, as given in Chapter 9.

Fault displacements can be investigated and confirmed by geodetic surveys, releveling for vertical movement and retriangulation for horizontal movements. This is perhaps the most reliable and effective way for scientific observation of displacement. This procedure implies a previous survey based on well-placed monuments, with survey lines extending out of the disturbed area. Inland changes of level are not easily established or studied, unless lines of precise leveling were previously carried far outside the area. Many large earthquakes have produced regional topographic distortions, often with displacement on several small faults. The Assam earthquake of 1897 is one such well-documented example (Oldham, 1899).

All faults may not be exposed on the surface; most faults may be subsurface and there may be no evidence of their existence on the surface. Faults hidden in the subsurface can be located by geophysical methods. These indirect methods help in finding density and magnetic anomalies associated with fault displacements. The method involving seismic waves, known as the seismic method, gives better estimates of fault characteristics. Interpretation of geophysical data helps in conceptual visualization and mathematical model of inaccessible faults (Hamzehloo et. al., 2002; Joshi et. al., 1995, 1999a, 1999b; Sinvhal and Srivastava, 1986, 1987; Sinvhal et. al. 1993, 1997, 1998). Interpretation of data recorded on seismological networks sometimes helps in estimating location and parameters of the causative fault for an earthquake. This is achieved through fault plane solutions. Data collected either from microearthquake networks, or from networks recording after shocks of an earthquake sometimes reveal the presence of currently active faults.

Faults that are caused during an earthquake can sometimes be mapped and identified in postearthquake field surveys. Trend of isoseismals also gives an indication of the presence of a causative fault for an earthquake. Elongation along higher isoseismals usually indicates a fault parallel to the elongation. For more on isoseismals, see Chapter 7 on earthquake intensity.

Since damage potential of earthquakes and faults is of such tremendous importance, these can be better understood if they are theoretically and computationally modeled. In the simplest case, a fault can be modeled as a plain rectangular surface, with a finite length, downward extension, dip, and strike. During an earthquake, rupture originates and propagates on this fault plane, and its seismic response is estimated at different locations on the surface of the earth. An example of this is given in Figure 4.5.

DAMAGE IMPLICATIONS

A fault can cause a myriad of earthquake effects that include topographic changes, surface distortions, regional warping of ground, uplift and submergence of coastlines, liquefaction in soft soil, fissures, water fountains, sand boils, offsets, land slides, rock falls, and many other associated effects. Some of these are given in Chapter 9. If the earthquake has a marine origin and the causative fault has vertical displacement, it can cause a destructive tsunami in coastal areas. The most recent example of this was provided by the Sumatra earthquake of December 26, 2004. This aspect is discussed in Chapter 10.

Relative displacement of two sides of a fault involves forces that can be very destructive to man-made structures. Casualties and injuries due to the primary effect of the earthquake alone, i.e., faulting, are rare, but the ground and the built environment located in the fault zone or close to it are susceptible to various kinds of damage. It is best to avoid any construction activity in the vicinity of a known fault, but in practice this luxury is not always possible. In that case, it is necessary to assess the hazard potential of known faults around the site and to design and construct a built environment accordingly, which will withstand seismic forces in its lifetime. Therefore, faults are of tremendous importance in the context of earthquake disasters. That there is an association between faults and earthquakes has been long established, but the nature of this association is becoming less obscure now. Therefore, when the location of important structures is under consideration, their proximity to known and active faults needs to be investigated thoroughly.

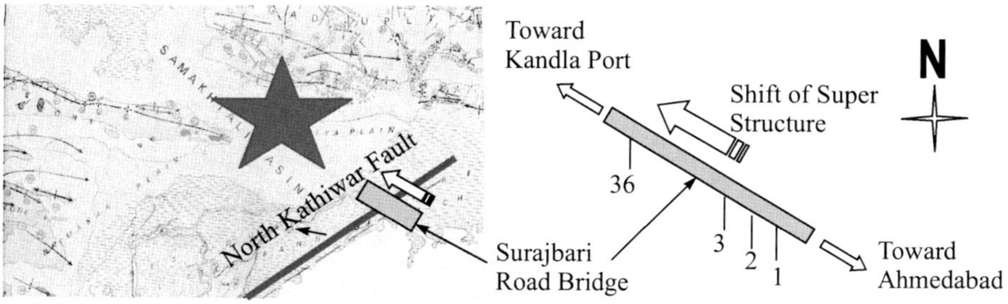

Fig. 4.6 The superstructure of Surajbari road bridge, located on the North Kathiawar fault, near Maliya Miyana in Kutch, was displaced by about 50 cm toward the north. Star depicts epicenter of the Kutch earthquake of 2001. (See color figure also.)

If a bridge crosses a fault line where there is displacement, the bridge may be severely damaged or it may even fail completely. The seismic performance of Surajbari Bridge in the Kutch earthquake of 2001, shown in Figure 4.6, (Sinvhal et al., 2001c), the Austen Bridge in the Sumatra earthquake of 2004 (Wason et al., 2006), and the bridge at Sarai Bandi in Baramulla district (Sinvhal et al., 2005, Pandey et al., 2006a, b), all in seismic zone V of the seismic zoning map of India, provide some appropriate examples. Tunnels, canals, and irrigation systems situated on a fault may be offset, shortened by displacement, may be damaged due to slumping, or emergence of ground water and sand.

CONCLUSION

This chapter discussed several aspects of faults and earthquakes, including damage potential of the two together. On the surface of the earth, damage can be in the form of topographic changes, surface distortions, regional warping of ground, uplift and submergence of coastlines, and many other associated effects. The built environment on this kind of damaged ground is liable to be adversely affected and sometimes claims thousands of human lives in a single earthquake. In the next chapter, we will see how the Indian plate evolved on the basis of the theory of plate tectonics, and how this gave rise to major tectonic units, especially several mega faults in the Himalayan tectonic zone, and how this affects current seismicity of the Indian subcontinent.

REFERENCES

Hamzehloo, H., A. Sinvhal and H. Sinvhal, 2002, Simulation of strong ground motion for the 1999 Kareh Bas (M_W 6.1), Iran Earthquake, in *Proceedings of the 12th Symposium on Earthquake Engineering*, Roorkee, p 215–223.

Hills, E. S., 1959, *Outlines of Structural Geology*, Methuen & Co. Ltd., London, 182 p.

Joshi, A., A. Sinvhal and H. Sinvhal, 1995, Modelling of rupture plane for Uttarkashi earthquake of 20th October 1991, in *Group Meeting on Seismo-tectonics and Geodynamics of the Himalaya*, Abstract volume, Roorkee, p 8–9.

Joshi, A., A. Sinvhal and H. Sinvhal, 1999a, A strong motion model for the Uttarkashi earthquake of October 20, 1991, in *Geodynamics of the NW Himalaya*, Eds. A. K. Jain and R. Manickavasagam, Memoir 6, p 329–334, Gondwana Research Group, Japan.

Joshi, A., B. Kumar, A. Sinvhal and H. Sinvhal, 1999b, Generation of synthetic accelerograms by modelling of rupture plane, *ISET Journal of Earthquake Technology*, **36**(1), p 43–60.

Oldham, R. D., 1899, Report on the Great Earthquake of 12th June 1897, in *Memoirs Geological Survey of India*, Volume 29, Geological Survey of India, 379 p.

Oldham, R. D., 1928, The Cutch (Kachh) Earthquake of 16th June 1819 with Revision of the Great Earthquake of 12th June 1897, in *Memoirs Geological Survey of India*, Volume 46, p 71–147, Geological Survey of India.

Pande, P., S. K. Gupta, N. V. Venkataraman and B. Venkataraman, 1995, Terrain changes consequent to the Killari earthquake of 30th Sept. 1993, in *Geol Survey of India Special Publication*, No. 27, p. 215–220, Geol Survey of India, Hyderabad.

Pandey, A. D., S. M. Pore and A. Sinvhal, 2006a, Damage to the engineered constructions due to Kashmir Earthquake of October 8, 2005, in *Proceedings of the 100th Anniversary Earthquake Conference*, April 18–22, 2006, San Francisco, California.

Pandey, A. D., A. Sinvhal and S. M. Pore, 2006b, Engineering Aspects of the Kashmir Earthquake of 8th October 2005 and the Need for a Blue Print for the Future, in *Proceedings of the Seminar on Seismic Protection of Structures*, Chief Engineer Chandigarh Zone Military Engineer Services, Chandigarh.

Sinvhal, A. and L. S. Srivastava, 1986, A note on simulation of ground motion due to quarry blasts, in *Proceedings of the Eight Symposium on Earthquake Engineering*, Roorkee, India, p 45–52.

Sinvhal, A. and L. S. Srivastava, 1987, Rupture model for simulation of near field earthquakes, in *Proceedings of the Sixth Indian Geological Congress*, Roorkee, India, p 209–211.

Sinvhal, A., A. Joshi and H. Sinvhal, 1993, Predicting strong ground motion by modelling the rupture at source, in *Proceedings of the 28th Annual Seminar on "Geophysics for Rural Development"*, Indian Geophysical Union, Hyderabad, India, p 68–74.

Sinvhal, A., H. Sinvhal, A. Joshi and P. R. Bose, 1997, Significance of Killari lineaments in the Latur earthquake, in *Proceedings of the Workshop on Earthquake Disaster Preparedness*, Roorkee, India, p 31–38.

Sinvhal, A., A. Joshi and H. Sinvhal, 1998, Rupture models using duration of strong motion records for three recent Himalayan earthquakes, in *Proceedings of the Eleventh Symposium on Earthquake Engineering*, Roorkee, India, p 255–262.

Sinvhal, A., P. R. Bose, A. Bose and V. Prakash, 2001, Damage observed to Surajbari Bridge due to the Kutch earthquake of 26th January 2001, in *Proceedings of the Workshop on Recent Earthquakes of Chamoli and Bhuj*, Roorkee, May 24–26, 2001, Indian Society of Earthquake Technology, p 423–431.

Sinvhal, A., V. Prakash, P. R. Bose, A. Bose, H. R. Wason, H. Sinvhal and A. D. Pandey, 2003, Ground damage observed in the Kutch earthquake of 26th January, 2001, in *Proceedings of the Indian Geotechnical Conference (IGC) 2003—-Geotechnical Engineering for Infrastructure Development*, Roorkee, India, p 273–276.

Sinvhal, A., A. D. Pandey and S. M. Pore, 2005, Preliminary Report on Kashmir Earthquake of 8th Oct. 2005, A damage survey report submitted to Department of Earthquake Engineering, Indian Institute of Technology Roorkee, Roorkee.

Stoffer, P. W., 2006, *Where's the San Andreas Fault? A Guidebook to Tracing the Fault on Public Lands in the San Francisco Bay Region*, USGS, California, 123 p.

Wason, H. R., A. Sinvhal, D. Shanker, A. Kumar and V. H. Joshi, 2006, Ground deformation observed due to the great Sumatra earthquake of December 26, 2004 and tsunami in and around Andaman and Nicobar Islands, in *Proceedings of the Thirteenth Symposium on Earthquake Engineering*, IIT Roorkee, December 18–20 2006, p 228–237.

http://www.eeri.org/lfe/clearinghouse/kashmir/reports/DEQ_IITR_KASHEQ05.pdf

http://www.iitr.ernet.in/EQ-Kashmir.pdf

http://www.iitr.ernet.in/news-system/files/58.pdf

5 CHAPTER

Tectonic Evolution of the Indian Plate

INTRODUCTION

The Indian plate is one of the major plates on this globe. Other major plates contiguous with the Indian plate are the plates of Africa, Antarctica, Eurasia, and Pacific (Figure 5.1). This chapter explains how the Indian plate evolved on the basis of the theory of plate tectonics. The present shape and position of

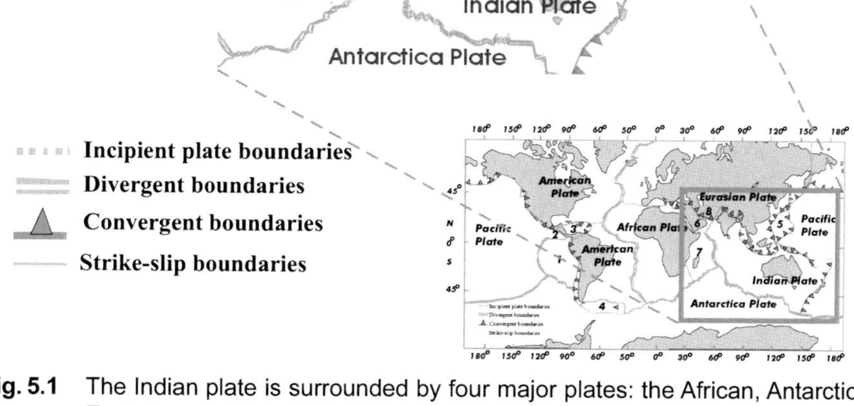

Fig. 5.1 The Indian plate is surrounded by four major plates: the African, Antarctica, Eurasian, and Pacific.

the Indian plate is very different from what it was earlier, in geological time. It has changed shape, traveled large distances to be where it is today, and continues to move northward. This has immense implications in terms of current seismicity.

BOUNDARIES OF THE INDIAN PLATE

India, Sri Lanka, Bangla Desh, Bhutan, parts of Pakistan, Nepal, and Afghanistan lie within the Indian plate. Australia and Tasmania are also part of the Indian plate but are located on a separate continental crust. Some countries contiguous with the Indian plate are Iran, Afghanistan, China, Tibet, Tajikistan, Kyrgystan, Myanmar, and New Zealand.

Diverse topographic features that characterize boundaries of the Indian plate include young mountain chains, trenches, island arcs, and mid-oceanic ridges. A large portion of the Indian plate is submerged below the Indian Ocean and the Pacific Ocean and consists of oceanic crust. The northern boundary of the Indian plate is defined by the Himalayan Mountains, which are part of the Alpine Himalayan seismic belt. The Himalayas stretch from Kashmir in the west to Arunachal Pradesh in the east and straddle Nepal and Bhutan in between. From Arunachal Pradesh, the boundary of the Indian plate swings sharply southward, from where it extends as the Arakan Yoma range of mountains. From there, it extends eastward toward the Andaman, Nicobar, and Indonesian Islands as a long continuous chain of Andaman, Sumatra, Java, Sunda Trenches. This intersects the Circum Pacific Belt near Philippines and enters the Pacific Ocean. Beyond this, it again turns southward, toward and through New Zealand, via the Kermadec-Tonga Trench, New Hebrides Trench, and Macquarie Ridge. Beyond this, it re-enters the Indian Ocean as South East Indian Ocean Ridge and swings toward the Arabian Sea as the South West Indian Ocean Ridge, Central Indian Ocean Ridge, and the Carlsberg Ridge. There it joins the Sulaiman and Kirthar ranges of Pakistan. Several major fracture zones are associated with the oceanic ridges.

The Indian plate is bound by all three kinds of plate boundaries, i.e., destructive, constructive and conservative. A destructive boundary indicates the presence of a subduction zone, which manifests as shortening of the crust, and its topographic manifestation is the trench and island arc system. The Andaman Sumatra Java Sunda trench represents the convergent boundary between the Indian plate and the Eurasian plate in the Bay of Bengal and likewise further east, in the Pacific Ocean, the New Hebrides trench, Tonga trench and the Kermadec trench represent the boundary between the Indian and the Pacific plate. A divergent, i.e., a creative plate boundary indicates sea floor spreading and is indicated by mid-oceanic ridges. A long chain of mid-oceanic ridges exists in the Indian Ocean. A conservative boundary indicates that the Indian plate is sliding past the adjacent plate in that region. The

Sulaiman and Kirthar ranges in Pakistan represent a conservative plate boundary. These ridges and trenches are shown in Figure 1.4.

The northern edge of the Indian Plate, represented by the Himalayan arc, has gone through all three kinds of destructive plate margins, i.e., from oceanic–oceanic, to oceanic–continental, and is currently of continent–continent type collision. To understand how this transformation happened, we go back several hundred million years in time on the geological time scale. This scale is given in Table 5.1.

GEOLOGICAL TIME SCALE

The earth is more than 4 billion years old. In the geological time scale, time is divided into eon, era, period, and epoch. Eon is the largest division of geologic time, and covers any span of one billion years. It embraces several eras (for example, the Phanerozoic, 600 million years ago (mya) to present; Proterozoic and Archaean). Era is a time period that includes several periods but is smaller than an eon. Commonly recognized eras are Precambrian, Palaeozoic, Mesozoic, and Cenozoic (or Cainozoic). Period is the most commonly used unit of geologic time, representing one subdivision of an era. Epoch is one subdivision of a geologic period, often chosen to correspond to a stratigraphic series. It is also used for a division of time corresponding to a Paleo-magnetic interval.

Cenozoic is a division of geological time that succeeds the Mesozoic and ends at the Quaternary. Pre-Quaternary refers to any time before 1.6 million years. The duration is approximately 65 million years. It is commonly used as a synonym for Tertiary. Quaternary is the period of geologic time starting 1.6 million years ago and continuing to the present day. It is divided into two epochs: Pleistocene and Holocene. Pleistocene is a name given to the geologic period between about 1.6 million years and 10,000 years before the present. Pleistocene is the earlier (older) epoch of the Quaternary period; the Holocene follows it.

EVOLUTION IN TIME

About 280 million years ago, which in geological time scale is known as the Permian age, there was a single supercontinent on this earth, Pangea, which means 'all earth'. Pangea was surrounded on all sides by the ocean Panthalassa, which means 'all seas'. Glacial deposits that were together at that time are spread in a wide geographical area today. This distribution is explained by postulating a single glacier flowing over the South Pole, before the breakup of Pangea. As time progressed, i.e., about 200 million years ago, at the end of the Jurassic period, the supercontinent split into two large continents. The northern one was called Laurasia and comprised of

Table 5.1 Geological Time Scale

Eon	*Era*	*Period*		*Epoch*	*Millions of Years ago*
Phanerozoic	Cenozoic	Quaternary		Holocene	10,000 years
				Pleistocene	1.8
		Tertiary	Neogene	Pliocene	5.3
				Miocene	24
			Paleogene	Oligocene	34
				Eocene	55
				Paleocene	65
	Mesozoic			Cretaceous	142
				Jurassic	206
				Triassic	248
	Paleozoic			Permian	290
			Pennsylvanian	Carboniferous	320
			Mississippian		354
				Devonian	417
				Silurian	443
				Ordovician	495
				Cambrian	545
Proterozoic				Pre-Cambrian	2500
Archean					4550

present-day North America, Europe, and most of north Asia. The southern one was called Gondwanaland, after the Gonds of central India, and comprised of Antarctica, South America, Africa, Madagascar, India, and Australia. The Tethys Sea separated these two large continents, and extended from present-day Spain in the west to Indonesia in the east, and supported abundant marine life. About 180 million years ago, India started to separate from Gondwanaland.

About 105 million years ago, in mid Cretaceous, after the eastern edge of India gradually separated from Gondwanaland and India became an island, India started a slow (10 cm/year) and long journey northward, toward Laurasia, as shown in Figures 5.2 and 5.3. To its north was the vast Tethys Sea. Leh, Dehra Dun, Kathmandu, and Darjeeling were then beaches on the northern edge of the Indian plate, beyond which was oceanic crust of the Tethys Sea. A convergent boundary in the north caused subduction of the oceanic crust of the Indian plate. This destructive boundary was the ocean–ocean type of convergence. As the oceanic crust of the Indian plate subducted, an arc of volcanic islands was formed in the Tethys Sea, between the Indian and the Laurasia plates. Simultaneously, a new ocean, the Indian Ocean gradually expanded along mid-oceanic ridges in the south, indicating a creative plate margin. Evidence of the extinct Tethys Sea is found at several places in the Himalayas as marine stratigraphy. These occur as fossils and see waves.

Fossils reveal the geological age at which the material was deposited, and also the type of environment and climate that existed at that time. Near Lhasa (capital of Tibet) and in Zanskar range of mountains, layers of sandstone was found to contain plant and animal fossils of marine origin, some of which lived

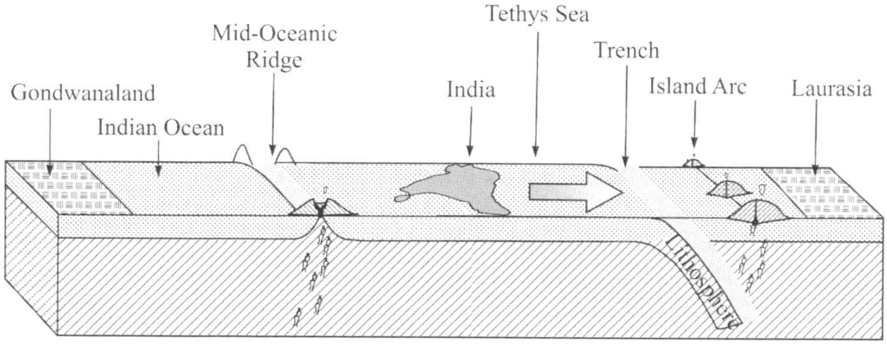

Fig. 5.2 Plate tectonic model showing movement of the Indian plate. Large arrow shows the direction of motion. India separated from Gondwanaland and moved northward toward Laurasia, through the Indian Ocean. The mid-oceanic ridge is a constructive plate margin and depicts a spreading center in the expanding Indian Ocean. The trench and arc of volcanic islands depict a destructive plate margin.

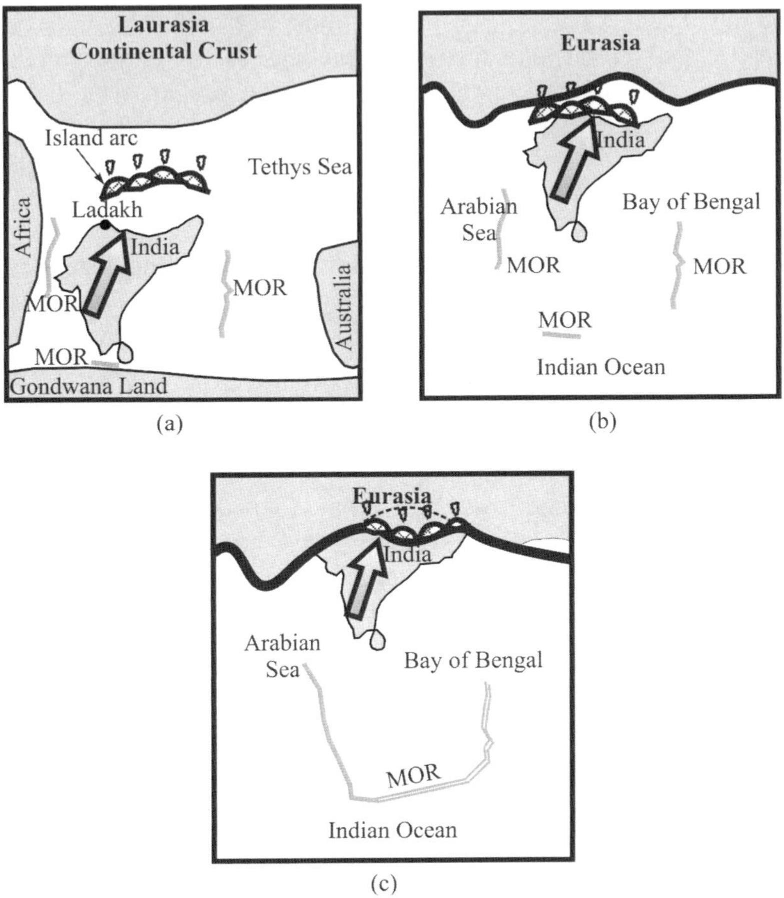

Fig. 5.3 Schematic illustration of possible stages in evolution of the Indian plate. India, originally joined to Antarctica, moved through the Indian Ocean to collide with Eurasia. (a) The picture as it was about 105 mya, (b) about 60 mya, (c) The position today. Gray areas indicate continental crust; large arrow shows direction of motion of Indian crust. MOR = Mid-oceanic ridge.

in a mild wet environment. This indicates that Tibet was once close to the equator, although today it has an arid and cold climate and that Tibet has moved 2000 km northward in the last 105 million years.

About 90 million years ago, as India continued to move northward, the island arc was pushed northward, it collided with Laurasia, and the intervening oceanic crust between Laurasia and the island arc folded and faulted. This was the first phase of orogeny. Orogeny means mountain building, particularly by folding and thrusting of rock layers. In the plate tectonic model, orogeny occurs primarily at boundaries of colliding plates, where the intervening material is crumpled and volcanism is initiated.

About 70–65 million years ago, at the end of the Cretaceous period, because of continuing northward movement of India, the Tethys Sea closed gradually and the island arc was squashed between crusts from two different plates. Further movements of the Indian plate caused further creasing of the intervening oceanic crust into a series of folded mountains and basins parallel to the zone of collision. This was the second phase of Himalayan orogeny. At the same time and elsewhere on the globe, South America separated from Gondwanaland and became a giant island. Australia and Africa may have only just begun to drift away from Gondwanaland. Continental crust of North America, Europe, and north Asia was perhaps still interconnected.

About 60–55 million years ago, i.e., in Paleocene times, northern tip of the Indian plate collided with the southern edge of Laurasia. The first contact took place near the present day Leh in Ladakh. The destructive plate boundary changed from collision between an oceanic (Indian plate) and a continental (Eurasian) plate, to continent–continent type collision. This started an era of collision between two continental plates. When the leading edge of two adjacent plates approach a sink and both are capped by lighter continental crust neither of the two plates sink at the subduction zone, and like two colliding icebergs resist downward motion. Because of this impediment, northward advance of India slowed down. Readjustment of continental crust took place on both plates and the crust thickened. After this continent-continent collision, because of the buoyancy factor, it became difficult for the Indian plate to subduct below the Eurasian plate. At the same time, another subduction zone developed along the eastern margin of the Indian plate. This gently inclined Benioff zone extends below the Islands of Andaman, Nicobar, Indonesia, and Philippines. This gave rise to the Andaman–Sumatra–Java–Sunda Trench system in the Indian Ocean and the Arakan Yoma range of mountains in Myanmar, in the lower Eocene times (50 mya). Also, eastern Myanmar was later uplifted into a high plateau, in a process similar to that of Tibetan Plateau. This was the third phase of Himalyan orogeny.

About 50–36 million years ago, i.e., in Upper Eocene and Oligocene times, after colliding with Eurasia, like a door slamming shut, India rotated anticlockwise. The intervening volcanic islands merged between the two large continental crusts. These islands are now part of Kohistan, Ladakh, and Tibet. The Tethys Sea closed altogether. This was the fourth phase of Himalayan orogeny. Almost 30 million years ago, a large migration of mammals occurred between Eurasia and India, indicating that the Himalayas was still a warm marshy zone, and the two continents were joined together.

About 20–23 million years ago, i.e., in Middle Miocene times, due to the strain produced by the collision, folding and faulting occurred along the boundary of the collision zone and the Tethys Sea disappeared altogether. This caused a rapid uplift of the Himalayan ranges. This was the fifth phase of Himalayan orogeny, and it may have been the most powerful one of all. Two to one million years ago further upheavals resulted in thickening of the crust, especially in Central Asia below the Pamir, Hindu Kush, Tibet, and the Pir Panjal ranges. Kumaon, Garhwal and several other ranges were formed. Tectonics of these areas has a very strong bearing on rivers, valleys, gorges, lakes, and hot springs in the area. The Mansarovar Lake and the Mediterranean Sea are a remnant of the Tethys Sea. The Alps evolved in a manner similar to that of the Himalayas, albeit with slower tectonic upheavals.

Due to the ongoing collision of the Indian plate with the Eurasian plate, the accumulated strain is released at several faults in the collision zone. Eighty percent of the strain resulting from this is absorbed in a 50-km wide region centered on the southern edge of the Tibet plateau. The remaining 20% of the strain is absorbed in the surrounding Himalayas. This results in a convergence of 2 m per century between the two plates, (Bilham et al., 2001). This manifests as a maximum horizontal southward velocity of 17.5 ± 2 mm between southern Tibet and India, and there is no relative motion between the Indo Gangetic plain and peninsular India. The GPS-derived horizontal strains are 226×10^{-9} for peninsular India and 2×10^{-7} for the Himalayas, which means that strains are higher in the Himalayas. A belt of strong gravity anomalies, indicating a lack of equilibrium, runs along the Himalayan arc.

In India, the main folding and thrusting into mountains began in the Cretaceous (144 mya), continued into the Eocene (55 mya), and climaxed in the mid Tertiary (5 mya). Northeastward movement of the Indian plate continues today. The effect of this is that the crust in the Himalayan collision zone is shortening, thickening, folding, and faulting. This renders the Himalaya geologically the youngest mountain chain and the most seismically active plate boundary in an intracontinental region on the globe.

This causes catastrophic earthquakes in the length and width of this destructive plate boundary, i.e., in the Himalayas and along the trenches in the Bay of Bengal. Epicenters of many great earthquakes lie in or near this belt. In the last 200 years alone, seven great earthquakes that caused heavy damage to life and property originated at or near the Indian plate boundary. These are shown in Figure 6.1.

Due to deep-seated tectonic forces, northward movement of the Indian plate is expected to continue into the future, therefore, destructive earthquakes in these regions will continue to occur. Population is dense in the foothill of the Himalayas, and the area is going through rapid construction activity in the form of dams, hydroelectric projects, bridges, houses, etc. Therefore these regions, vulnerable to earthquake hazards and risks, are of particular seismotectonic interest, and require an appropriate understanding of seismotectonics of the region, seismic monitoring, and special engineering interventions to mitigate future earthquake disasters.

MAIN TECTONIC UNITS

As the Indian plate moved northward, the crust deformed. This formed the shape of several tectonic features of various ages, shapes, sizes, and seismic implications. The continental crust of India can be divided into three broad tectonic units: the Himalayan tectonic zone, the Indo Gangetic plains, and the peninsular region. These are shown in Figure 5.4.

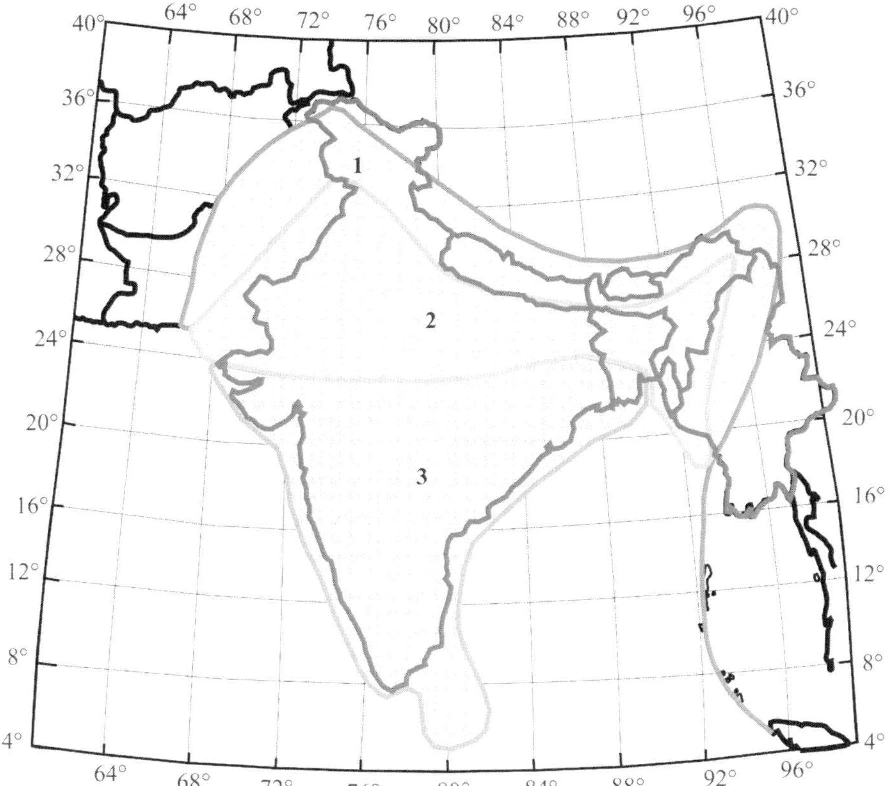

Fig. 5.4 Simplified tectonic divisions of India are as follows: (1) Himalayan Tectonic Zone, (2) Indo Gangetic Plains, and (3) Peninsular Region.

The Himalayan Tectonic Zone

The Himalayan tectonic zone, which is coincident with the Himalayan region, occupies the entire northern boundary of the Indian plate. With a linear arcuate trend, about 2500 km long, the Himalayan arc is convex toward the south, i.e., the Indian peninsular. So is the Baluchistan arc, which comprises the Sulaiman and Kirthar ranges in the west, and also the arc containing the Arakan Yoma folded ranges in the east. These three arcs, convex toward the Indian peninsular, are an indication of the enormous stresses existing around the edges of the Indian plate, and this manifests as high seismicity. The region where the two arcs meet, in a knee-bend type of situation, is referred to as a syntaxis. These regions are seismically more active than the three arcs and are characterized by a knot of rugged mountain terrain. At both these places, the rivers take a sharp turn to enter the Indian plate. The three arcs and the two syntaxis are shown in Figure 5.5.

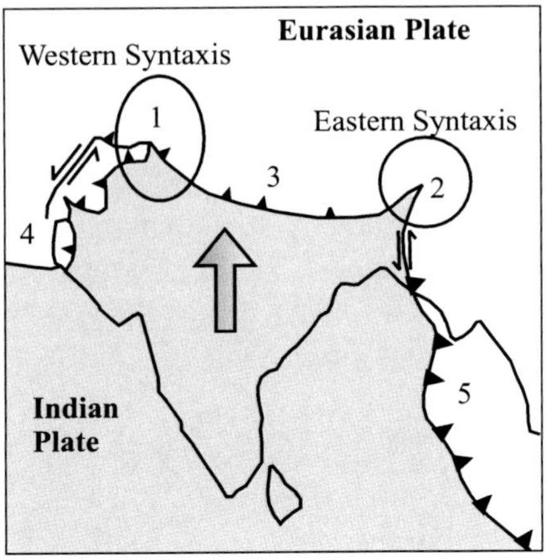

Thrust Fault
Subduction Zone
Direction of motion of Indian plate
Strike-slip fault

Fig. 5.5 Simplified tectonic map showing the collision zone between the Indian and the Eurasian plates. Large circles indicate the: (1) Western Syntaxis, (2) Eastern Syntaxis, (3) Himalayan arc, (4) Baluchistan arc, and (5) Arakan Yoma arc.

The western syntaxis is defined by the intersection of the Baluchistan arc with the Himalayan arc, in the Hindu Kush, and Pamir regions. Its center is approximated by the Nanga Parbat, and the Indus winds around the Nanga Parbat. Several mountain ranges radiate from the Pamir knot; some of the prominent ones are Karakoram, Kailash, Koh-i-baba, Kunlun, Altyn Tagh, Himalayas, Kirthar, Sulaiman, Tangla, and Nyenchentangla (in Tibet), Alinkangra, Mekran, Saman, Pir Panjal, and Ladakh. One of the branches stretches towards Tien Shan and Mongolia. The western syntaxis encompasses parts of Kashmir. The Kashmir earthquake of October 8, 2005, magnitude 7.6, had its epicenter within the western syntaxis. The eastern syntaxis is defined by the intersection of the Himalayan arc with the Arakan Yoma arc. It is in the Tibet-northeast India region. The Namcha Barwa approximates its center, and the Brahmaputra winds around this high mountain peak to enter the Indian plate.

The mountain ranges in the east and west of the Himalyan arc are different from each other. The eastern Himalayas are higher, narrower, about 150 km wide, the ranges are longer, are almost parallel to the Himalayan arc, and are more uniform in their composition compared with the western ranges. The western Himalayas are broader, about 400 km wide.

Mega Thrusts

Several distinct parallel features exist in the Himalayan tectonic zone. From north to south, these are the Tethys Himalaya (sometimes also referred to as Trans Himalayas), the Greater Himalaya, the Lesser Himalaya, and the Outer

Himalaya. Mega thrust sheets separate these: the Indus Tsangpo Suture Zone (ITSZ), Main Central Thrust (MCT), Main Boundary Thrust (MBT), and Frontal Foothill Thrust (FFT). These are shown in Table 5.2 and Figure 5.6. These are prone to frequent earthquakes and landslides.

The Tethys Himalayas involve a stratigraphic column from late Precambrian to Eocene. The Indus Tsangpo Suture Zone (ITSZ) is a major

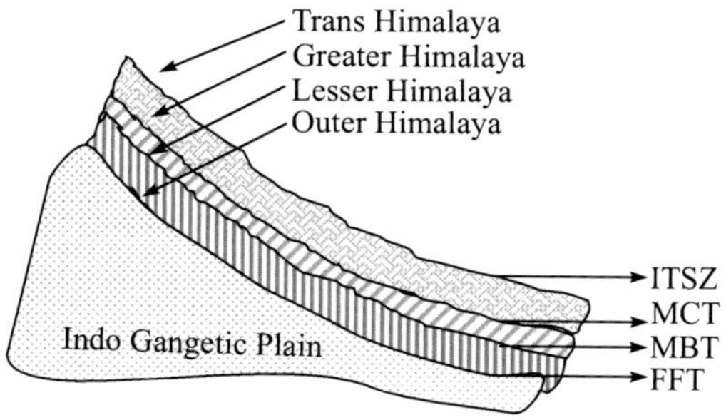

Fig. 5.6 Simplified version of the four subdivisions of the Himalayan tectonic zone, which are separated by mega faults, e.g., ITSZ—Indian Tsangpo Suture Zone, MCT—Main Central Thrust, MBT—Main Boundary Thrust, and FFT—Frontal Foothill Thrust.

Table 5.2 Column 1 shows mega faults along the Himalayan arc from north to south, column 2 shows the different names given to subdivisions, column 3 shows main characteristics of each division, and column 4 shows other features, e.g., mountain ranges. ITSZ = Indus Tsangpo Suture Zone, MCT = Main Central Thrust, MBT = Main Boundary Thrust, FFT = Frontal Foothill Thrust.

1 *Mega faults*	*2* *Nomenclature*	*3* *Characteristic*	*4* *Mountain ranges*
	Trans Himalaya or Tethys Himalaya	Tethys sediments Fossiliferous	Ladakh Harmosh Kailash Mahabharat Nag Tibba
ITSZ	-----------------	-----------------	-----------------
	Greater Himalaya or Higher Himalaya or Central Himalaya or Himadri	Fossils Metamorphic rocks	Karakoram Zanskar Kailash Patkai
MCT	-----------------	-----------------	-----------------
	Lesser Himalaya or Lower Himalaya	Metamorphic rocks Early Tertiary	Pir Panjal Dhaula Dhar Mahabharat Barail, Khasi Jaintia, Garo
MBT	-----------------	-----------------	-----------------
	Outer Himalaya or sub-Himalaya	Siwalik Miocene	Siwaliks
FFT	-----------------	-----------------	-----------------
	Indo Gangetic Plain Peninsular region		

suture zone and is characterized by oroganic sediments and large thrust sheets. Formed in late Cretaceous, it indicates the zone of initial collision of oceanic crust of the Indian plate with oceanic crust of the Eurasian plate. It separates the Tethys Himalayas from the Greater Himalayas.

The Greater Himalayas consist of thick crystalline thrust sheets. Several mountain ranges like the Karakoram, Zanskar, Kailash, the Great Himalayan Range in the west, and Patkai Mountains in eastern Himalayas are part of the Greater Himalayas. The highest mountain peaks are within this subdivision, e.g., Mt Everest (8848 m) in Nepal, Kanchenjunga in Sikkim, Nanga Parbat in Kashmir, Nanda Devi in Uttarakhand and Namcha Barwa in Tibet. Later, south and parallel to the ITSZ the MCT was formed, which vaguely defines the present collision zone. MCT is the boundary between the Greater and the Lesser Himalayas.

The Lesser Himalayas have huge sedimentary sections. These are of late Precambrian age and are covered by Gondwana-type rocks and by crystalline thrust sheets. The Pir Panjal Mountain ranges in Jammu and Kashmir, Dhauladhar in Jammu and Kashmir and Himachal Pradesh, Mahabharat in Nepal, and Barail, Khasi, Jaintia and Garo hills in Eastern Himalayas are all part of the Lesser Himalayas. Several popular hill stations like Dalhousie, Simla, Dharamsala, Mussoorie, Naini Tal and Darjeeling are located within these ranges. South of this is the Main Boundary Thrust (MBT), which is the boundary between the Lesser Himalayas and the Outer Himalayas.

The Outer Himalayas approximate the area between the MBT and the FFT. These consist of the youngest mountain ranges, the Siwaliks, which form the foothill of the Himalayas, have Tertiary sediments, and border the basins of Indo Gangetic plains. These are more prominent in western Himalaya and consist of a large amount of unconsolidated river deposits. These too are prone to earthquakes and landslides. The southern most mega Himalayan thrust is the FFT. It marks the southern boundary of the Himalayas, and south of these are the Indo Gangetic plains. These are shown in Figure 5.6.

Peninsular India

The peninsular region of India consists of continental crust. Geologically very old rocks, of Archaean and Proterozoic age, are exposed over more than half the Indian peninsular. The rest is covered by thick lava flows, which were extruded due to profuse volcanic activity in the Cretaceous–Eocene interval, i.e., these are about 65 million years old. These basalts constitute the Deccan traps. Peninsular India consists of elongated basins, which are filled with thick sediments. Volcanic rocks are found intercalated with sediments. Sedimentary deposits were subsequently deformed into folded mountains due to orogenic forces. During the Pre-Cambrian age, folding occurred in different stages. These correspond to Dharwar folding, Aravalli folding, Eastern Ghat folding,

Satpura folding, and Delhi folding. Dharwar folding is the oldest, followed by Aravalli, Eastern Ghat, Satpura, and Delhi orogeny.

The Indo Gangetic Plain

The Indo Gangetic plain is a depression that separates the Himalayan tectonic zone from the peninsular region. It is covered by thick alluvium, which may be as thick as 6 km in places.

TECTONIC EVOLUTION

Tectonic evolution of the continental crust in India occurred in six sequences, ranging in time from Proterozoic to Neogene. These are Neogene sequence, Palaeogene sequence, Mesozoic sequence, Gondwana sequence (Upper Carboniferous to Lower Cretaceous), Vindhyan sequence (Upper Proterozoic to Lower Paleozoic), and Cuddapah sequence (Proterozoic) (Krishnan, 1953, 1982; Mathur and Evans, 1964). Each sequence is limited by an unconformity in a wide geographical area. Several sedimentary basins and tectonic units subsequently developed within each sequence, and are classified into four groups on the basis of tectonics and area (Eremenko and Negi, 1968).

Structures of superorder represent subsided areas of more than 60,000 km^2 and are represented by the Deccan syneclise and the Vindhyan syneclise. Structures of first order represent areas between 6000 and 60,000 km^2 and include shelf, depression, graben and ridge elements. Examples of this are: the Bastar depression, Cambay graben, Chattisgarh depression, Cuddapah depression, East Uttar Pradesh shelf, Faizabad ridge, Gandak depression, Godavari graben, Indo-Ceylon graben, Laccadive-Kerala graben, Lahore-Delhi ridge, Mahanadi graben, Malwa ridge, Monghyr-Saharsa ridge, Narmada-Son-Damodar graben, Northern Shillong shelf, Punjab shelf, Rajasthan shelf, Sarda depression, Saurashtra-Kutch shelf, Southern Shillong Shelf, Upper Assam shelf, West Bengal shelf, and West Uttar Pradesh shelf. Bundelkhand massif, Shillong massif, and Mikir Hill massif form isolated outcrops of folded basement.

Structures of second order have an area less than 6000 km^2 and are located within the first-order structures. These include smaller depressions, ridges, arches, and hinges. Examples of this are the Ariyalur-Pondicherry depression, Banni depression, Bapatla ridge, Bhimavaram-Tanuku ridge, Bhubaneswar ridge, Broach depression, Cuttack depression, Delhi-Hardwar ridge, Devakkottai-Mannargudi ridge, East Godavari depression, Hinge zone (West Bengal shelf), Island Belt ridge, Krishna depression, Kumbakonam-Shiyali ridge, Mainland ridge, Mari-Jaisalmer arch, Nagapattinam depression, Puri depression, Ramnad-Palk strait depression, Sanchor depression, Shahgarh depression, Tarapur depression, Thanjavur depression, Tranquebar depression, Upper Assam depression, Wagad ridge, and West Godavari depression. Structures of third order represent local structures of limited extent like anticlines, noses, etc.

TECTONICS AND SEISMICITY

As the entire Indian plate continues to drag northward, the Himalayan edge, in a continent–continent collision zone, is affected the most, (Balakrishnan, 1997; Sinvhal, 1996). This causes Himalayan seismicity. A similar explanation exists for high seismicity in the Bay of Bengal. Preexisting faults may sometimes get reactivated or new faults may be formed due to the drag. The rest of the plate is also affected, but to a lesser extent. This causes intraplate seismicity in the form of scattered earthquakes away from the plate boundary, in the valleys of Krishna, Godavari, Narmada, Son, and Damodar Rivers. This gives rise to scattered earthquakes in the Indian peninsular, e.g., the Latur earthquake of 1993 and the Jabalpur earthquake of 1997 occurred in the Narmada–Sone–Damodar graben. Although it is now well-recognized that an association exists between earthquakes and tectonic units, yet the nature of this association is still obscure.

CONCLUSION

This chapter explained how the Indian plate evolved on the basis of the theory of plate tectonics and what it means in terms of thrusts, faults, fault zones, tectonic units and current seismicity. The next chapter will deal with seismicity of India and contiguous regions.

REFERENCES

Balakrishnan, T. S., 1997, *Major Tectonic Elements of the Indian Subcontinent and Contiguous Areas: A Geophysical View*, Memoir 38, Geological Society of India, 155 p.

Bilham, R., V. K. Gaur and P. Molnar, 2001, Himalayan seismic hazard, *Science*, **293**, p 1442–1444.

Eremenko, N. A. and B. S. Negi, 1968, *Tectonic Map of India, 1 : 2,000,000 scale, and Tectonic Guide*, Oil and Natural Gas Commission, Dehradun.

Evans, P., 1964, The tectonic framework of Assam, *Journal of the Geological Society of India*, **5**, p 4–34.

Krishnan, M. S., 1953, *The Structural and Tectonic History of India*, Memoir 81, Geological Society of India.

Krishnan, M. S., 1982, *Geology of India and Burma* (Sixth Edition), CBS Publishers and Distributors, Delhi, 536 p.

Mathur, L. P. and P. Evans, 1964, Oil in India—Special Brochure, in *Proceedings of the XXIIth International Geological Congress*, New Delhi, 86p.

Sinvhal, A., 1996, Evolution of Himalayas, in *Proceedings of the VIIth All India Meeting of Women in Science (IWSA)—-Role of Women in Science Society Interaction*, Roorkee, India, 125 p.

6 CHAPTER

Seismicity of India

INTRODUCTION

Earthquakes have claimed, and continue to claim, thousands of human lives. The larger and more frequent ones are associated with interplate environments. Most large and destructive earthquakes in India occur along and close to margins of the Indian plate. Regions of high seismicity can be identified as the Himalayan arc, with a dense concentration of epicenters in the eastern and western syntaxis. This trend continues along the trench systems in the Bay of Bengal, the Arakan Yoma and Andaman and Nicobar region, and in the Kutch region. The three arc systems, as shown in Figure 5.5, are as follows: the Himalayan arc, the Baluchistan arc, and the Arakan Yoma arc. Most epicenters are confined in these regions. Moderate-sized earthquakes and microearthquakes are even more frequent in these regions, but are less damaging than large earthquakes. Earthquakes occur in other parts of the country too but with reduced magnitude and frequency. Intraplate earthquakes are usually smaller and occur less frequently, like in the Indo Gangetic plains and in peninsular India.

GREAT EARTHQUAKES IN INDIA

Seven great earthquakes have devastated the Indian subcontinent in the last two centuries, together with numerous other events. An earthquake of magnitude 8+ is catastrophic in a very large area and is referred to as a great earthquake. These are plotted in Figure 6.1, and are listed in Table 6.1 with their salient features. These great earthquakes originated at or near boundaries of the Indian plate and caused immense destruction of life and property in large geographical areas. These were the great earthquakes of Kutch in the year 1819, Assam in 1897 and again in 1950, Kangra in 1905,

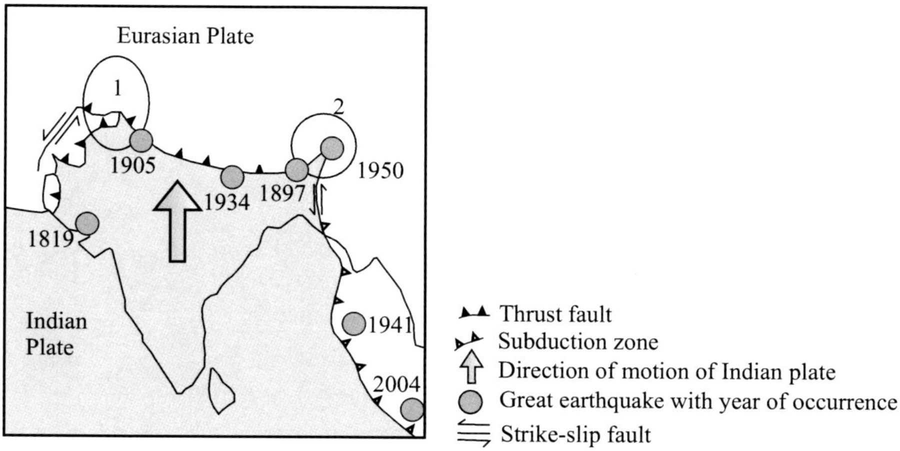

Fig. 6.1 Epicenters of seven great earthquakes and simplified boundary of the Indian plate. Western syntaxis is shown by 1 and eastern syntaxis by 2.

Nepal–Bihar region in 1934, and in the Bay of Bengal in 1941 and again in 2004. Meizo seismal area of these earthquakes is shown within the Himalayan tectonic zone in Figure 6.2.

A single great earthquake not only covers almost all damaging effects that can occur in any earthquake, which includes damage to ground, the built environment, and the human tragedy, but is also the place where damaging earthquakes occur later also. Due to the Kutch earthquake in 1819, an 80-km long fault was formed on the surface, and damaging effects produced by this earthquake are given in Chapter 4, on faults.

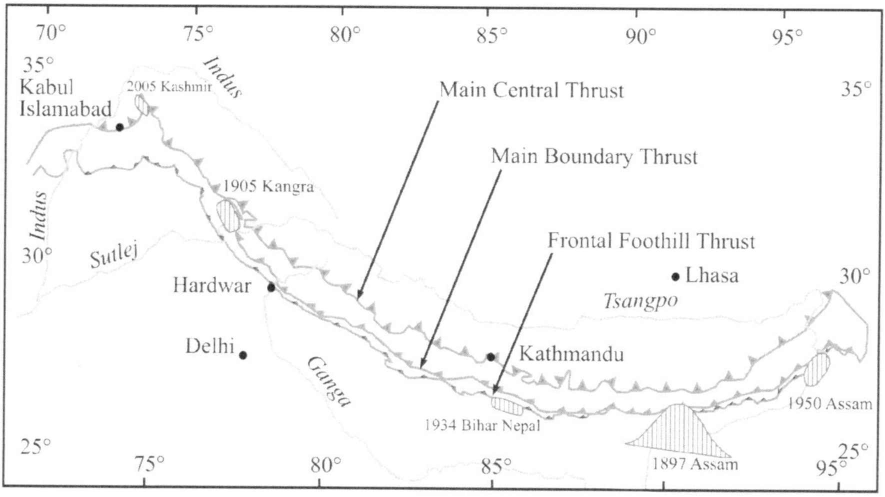

Fig. 6.2 Meizoseismal area of four great Indian earthquakes, and the Kashmir earthquake of 2005, plotted on map showing mega faults within the Himalayan tectonic zone.

Table 6.1 Earthquake parameters, maximum intensity, casualties, and places severely affected for seven great Indian earthquakes. In a large earthquake the rupture length may sometimes be as large as 250 km. Thus, the position of focus and epicenter becomes uncertain. Even with this uncertainty, an epicentral location is useful as it gives a broad picture of seismicity.

S. No.	*Earth quake*	*Date D M Y*	*OT (IST)*	*Place*	*Epicenter*		*Magnitude*	*Max Intensity*	*Depth of Focus (km)*	*Casual-ties*	*Places severely affected*	*Reference*
					Lat (°N)	*Long (°E)*						
1	Kutch	16.06.1819	Evening ~6.50 pm	Kutch	24.1 23.5	69.1 69.5	> 8	XI MMI	Not known	>10,500	Kutch	Oldham (1928)
2	Assam	12.06.1897	5.15 pm	Shillong	26.0	91.0	8.7	XII MMI	8	>1542	Shillong, Goalpara, Guwahati, Nowgong, Sylhet, Tura, Dhubri, Kuch-Bihar, Nalbari	Oldham (1899)
3	Kangra	04.04.1905	06.20 am	Kangra Dharamsala	32.5	76.5	8.6/8.4	X RF	33	>19,000	Kangra, Kulu, Dharamsala, Palampur, Mussoorie, Dehra Dun, Lahore	Middlemiss (1910)
4	Bihar-Nepal	15.01.1934	Afternoon 14 h 21min 18 sec 2.21.18 pm	Motihari-Madhubani Kathmandu Monghy	26.50	86.50	8.4/8.3	X MI	Not known	>12,000	Motihari, Madhubani, Bhatgaon, Darjeeling, Kathmandu, Monghyr, Muzaffarpur, Patna, Purnea, Sitamarhi	Auden (1939)
5	North Andaman	26.06.1941		Middle Andaman	12.50	92.50	8.1 M_W 7.7 M_o 4.27x 10^{30} Nm	VIII^+ MMI	60		Middle Andaman, S. Andaman, Baratang	IMD
6	Assam	15.08.1950	19 h 39.5 min evening 7.39.30 pm	Rima on India-Tibet border	28.0 28.5	95.8 96.5	8.7	XII MMI	15-25 14	>1526	Lakhimpur, Sadiya, Sibsagar, Dibrugarh, Jorhat	Banerji 1953 Pramanik and Mukherjee 1953
7	Sumatra	26.12.2004	06:29 am	Indian Ocean	3.33 3.27	96.13 95.82	M_s 8.6 M_s 9.0 M_w 9.3		- 30	~2,30,000	Rim of 12 countries in the Indian Ocean	IMD USGS

Note: D—Date; M—Month; Y—Year; OT—Origin Time; IST—Indian Standard Time; Lat—Latitude; Long—Longitude; M—Magnitude; MI—Mercalli Intensity Scale; MMI—Modified Mercalli Intensity Scale; RF—Rossi Forel Scale.

THE ASSAM EARTHQUAKE OF 12TH JUNE 1897

The epicenter of this great earthquake was in Shillong, in present-day Meghalaya. It occurred in the evening, when most people were awake, many were outdoors, and most had a chance to escape to safety. It was assigned magnitude 8.7 by Gutenberg (1956), and it was placed among the largest known earthquakes in the world at that time. For such a great earthquake, casualty figures were mercifully low, 1542. Oldham (1899) authored a valuable scientific memoir on this earthquake, believed to be the first book ever written totally devoted to a single earthquake.

This earthquake was destructive in a very large area, of approximate radius 500 km, and devastated the Shillong plateau and the Assam hills. As a consequence of this earthquake, the Chedrang fault and Samin faults were exposed on the surface, details of which are given in Chapter 4. Ground fissures were numerous and large, and earthquake fountains emerged from some of these. Several new ponds and waterfalls were formed, indicating widespread alteration of the drainage system. Gigantic landslides denuded the Assam hills. This blocked the Brahmaputra River in several places and caused floods all along the Brahmaputra River. Flooding was maximized around Shillong, Garo hills, Goalpara, and Sylhet. Borpeta (26° 20° N, 91′ 03′ E) was the worst affected region. More details are given in Chapter 9.

Topographic distortions occurred in the form of shifting of hills on either side of the Brahmaputra River, where it took a sharp turn, near Tura. Before the earthquake, it was possible to regularly exchange heliograph signals between Rowamari and Tura from a certain spot by a ray over an intervening hill. After the earthquake, instead of Rowamari being just visible over the hilltops, a broad stretch of plains east of the Brahmaputra was visible. Resurveys after the earthquake confirmed extensive change. Since all monuments were in the disturbed area, all reference points were disturbed and details of warping was not derivable. Resurveys after the earthquake indicated a change in height of hills.

Oldham designed a seven-point intensity scale to map the extensive damage caused by this earthquake. This scale is given in Appendix II. Later, Richter (958) modified and extended the then popular ten-point Mercalli scale at the higher end to a twelve-point scale, to account for the immense devastation caused by this earthquake. Therefore, on the Modified Mercalli Intensity Scale, this earthquake was later assigned the highest intensity possible, i.e., MMI XII.

The built environment, including masonry houses, in several major towns like Goalpara, Guwahati, Nowgong, Sylhet, and Shillong, and all villages around these was devastated. Several boulders lifted out of the ground vertically upward, without cutting edges of their former seats, indicating vertical accelerations exceeding that of gravity. Eyewitness reports describe pebbles bouncing on the ground 'like peas on a drumhead'. Oldham estimated peak ground accelerations at several places in the meizoseismal area; these are given in Table 6.2.

Table 6.2 Peak ground acceleration for different places as estimated by Oldham (1899).

S No.	*Place*	*Maximum horizontal acceleration (mm/sec^2)*
1.	Cherrapunji	3,000
2.	Dhubri	2,700
3.	Guwahati	2,600
4.	Shillong	4,200
5.	Silchar	1,200
6.	Sylhet	4,200

Aftershocks occurred in a large area with approximate dimensions 300 × 80 km. An earthquake generally does not occur as a single event but comes as a series of events. The event with the largest magnitude in this series is called the main event. A large earthquake occurs due to fracture of rocks under strain. The strained blocks eventually regain equilibrium. Events preceding the main shock in a restricted volume are called foreshocks, and events that occur after the main shock are called after shocks. Aftershocks indicate that readjustment of equilibrium is not completed at once. Usually, the number of aftershocks is much larger than the number of foreshocks. It has been observed that large earthquakes are usually preceded by foreshocks and are followed by numerous aftershocks. In a large earthquake, aftershock activity may continue for weeks or even months. The frequency of occurrence and magnitude of aftershocks generally reduce with time.

Paleo seismological studies based on carbon–14 dating revealed liquefaction and deformed features in trenches, indicating three earlier events of magnitude comparable to the 1897 event. This gives the return period of great earthquakes along the Chedrang fault in Shillong plateau, responsible for the 1897 earthquake, as 400–600 years (Sukhija et al, 1999).

THE KANGRA EARTHQUAKE OF APRIL 4, 1905

This early-morning earthquake had twin epicenters 160 km apart, in the Kangra–Kulu region (32.5° N, 76.5° E) and in the Dehradun–Mussoorie region. It is one of the earliest great Indian earthquakes for which instrumental magnitude, 8.6, is available. This earthquake witnessed a very large casualty figure of more than 19,000. Kangra, Dharamsala, and Palampur were devastated, and Mussoorie, Dehradun, and Lahore were severely affected. The earthquake was felt in large parts of northern India; in Quetta and Sind in the west, in the south in the Tapti valley and in the east the Ganga delta. Middlemiss (1910) documented effects of this great earthquake in another Geological Survey of India memoir.

A complex network of faults such as the Main Central Thrust, the Main Boundary Thrust, and the Krol Thrust exists in this region. Extensive ground damage was reported like faulting and fractures. Height and level of stations and hilltops was altered. Dehradun and Siwalik hills showed a rise of 30 cm relative to Mussoorie, indicating topographic changes across the Main Boundary Thrust. Numerous landslides and rock falls were spread in a very wide area. The earthquake altered the drainage system, and disturbed springs, streams, and canals. Maximum intensity assigned to this earthquake was X on the Rossi Forel scale, which is the top of this scale. Several hundred aftershocks continued for months after the main event. This earthquake occurred due to slip at two points, in a fault parallel to the Main Boundary Thrust, at the foothill of the Himalayas. The linear extent of fault was large as is evident by the two epicenters 160 km apart.

THE BIHAR–NEPAL EARTHQUAKE OF JANUARY 15, 1934

This earthquake occurred in the winter afternoon, when most people were awake and outdoors. The instrumentally determined epicenter, 26° 50′ N, 86° 50′ E, was near the eastern edge of the meizoseismal area. Instrumental magnitude, 8.4 (M_w = 7.9, m_b = 7.8, M_s = 8.3...) was assigned to this earthquake. For the main event, most recording instruments were thrown out of action. However, it was recorded at distant stations such as Pasadena, Leningrad, and Tokyo. Auden et. al. (1939) documented this earthquake in another monumental memoir brought out by the Geological Survey of India. The earthquake claimed more than 12,000 lives, 3400 of which were in Nepal and 1260 in Monghyr alone.

Ground effects such as slumping, liquefaction, fissures, alteration of levels, landslides, fault scarps, surface distortions, and disturbance of drainage were reported in a wide geographical region. This earthquake caused soil liquefaction effects in a very large area, which was named as the slump belt. Seismic effects in the slump belt are given in Chapter 9 on ground damage. Landslides occurred in the north of the epicenter, along the Kosi River and in hilly regions of Nepal and Darjeeling. In Nepal, these were confined to highly weathered and metamorphosed rocks (pegmatite, gneiss, and schist of Mahabharat range). In Dharan Dhankutta (26° 59′, 87° 21′), the earthquake claimed more than 80 human lives. Near Muksar (26° 52′, 86° 23′), it blocked the nala in four places to form lakes. Two of these lakes emptied after few weeks.

This earthquake had three separate regions where intensity was X, the highest on the Mercalli Intensity Scale. The largest of these regions was about

80 miles long and 20 miles wide (128 × 32 km) and consisted of parts of the districts of Darbhanga (defined by the Motihari-Madhubani, Rajnagar, Mirzapur region) and Muzaffarpur (Riga, Sitamarhi). Two other spots, almost 100 miles (160 km) on opposite sides of this east–west trending slump belt, were centered at Monghyr, south of the Ganga, and Kathmandu in the north. In all these places, brick masonry houses were ruined and collapsed.

Intensity IX included an area that was about 190 miles (304 km) long and was of irregular width that exceeded 40 miles (64 km) at places. It included districts Saan (Gopalganj, Chapra), Champaran (Motihari, Sagauli, Kesariya), Muzaffarpur (Muzaffarpur, Sitamarhi, Musahari, Sheohar, Belsand, Motipur, Riga,), Darbhanga (Darbhanga, Lakheria Sarai, Sakri, Lohat, Pandaul, Pusa, Bahera, Berhampur, Jaynagar, Supaul, Madhepura, Pratapganj, Murliganj), Purnea (Purnea, Forbesganj, Jogbani, Raniganj, Champanagar), and Patna (South of Ganga, Patna, Barh, Mokameh, Monghyr, Jamalpur, Dharhara).

The earthquake-affected area in Bihar is endowed with thick alluvium of the Ganga basin. Ridges are exposed through this alluvium in several places. Monghyr is located on thick alluvium and a ridge of Archaean quartzite emerges through it. When a ridge emerges through thick alluvium, the former resists severe shaking and heavy damage is confined to the surrounding unconsolidated sediments and alluvium of geologically recent age. Unconsolidated soil absorbs seismic energy and is prone to slumping, liquefaction, compaction, subsidence, ground fissures, and causes sinking of heavy structures. Therefore different parts of Monghyr suffered different kinds of damage. A similar explanation is applicable to Nepal valley, which includes Kathmandu, that rests on weathered metamorphic rocks.

THE ANDAMAN EARTHQUAKE OF JUNE 26, 1941

This great earthquake originated in the west of Middle Andaman Island (12.50°N, 92.50°E), and was assigned magnitude M = 8.1, $M_w = 7.7$, IMD, $M_0 = 4.25 \times 10^{30}$ Nm. This thrust type convergent margin event caused extensive damage, VIII$^+$ on MMI scale, to masonry buildings in Middle Andaman, South Andaman and Baratung Islands. This tsunami genic earthquake flooded and damaged masonry structures in Port Blair and on the east coast of mainland India.

THE ASSAM EARTHQUAKE OF AUGUST 15, 1950

This Independence Day great earthquake had its epicenter in the Dibang valley. Poddar (1953) assigned it epicenter, 96°E, 28°40′ N, based on instrumental data recorded in India, Pasadena, and Strasbourg. According to other workers, the epicenter was outside the Indian border, at Rima, which is on the northeastern border of India, with China and Tibet, 200 miles north of

Sadiya. The epicentral region lies in the eastern syntaxis and includes the Mishmi and Lohit thrust zone. The affected region was sparsely populated and the earthquake occurred in the evening at 19 h 39.5 m IST, when most people were awake. Fifteen hundred and twenty six (1526) people lost their lives, of these 952 were in Mishmi and Abor hills, 80% of the rest were along the Subansiri River.

This earthquake was more devastating than the 1897 great Assam earthquake. It had profound effects around the valleys of Brahmaputra River and its many tributaries and in upper Assam, which is now Arunachal Pradesh. Approximately 15,000 sq miles of area, which included Lakhimpur, Sibsagar, Dibrugarh, Jorhat, and Sadiya districts, suffered extensive damage to life and property. All possible damaging ground effects associated with great earthquakes were reported for this earthquake. This included topographic changes, formation of huge fissures that gave rise to sand and water fountains and oozes; subsidence and elevation of ground, gigantic landslides along the Brahmaputra and its many tributaries, damming of rivers and postearthquake floods. Drainage pattern was altered along the Brahmaputra and its many tributaries such as Burhi Dihing, Dibang, Dihang, Lohit, Subansiri, Tidding, and others.

Several isoseismal maps published for this great earthquake showed that all isoseismals were elongated along the Brahmaputra valley. Ray (1953) used the MMI scale and based his map on media reports published between August 15 and September 3, 1950. Poddar (1953) and Tandon (1953) used the Rossi Forel (RF) scale. Dibrugarh showed maximum damage of grade XII on the MMI and X on the RF scale, i.e., top of whichever scale was used. Isoseismal VIII was approximately 75,000 sq miles in area. The entire built environment including buildings, roads, bridges, and telephone lines collapsed. Pramanik and Mukherjee (1953) estimated epicentral accelerations to be 0.2 g for rock formations, 0.4 g for alluvium, and 1.0 g for top of very tall structures.

This earthquake had numerous aftershocks between longitude 90° E and 97° E, most of which were well recorded instrumentally, had good epicentral locations, and for some magnitude exceeded 6.0. Four aftershocks of magnitude 7.0 occurred within 1 month of the earthquake. The probable cause of this earthquake is attributed to the complex fault-system and tectonics in the eastern syntaxis, and rupture on a NE–SW trending, 150-km long fault.

THE SUMATRA EARTHQUAKE OF DECEMBER 26, 2004

No great earthquake occurred in India or its plate margins after 1950. Then the great Sumatra earthquake of December 26, 2004 brought with it the tsunami, which caused unprecedented human casualties on the rim of the Indian Ocean. For more on this earthquake, see Chapter 10 on tsunami.

FUTURE IMPLICATIONS

The four great earthquakes in the Himalayas are not evenly distributed on the 2500-km long Himalayan arc. Probability of occurrence of an earthquake is large in the several seismic gaps that exist between the epicenters of great earthquakes, as shown in Fig. 6.3. The gap between the great earthquakes of Kangra (1905) and Bihar (1934) is approximately 1000 km long; between Bihar (1934) and Assam (1897) earthquakes it is 400 km; between the two earthquakes in Assam (1897 and 1950) it is approximately 600 km long; between Arunachal Pradesh (1950) and Andaman (1941) it is almost 1500 km; and (e) between Andaman (1941) and Sumatra (2004) it is approximately 1000 km. These are the places where the epicenter of a future great earthquake cannot be ruled out, together with the eastern and western syntaxis.

Four great earthquakes occurred within the Himalayan arc between 1897 and 1950, that is in a time span of 53 years. There has been no great earthquake in this arc after 1950. If a great earthquake were to originate now in the Himalayan tectonic zone, it will be more devastating than any of the earthquakes discussed in this chapter. Not only this it will also have a larger geographical spread of destructive influence. Such an earthquake will cover an area that will be defined by an arc parallel to the Himalayan arc, of at least a width of 400 km, and would include a large part of the Indo Gangetic and Brahmaputra basins, i.e., sedimentary basins with soft sediments, on which several states and major cities are founded. Jammu and Kashmir, Himachal Pradesh, Punjab, Haryana, Uttar Pradesh, Uttarakhand, Bihar, West Bengal, all the seven states of North East India and large portions of Pakistan, Nepal, and Bangladesh lie within this arc. This region supports more than half the population of the country. Population has trebled since the last great Himalayan earthquake occurred in 1950. Moreover, several large and densely populated urban centers exist and new ones are coming up within this threatened area, even within the meizo seismal area of the great Himalayan earthquakes, most of which, at best, are only partially planned. This has increased seismic risk several folds.

Therefore, the human habitat in the Himalayan tectonic zone and the Indo Gangetic plains require immediate preparedness to meet an imminent disaster from a great earthquake. The dismal seismic performance of stone houses is well known and is dealt in Chapter 11. Chapter 12 deals with seismic response of tall buildings at large epicentral distances and the several factors that come into play, especially site effects. Examples in that chapter are taken for Ahmedabad and for Surat, which were 250 and 350 km away from the epicenter of the Kutch earthquake of 2001, magnitude 6.9. Similarly lifelines and infrastructure also stand threatened, and their seismic response is given in Chapter 13. The destructive reach of the Kutch earthquake of 2001 extended beyond 300 km. By extrapolation, for a great earthquake this is expected to be much larger, say 400 km. This is shown in Fig. 6.3.

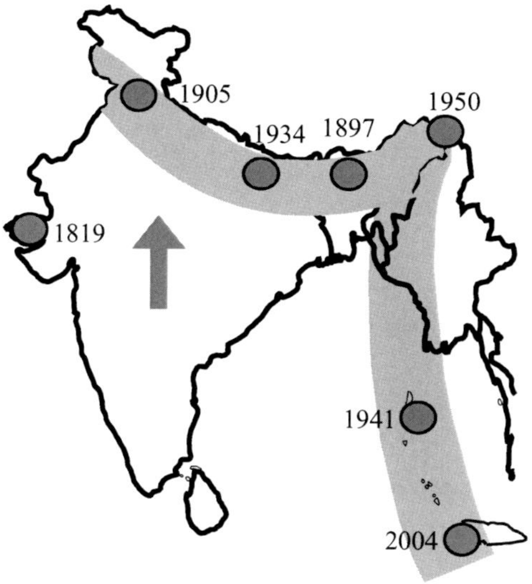

Fig. 6.3 Epicenters of the seven great earthquakes that caused extensive damage in India in the last 200 years are shown on map of India. Areas at higher seismic risk are shown within the shaded arc. This includes several densely populated cities.

OTHER IMPORTANT DISASTROUS EARTHQUAKES IN INDIA

Earthquakes are repeated frequently in an area where a great earthquake occurred earlier, even though the magnitude may be smaller. This is evident from several earthquake catalogues, e.g., Oldham (1870), Tandon (1974) and Bapat (1983). The Kutch earthquake of 2001 and the Anjar earthquake of 1956 occurred in the meizoseismal area of the great Kutch earthquake of 1819. Similarly the Dharamsala earthquake of 1986 occurred within the meizoseismal area of the great Kangra earthquake of 1905, and the Bihar earthquake of 1988 occurred within the meizoseismal area of the great Bihar–Nepal earthquake of 1934. In the case of the two Assam earthquakes, and in the Andaman Nicobar region the list of subsequent damaging earthquakes is very long.

Damaging earthquakes have occurred in other parts of the country too; notable among these were the earthquakes of Quetta, (1935), Koyna (1967), Uttarkashi (1991), Latur (1993), Jabalpur (1997), Chamoli (1999), Kutch (2001), and Kashmir (2005). These earthquakes caused heavy damage in densely populated areas and took a heavy death toll. In addition to these, there have been hundreds of other damaging earthquakes. The filling of the

Table 6.3 List of some damaging earthquakes in India. Great earthquakes for almost the same period are listed in Table 6.1.

S. No.	*Earth-quake*	*Date (D M Y)*	*Origin Time IST H M S*	*Epicenter*		*Magni-tude*	*Max Intensity*	*Depth (km)*	*Casual-ties*	*Places Affected*	*References*
				Lat (°N)	*Long (°E)*						
1.	Delhi	15.07.1720				6.5					Tandon (1974)
2.	Bellary	01.04.1843				5.5–6				Felt in Deccan Plateau	Tandon (1974)
3.	Kashmir	30.05.1885		34.6	74.38	7			6000		Tandon (1974)
4.	Coiambatore	08.02.1900		10.7	76.7	6.0	VII MMI				Basu (1964)
5.	Calcutta	29.09.1906		21	88		VII / VI MMI				Middlemiss (1907)
6.	Baluchistan	21.10.1909		30	68	7.2	IX MMI				Haron (1912)
7.	Srimangal	08.07.1918	10 22 07	24.5	91	7.6	X MMI			Tea Estates Ruined	Stuar (1920)
8.	N W Himalaya	01.02.1929		36.5	70.5	7.1	IX MMI				Mukherjee (1950)
9.	Dhubri	02.07.1930	21 03 34	25.8	90.2	7.1	IX MMI			Gauhati, Rangpur, Berhampur, Mymensingh	Gee (1934)
10.	Sharigh	25.08.1931		30.2	67.7	7.0	VIII MMI				West (1934)
11.	Mach	27.08.1931		29.8	67.3	7.4	VIII MMI				
12.	Great Pamir	21.11.1934		36.5	70.5	6.9	IX MMI				West (1934)
13.	Quetta (Baluchistan)	31.05.1935		29.5 30.2	66.8 67.03	7.5 7.6	X		> 50,000		West (1934)
14.	Hindu kush	14.11.1937	21 37 22	35.1	78.1	6	IX				Coulson (1938)
15.	Pamir	21.11.1939				6.9					Coulson (1940)
16.	Anjar	21.07.1956	15 32 26	23.3	70.0	7.0	VIII MMI				Tandon (1959)
17.	Delhi	27.08.1960	15 58 59	28.2	77.4	6.0	VII MMI				Mukti Nath (1969)
18.	Calcutta	15.04.1964		21.5	88.1	5.5	VII MMI				Jhingran (1969)
19.	Ongole, A.P.,	27.03.1967		17.7	73.9	5.0	VI		8		

(Contd.)

Table 6.3 (Contd.)

S. No.	Earth-quake	Date (D M Y)	Origin Time IST H M S	Epicenter		Magni-tude	Max Intensity	Depth (km)	Casual-ties	Places Affected	References
				Lat (°N)	Long (°E)						
20.	Koyna	10.12.1967	22 51 24	17.4	73.9	6.5	IX MMI	15-20			Chatterjee, et al., (1969)
21.	Bhadrachalam	13.04.1969	15 34 55	17.9	80.6	5.7 6.5 (NDI)	VII MMI				Mukherjee, (1971)
22.	Broach	23.03.1970		21.7	73	5.8 6.0 5.4	VII MMI		30		Chaudhury et al., (1970)
23.	Kinnaur	19.01.1975		32.5	78.4	6.8 Ms	IX MMI				Singh et al., (1977)
24.	Shimoga	12.05.1975		13.8	75.1	5.1	V MMI				Gosavi et al., (1977)
25.	Roorkee	06.11.1975	05 41 30	29.5	78.1	4.7	VI MMI				Arya et al., (1977)
26.	Dharamsala	04.1986		32.75	76.86	5.5					Arya et al., (1986)
27.	Assam	06.08.1988									IMD
28.	Bihar	21.08.1988	Predawn 04.54	26.5	86.5	6.6	IX		282 India 704 Nepal		IMD
29.	Uttarkashi	20.10.1991	02 53 15	30.75	78.86	7.0 Ms 6.5 mb	IX + MMI	12 km	>770	Uttarkashi, Chamoli, Tehri (Uttarakhand)	USGS PDE, (1991)
30.	Latur	30.09.1993	03.56	76°35' 76.5	18°03' 18.1	6.4 Ms 6.3 mb 6.1 Mw	VIII + MMI	8 5	>10,000	Latur, Osmanabad Maharashtra	Sinvhal et al., (1994a,) USGS
31.	Jabalpur	22.05.1997		23.18	80.02 5.7 Ms 5.8 Mw	6.1 mb	VIII	32 NEIC	37	Kosamghat	Rai et al., (1997)

(Contd.)

Table 6.3 (Contd.)

S. No.	Earthquake	Date (D M Y)	Origin Time IST H M S	Epicenter		Magnitude	Max Intensity	Depth (km)	Casualties	Places Affected	References
				Lat (°N)	Long (°E)						
32.	Chamoli	23.03.1999	00 35 11	30.4	79.42	6.8	VIII MSK	20	100	Chamoli Distt. Rudraprayag Distt.	IMD
33.	Kutch	26.01.2001	08 46 39.32	23.6 23.36 23.4	69.8 70.34 70.27	M_L 6.9 M 7.7 M_S 7.6 M_W 7.6	X	15 22	> 10,000	Kutch, Surat Ahmedabad	IMD USGS Sinvhal, et al., 2003
34.	Kashmir	08.10.2005	09 20 38	34.402	73.560	7.6 m_b 7.3 M_s 6.8	XI+	10	> 86,000	Baramulla Distt.	IMD USGS GSI Sinvhal, et. al., (2005)

D—date; M—month; Y—year; IST—Indian Standard Time; MI—Mercalli Intensity Scale; MMI—Modified Mercalli Intensity Scale.

reservoir of the Koyna dam triggered the most destructive earthquake in peninsular India, with epicentral intensity MMI IX. Isoseismals were elongated in a N – S direction, probably indicating the trend of a subsurface fault. Peninsular India experienced several intraplate earthquakes of magnitude almost 6 on the Richter scale (Rao and Rao, 1984). This also brings home the point that no place in India can be considered free from earthquakes; it is only a variation of risk which may be greater in one region than in another. A list of some important damaging earthquakes is given in Table 6.3.

CONCLUSION

The larger, more frequent, and destructive earthquakes in India are associated with interplate environments. The three arc systems are the Himalayan arc, the Baluchistan, arc and the Arakan Yoma arc, which confine most epicenters in these regions. The Himalayan arc, together with the eastern and western syntaxis has a dense concentration of epicenters. This trend continues along the trench systems in the Bay of Bengal, and in the Andaman and Nicobar region. Moderate sized and micro earthquakes are even more frequent in all these regions. Intra plate earthquakes are usually smaller and occur less frequently, like in the Indo Gangetic Plains and in peninsular India, but can sometimes be equally devastating. This has led to the formulation of a widely understood disaster mitigation strategy, which starts with the seismic zoning map of the country. Chapter 8 deals with this aspect, but before we go to that we will take a small digression to understand what is magnitude and intensity of an earthquake, in the next chapter.

REFERENCES

Arya, A. S., S. Singh, H. Sinvhal, R. Prakash, P. N. Agrawal, K. N. Khattri, B. Prakash and A. Sinvhal, 1977, A macro seismic study of November 6, 1975 Roorkee earthquake, Roorkee, India, in *Proceedings of the Sixth World Conference on Earthquake Engineering (6WCEE)*, New Delhi, Volume 1, p 255–261.

Arya, A. S., B. V. K. Lavania, S. P. Gupta and A. Kumar, 1986, Dharamsala earthquake of 26 April 1986, in *Proceedings of the 8th Symposium on Earthquake Engineering*, p 73–91.

Auden, J. B., J. A. Dunn, A. M. N. Ghosh, D. N. Wadia and S. C. Roy, 1939, *The Bihar–Nepal Earthquake of 1934*, Memoirs of GSI, Volume 73, 391 p.

Banerji, S. K., 1953, The origin and the nature of the disturbance produced by the Assam Earthquake of August 15, 1950 and its aftershocks, in *A Compilation of Papers on the Assam Earthquake of August 15, 1950*,

Ed. M. B. R Rao, Pub No. 1, The Central Board of Geophysics, Calcutta, Government of India, p 11–15.

Basu, K. L. 1964, A note on the Coiambatore earthquake of 8 February 1900, *Indian Journal of Meteorology and Geophysics*, **15**, p 281–286.

Chatterjee, G. C. *et al.*, 1969, Geological report on the Koyna earthquake of 11th December, 1967, Satara District, Maharashtra, GSI open file Report No. 5, GSI.

Chaudhary, H. M., S. N. Bhattacharya and S. R. Basu, 1970, Recent earthquake activity in India, in *Proceedings of the Fourth Symposium on Earthquake Engineering*, November 14–16, 1970, Roorkee, p 382–388.

Coulson, A. L., 1938, The Hindu Kush earthquake of the 14th November 1937, *Rec GSI*, **73**, p 135–144.

Coulson, A. L., 1940, An earthquake in the Great Pamir, *Rec. GSI*, **75**, Professional Paper No. 12.

Gee, E. R., 1934, *Dhubri Earthquake of 3rd July, 1930*, Memoirs of GSI, Volume 65, Part 1.

Gosavi, P. D., A. V. Bapat and S. K. Gupta, 1977, Macroseismic studies of four Indian earthquakes, in *Proceedings of the Sixth World Conference in Earthquake Enginering*, January 10–14, 1977, p 49–54.

Gutenberg, B., 1956, Great earthquakes 1896–1903, *Transactions of American Geopysical Union*, **37**, p 608–614.

Haron, A. M., 1912, The Baluchistan earthquake of 21st October 1909, *Rec. GSI*, **41**, Part 1.

Jhingran, A. G., C. Karunakaran and J.G. Krishna Murthy, 1969, The Calcutta earthquakes of 15th April and 9th June, 1964, *Rec. GSI*, **97**, Part 2, p 1–29.

Middlemiss, C. S., 1907, Two Calcutta earthquakes of 1906, *Rec GSI*, **36**, Part 3, p 214–232.

Middlemiss, C. S., 1910, *The Kangra Earthquake of 4th April 1905*, Memoirs of Geological Survey of India, Volume 38, 409 p.

Mukherjee, S. M., 1950, Remarks on two Hindu Kush earthquake shocks, *Indian Journal of Meteorology and Geophysics*, **1**, p 297–302.

Mukherjee, S. M., 1971, On two recent earthquakes in Deccan, *Indian Journal of Meteorology and Geophysics*, **22**, 589–594.

Narula, P. L., S. K. Shome, S. Kumar and P. Pande, 1995, Damage patterns and delineation of isoseismals of Uttarkashi earthquake of 20th October 1991, in *Uttarkashi Earthquake*, Eds. H. K. Gupta and G. D. Gupta, Memoir 30, Geological Society of India, 233 p.

Nath, M. *et. al.*, 1969, Delhi earthquake of 27th August, 1960, *Rec GSI*, **95**, Part 2.

Oldham, R. D., 1870, A Catalogue of Indian Earthquakes, from the earliest times to the end of 1869 A D, in *Memoirs of Geological Survey of India*, Geological Society of India, 63 p.

Oldham, R. D., 1884, Note on the earthquake of 31 December 1881, *Rec. GSI*, **XVII**(2), p 47–53.

Oldham, R. D., 1899, Report on the Great Earthquake of 12th June 1897, in *Memoirs Geological Survey of India*, Volume 29, 379 p.

Oldham R. D., 1928, The Cutch (Kachh) earthquake of 16th June 1819 with revision of the great earthquake of 12th June 1897, in *Memoirs Geological Survey of India*, Volume 46, p 71–147.

Poddar, M. C., 1953, A short note on the Assam earthquake of Aug 15, 1950, p 38–42, in *A Compilation of Papers on the Assam Earthquake of August 15, 1950*, Ed. M. B. R. Rao, Publication No. 1, The Central Board of Geophysics, Govt. of India, Calcutta.

Pramanik, S. K., and S. M. Mukherjee, 1953, The Assam Earthquake of 1950, in *A Compilation of Papers on the Assam Earthquake of August 15*, Ed. M. B. R. Rao, Publication No. 1, p 26–34, The Central Board of Geophysics, Govt. of India, Calcutta.

Rai, D. C. and C. V. R. Murty, 2003, Reconnaissance report, North Andaman (Diglipur) earthquake of 14 September 2002, 38 p.

Rao, B. R., and P. S. Rao, 1984, Historical seismicity of peninsular India, *BSSA*, **74**(6), p 2519–2533.

Ray, S., 1953, Isoseismals for the great Assam earthquake of August 15, 1950, in *A compilation of papers on the Assam earthquake of August 15, 1950*, Ed. M. B. R. Rao, Publication No. 1, p 35–37, The Central Board of Geophysics, Govt. of India, Calcutta.

Richter, C. F., 1958, *Elementary Seismology*, W. H. Freeman and Co., San Francisco, 768 p.

Singh, S., A. K. Jain, V. N. Singh and L. S. Srivastava, 1977, Damage during Kinnaur earthquake of January 19, 1975 in Himachal Pradesh, India, in *Proceedings of the Sixth World Conference in Earthquake Engineering*, Delhi.

Sinvhal, A., P. R. Bose and R. N. Dubey, 1994, Damage report for the Latur Osmanabad earthquake of September 30, 1993, *Bull. Ind. Soc. Earthquake Tech.*, **31**(1), p 15–54.

Sinvhal, A., P. R. Bose, V. Prakash, A. Bose, A. K. Saraf and H. Sinvhal, 2003, Isoseismals for the Kutch earthquake of 26th January 2001, *Earth and Planetary Sciences*, **112**(3), p 1–8.

Sinvhal, A., A. D. Pandey and S. M. Pore, 2005, Preliminary report on the 8th October 2005 Kashmir earthquake, Department of Earthquake Engineering, IIT Roorkee, 60 p.

Stuart, M., 1920, The Srimangal Earthquake of 8th July 1918, *Mem GSI*, Volume 46, Part I.

Sukhija, B. S., M.N. Rao, D. V. Reddy, P. Nagabhushanam, S. Hussain, R. K. Chadha and H. K. Gupta, 1999, Paleo-liquefaction evidence and periodicity of large prehistoric earthquakes in Shillong Plateau, India, *Earth and Planetary Science Letters*, **167**, p 269–282.

Tandon, A. N., 1953, The very great earthquake of Aug 15, 1950, in A compilation of papers on the Assam earthquake of August 15, 1950, M. B. R. Rao,Publication No. 1, p 80–89, The Central Board of Geophysics, Govt. of India, Calcutta.

Tandon, A.N., 1959, The Rann of Kutch earthquake of 21 July, 1956, *Ind. J. Met. Geophys*., **10**, p 137–146.

Tandon, A. N. and H. N. Srivastava, 1974, Earthquake occurrence in India, in *Earthquake Engineering*, Jai Krishna Sixtieth Birth Anniversary Commemoration Volume, p 1–49, Sarita Prakashan, Meerut.

West, W. D., 1934, Baluchistan earthquakes of August 25th and 27th, 1931, *Mem GSI*, Volume 67, Part I.

West, W. D., 1938, Preliminary Geological report on the Baluchistan (Quetta) earthquake of May 31st, 1935, *Rec. GSI*, **69**, Part 2, p 203–240.

http://www.eeri.org/lfe/clearinghouse/kashmir/reports/DEQ_IITR_KASHEQ05.pdf

http://www.iitr.ernet.in/EQ-Kashmir.pdf

http://www.iitr.ernet.in/news-system/files/58.pdf

7 CHAPTER

Measures of an Earthquake, Magnitude, and Intensity

INTRODUCTION

After every earthquake, one question that is always asked is 'how big was it?' The answer to this question is best given by the most often used term associated with earthquakes, magnitude. Earthquake magnitude is a fundamental parameter used to quantify and compare the size of large and small earthquakes. It is very common to confuse between the two commonly used measures of an earthquake—intensity and magnitude. In many instances, the two terms are erroneously used interchangeably. Intensity is based on postearthquake damage surveys, is a descriptive scale, is indicative of shaking at that place, is written in Roman numerals, is space-dependent, and varies from point to point in the affected area.

MAGNITUDE

Magnitude is expressed numerically; it is a definite Arabic number for any given earthquake and is estimated from instrumentally recorded seismograms. It is indicative of the energy released at the source during an earthquake. It is a unique value for a specific earthquake event and does not change with change of observation or change of place. The magnitude scale is open-ended on either side. For very small earthquakes, magnitude can be 3, 2, 1, 0, −1, −2, −3, and so on at the lower end of the scale and can be 7, 8, and so on at the higher end. Magnitude can be almost 9 for local or surface wave magnitude, and just below 10 for moment magnitude. The Sumatra earthquake of 2004 was one such rare event and was assigned magnitude 9.3. The magnitude scale has no theoretical upper limit, but a practical limit that depends on the strength of materials in rocks.

SOME COMMON MAGNITUDE SCALES

Several methods of estimating earthquake magnitude are currently in practice. However, Richter originally defined the concept in 1935. It is the most often quoted magnitude scale.

Richter Magnitude

The simplest definition of Richter magnitude is that it is the logarithm to base 10 of the maximum amplitude traced on a seismogram by a standard instrument placed at a distance of 100 km from the epicenter. Amplitude is measured in microns, 1 micron being 10^{-4} cm. Amplitude can vary enormously, with the size of the earthquake and the epicentral distance at which it is recorded. Therefore, the logarithmic scale is more manageable than a linear scale. Because the scale is logarithmic, every upward step of one magnitude unit means multiplying the recorded amplitude by an order of 10. The standard instrument is a short period Wood Anderson seismograph, which has magnification 2800, time period 0.8 sec, and damping coefficient 0.8.

Reduction of observed amplitudes at various distances to the expected amplitudes at the standard distance of 100 km is made by the use of empirical tables. The scale applies to earthquakes of normal focal depth. To achieve this, Richter gave the following expression in 1935.

$$M = \log[A(\Delta)/A_0(\Delta)]$$
$$= \log A(\Delta) - \log A_0(\Delta) \qquad (7.1)$$

M is Richter magnitude, Δ is epicentral distance and is given in km, A is maximum trace amplitude of the event to be measured, and A_0 is the maximum trace amplitude of a zero magnitude earthquake. This standard earthquake gives amplitude of one micron on the standard instrument at an epicentral distance of 100 km (in this case $A = A_0$). It rates other earthquakes in a relative manner under identical observational conditions. The range of energy released in different earthquakes is very large; therefore, amplitudes can vary from 0.1 mm to up to 12 cm and more. This large variation is taken care of by the logarithmic scale. Richter's (1935) formula gives local magnitude, i.e., for an epicentral distance less than 600 km and for shallow focus earthquakes in California. The empirical relation is given by

$$\log A = 6.37 - 3\log\Delta \qquad (7.2)$$

A is the maximum amplitude in microns. Since magnification of the standard instrument is 2800, and if ground amplitude 'a' is in microns, then the measured amplitude can be written as

$$\log A = \log(2800a). \qquad (7.3)$$

By substituting Equations (7.2) and (7.3) in Equation (7.1), magnitude can be determined in the following way

$$M = \log(2800a) - (6.37 - 3\log\Delta)$$
$$= \log 2800 + \log a - 6.37 + 3\log\Delta$$

$$\boxed{\mathbf{M = \log a + 3\log\Delta - 2.92}} \tag{7.4}$$

This is the Richter magnitude. It is also known as local magnitude, and is denoted by the symbol M_L. Equation (7.4) can be used for any type of seismograph, if ground amplitude and epicentral distance are known. An example for this is illustrated in Figure 7.1.

Several modifications were made to the original concept of magnitude by considering observations made at epicentral distance other than 100 km and for different types of instrument. Gutenberg and Richter (1945) gave empirical tables for these modifications, assuming that amplitude of ground

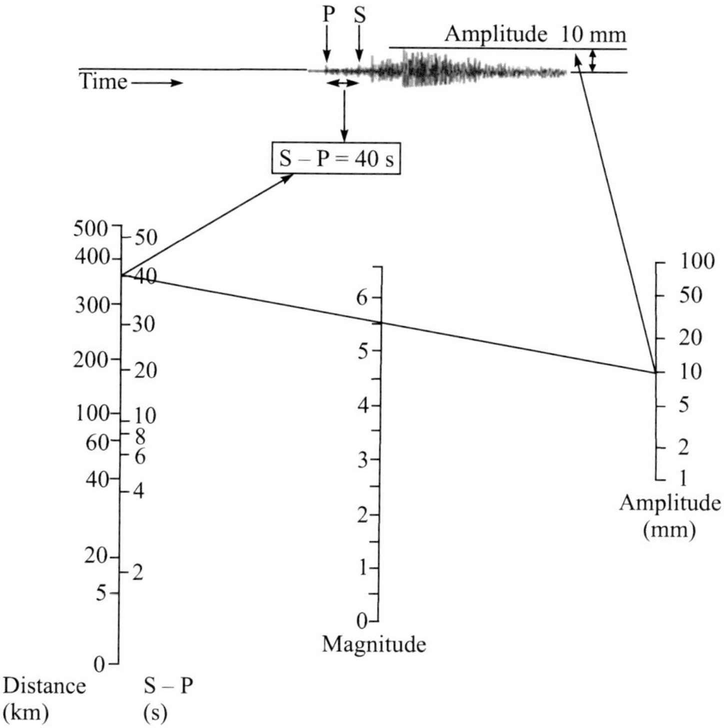

Fig. 7.1 Graphic procedure for calculating magnitude of a local earthquake using Richter's method is shown here. Maximum amplitude measured on the seismogram is 10 mm. This amplitude is suitably scaled to account for magnification, time period, and damping of the recording instrument. Hypocentral distance is estimated using the difference in the arrival time of *S*- and *P*-waves (example $S - P = 40$ s). If a straight edge is placed between appropriate points on the distance (left) and amplitude (right) scales, then the point where the line joining the two values calculated above cuts the magnitude scale gives the magnitude of the event as $M_L = 5.5$.

motion is proportional to the amount of energy released at the time of the earthquake. These empirical tables were subsequently extended for body and surface wave observations, for tele seismic events, i.e., for the case when epicentral distance was larger than 600 km, and also for deep focus earthquakes. Therefore, Richter's original concept of magnitude kept on expanding and several magnitude scales were in use in due course of time. Some magnitude scales that are in common use, such as the surface wave magnitude, body wave magnitude, and moment magnitude are given in this chapter.

Surface Wave Magnitude (M_s)

Seismograms of shallow focus tele-seismic events are dominated by long-period surface waves that have periods of the order of 20 sec. Magnitude of such events is calculated by the following expression

$$M_s = \log a + \alpha \log \Delta + \beta \tag{7.5}$$

Where M_s is the surface wave magnitude, a is maximum amplitude (in microns) of horizontal ground displacement for surface waves of 20 sec period, and Δ is epicentral distance in km. α and β are constants, empirically determined by taking into account several reference earthquakes whose magnitude is known. The values of α and β are chosen such that the magnitude calculated using Equation (7.5) gives values that are consistent with the values calculated using Equations (7.1) and (7.4). Gutenberg and Richter (1956a, b) assigned numerical values to α and β as $\alpha = 1.656$ and $\beta = (1.818 + C)$. C, a station constant, is a function of local conditions. Surface wave magnitude enables one to measure the size of large earthquakes even though it tends to saturate for very large earthquakes beyond magnitude 8. It is a widely used magnitude scale for tele-seismic earthquakes and is independent of the instrument used. The Sumatra earthquake of December 26, 2004 was assigned surface wave magnitude, M_s, 9.3.

Body Wave Magnitude (m_b)

Gutenberg and Richter (1945, 1956a) investigated *P*, *S*, and other phases of body waves, with respect to distance, for shallow and deep focus earthquakes. For *P*-waves with periods of 1 s the body wave magnitude m_b is given by

$$m_b = \log (a/T) + Q(h, \Delta). \tag{7.6}$$

Where 'a' is ground amplitude in microns, T is period in seconds of the measured wave; $Q(h, \Delta)$ is a function of depth of focus, h, and epicentral distance Δ. Occasionally, long-period instruments are used to determine body wave magnitude for periods from 5 to 15 s and these are usually for *PP*, *SP* phases. Different phases of seismic waves are given in Chapter 3 on seismic waves.

Richter's (1958) relations between body wave magnitude and surface wave magnitude are given below.

$$M_s = 1.59\, m_b - 3.97$$
$$m_b = 2.5 + 0.63\, M_s$$

The numerical value of body and surface wave magnitudes are same at $6^{3/4}$. Above this value surface wave magnitude is larger than the body wave magnitude, for $M_s = 8.9$, m_b is 8.1, and below this the reverse case holds, for $M_s = 0$, m_b is 2.5.

Seismic Moment Magnitude (M_w)

A better measure of the size of a large earthquake is the seismic moment, M_o. It takes into account rupture along a fault, which involves forces that are equal and opposite and produce a couple, dimensions of fault rupture and energy released at the source. Seismic moment M_o is given by μUA, where μ is the modulus of rigidity (= 3×10^{10} Nm^{-2} for crust and 7×10^{10} Nm^{-2} for mantle); U = average offset or longitudinal displacement of the fault, and A is the area of the fault (length × depth). Work done is given by $\Delta W = (\sigma/\mu)\, M_o$. Drop in strain energy in an event is expressed as work done at the fault surface. This is given by $\Delta W = \sigma UA$, where $\sigma = (\sigma_1 + \sigma_2)/2$; and σ_1 and σ_2 are stresses at the fault before and after an event. If it is assumed that after slip at a fault surface stress is equal to frictional stress, σ_f, i.e., $\sigma_2 = \sigma_f$, then the above equation becomes $\Delta W - \sigma_f UA = (\Delta\sigma/2\mu)\, M_o$. For large earthquakes, change in strain energy, $\Delta\sigma$, is almost 30 bars and $(\Delta\sigma/2\mu) \approx 1/(2 \times 10^4)$. Then energy in an earthquake comes out to be $M\sigma/(2 \times 10^4)$. On the basis of this relation and the conventional energy magnitude relation ($\log E = 11.8 + 1.5\, M$), given as Equation (7.9), Kanamori (1977) proposed the moment magnitude scale, M_w, as

$$M_w = (\log M_o/1.5) - 10.7. \qquad (7.7)$$

The Kanamori scale, M_w, has the added feature (over M_s) that it introduces quantification of very large earthquakes and involves the concept of earthquake-related fault. Seismic moment is measured from seismograms using long-period seismic waves.

Values obtained for the same earthquake using different inputs and methods give slightly different magnitude values, as can be seen for some earthquakes in Table 6.3. The Kashmir earthquake of 2005 and Kutch earthquake of January 26, 2001 were assigned magnitude on different scales by various agencies. These are given in Tables 14.1 and 14.2. The initial value of magnitude given immediately after an earthquake is sometimes modified slightly if more data from other recording instruments are incorporated. Therefore, even under the most favorable conditions uncertainties creep in between the ranges 0.5–0.8. Also it has to be borne in mind that the formulae used for determining magnitude are derived empirically and the complex process at the source is theoretically oversimplified by assuming a simple seismic source.

RELATION BETWEEN MAGNITUDE AND OTHER ASPECTS OF AN EARTHQUAKE

Repeated attempts have been made to find a relation between magnitude and other quantities such as earthquake damage, energy released in an earthquake, strain energy, frequency of occurrence, aftershock area, source volume, acceleration, intensity, fault length, time period, etc. Some of the frequently used relationships are given here.

Magnitude and Damage

Earthquakes with magnitude 3 or less are referred to as micro earthquakes and are barely perceptible to human beings even at the epicenter. About a 100 micro earthquakes are recorded annually by the micro earthquake network that operates around the Tehri region (EQ 87-16 and other reports). Earthquakes of magnitude about 4.5 cause slight damage near the epicenter. The 4.7 magnitude Roorkee earthquake of 1975 damaged the 125-year-old brick masonry building on the IIT Roorkee campus (Arya et al., 1977). In the main administrative building, plaster peeled off in several places but structural integrity of the building was intact. Earthquakes that have magnitude greater than 6 can usually damage life and property within a small area. The Latur earthquake of 1993, magnitude 6.4, claimed more than 10,000 lives in an area which was barely 12 km long (Sinvhal et al., 1994). An earthquake that has magnitude greater than 8.0 is referred to as a great earthquake as it can cause immense devastation in a very large area. Damaging effects of the seven great earthquakes that occurred in India in the last two centuries is given in Chapter 6.

Magnitude and Energy

Since an earthquake is associated with sudden release of energy at the source, there is a need to quantify this energy, despite the many practical difficulties involved in its estimation. Energy released in an earthquake of magnitude M is given by the expression

$$aM = \log_{10}(E_M/E_0). \tag{7.8}$$

Where a is a constant and is 1.5, and E_M is the energy released in an earthquake of magnitude M. E_0 is the energy released in an earthquake of zero magnitude, and Gutenberg and Richter (1956a, b) have estimated this to be 2.5×10^{11} ergs. A unit change in magnitude M changes the energy E by a factor of $10^{1.5}$, i.e., approximately 31 times. In other words, the energy in an earthquake of magnitude 6 is about 31 times as large as that for an earthquake of magnitude 5, and about 1000 times (31 × 31) that for an earthquake of magnitude 4. The energy released in a large earthquake, e.g., 8.9 magnitude, is 5.6×10^{24} ergs. This is equivalent to about 100 nuclear explosions, each with strength of 1 megaton (1 million tons) of TNT (Bolt, 2004). Energy released in all earthquakes annually sums to about 10^{25} ergs. Large earthquakes contribute a major portion of the total seismic energy released.

To arrive at a relation between magnitude and energy requires several assumptions. A simple harmonic plane wave starts at the source, spherical wave fronts develop, and the wave travels without distortion in a homogeneous, elastic, and isotropic medium. Ground amplitude depends on epicentral distance, focal depth, and time period of body and surface waves. The relation, as deduced in Appendix I, is given by

$$\log E = 11.8 + 1.5\,M \tag{7.9}$$

The accumulated strain energy in the crust is a possible source of seismic energy. The density of energy released from the source is uniform, and is approximately 10^3 erg cm^{-3}. This is similar to potential energy in rocks at ultimate strain. The volume where strain is released helps to define the quantity of seismic energy released and, therefore, the magnitude of the resultant earthquake. The earth's crust may break if it is strained beyond a certain limit. With the exception of faulted zones, strains in the crust are of the order of 10^{-4} or less. The earth's crust may be strained up to this level elastically, but beyond this it is liable to break.

Frequency of Occurrence

Earthquakes occur more often than one might tend to believe. However, large earthquakes are observed less often and are usually confined within well-known seismic belts. This is discussed in Chapter 1, on global seismicity. A large earthquake, of magnitude equal to or greater than 8.0, may occur once in a decade, whereas almost 10 earthquakes in a lower magnitude range, 7.0–7.9, may occur once in a year. The frequency of occurrence of earthquake events decreases exponentially as their magnitude increases, i.e., small magnitude earthquakes occur in large numbers. Almost 800,000 earthquakes may occur annually, in the magnitude range 2.0–3.4.

If N is the average number of earthquakes per year for which magnitude lies in the range M and $M + 0.1$, then for a wide range of magnitudes Gutenberg and Richter (1954) gave empirical relations of the form

$$\text{Log}_{10} N = \text{a} - \text{b}M \tag{7.10}$$

In this equation a and b are constants and help to define seismicity of a region. This relation holds for the entire world and also for particular regions for shallow focus earthquakes. For the whole world for shallow focus earthquakes a = 8.2 and b = 1.1 for magnitude greater than 7.3; and a = 4.6 and b = 0.6 for $5.8 \le M \le 7.3$, Richter (1958).

Fault Length and Magnitude

Active faults indicate future earthquake potential in a region. Length of a fault has been empirically related to the amount of energy that can be released, i.e., earthquake magnitude. The relation given by Kasahara (1981) is $\log L = p + qM$, where L denotes the length of fault, M is magnitude, and p and q are constants; q is generally within the range 0.5–1.2, and depends on regional structure. Otsuka (1965) gave a formula to relate magnitude M with the upper limit of fault length as $\log L_m = 3.2 + 0.5\,M$, where L_m is in centimeters. Wells and Coppersmith (1994) gave a relationship between moment magnitude M_w and length of surface rupture as $M_w = 1.16 \log (L) + 5.08 \pm 0.28$.

Aftershock Area and Magnitude

Utsu and Seki (1955) studied major earthquakes in the Japanese area and found an empirical relation between aftershock area and magnitude as

$$\log A = 1.02M + 6.0, \tag{7.11}$$

where A is measured in cm^2. Large shallow focus earthquakes tend to produce surface effects such as fault off sets and surface deformations. Dambara (1966) approximated the area of deformation as a circle with radius r. On the basis of Japanese data, he gave the following formula, $\log A = 1.02M + 6$; $\log (\pi r^2) = 1.02M + 6$; $\log \pi + 2\log r = 1.02M + 6$

$$\log r = 0.51M + 2.73, \tag{7.12}$$

where r is measured in centimeters. Tsuboi (1956) converted this as $\log A' = \log \pi + 2\log (0.51M + 2.73)$

$$\log A' = 1.02\,M + 5.96. \tag{7.13}$$

A' is the area of land deformation. This suggests that $A' \approx A$ and that the area of aftershocks is approximately the same as area of land deformation around the epicenter. For an earthquake of magnitude 6, the aftershock area is approximately 100 km^2 and the radius is about 10 km. For the largest historical earthquake, of magnitude $M = 8.6$, $A' = 50{,}000$ km^2, with effective radius of approximately 120 km.

Time Period and Magnitude

The period (T) of spectral peak for body and surface waves increases with magnitude. Kasahara (1981) has given a formula for P wave spectra of large earthquakes ($M > 5$), assuming a spherical source.

$$\log T = 0.51M - 2.59. \tag{7.14}$$

For smaller earthquakes ($M < 3$), Terashima (1968) has given the relation

$$\log T = 0.47M - 1.79. \tag{7.15}$$

Magnitude and Acceleration

Acceleration is one of the parameters that is considered while designing structures that are expected to show a desirable seismic performance. Several studies have been carried out to link acceleration with magnitude of an expected earthquake in the region around a proposed site, its hypocentral distance, closest distance to a causative fault, type of fault, attenuation characteristics, tectonic environment, i.e., whether it is inter- or intra-plate, and site conditions, etc. Numerous empirical formulae exist to link magnitude with acceleration. Depending on the available data, a choice is made from these to link magnitude with acceleration.

INTENSITY

When seismic waves reach the free surface of the earth, the ground shakes. When this shaking is severe, ground is damaged and all structures founded on it shake and some are damaged. After a devastating earthquake, it is relevant to know about the kind of damage that took place and its geographical extent. The answer to this is provided by the earthquake intensity. Intensity is a space-dependent descriptive rating of changes observed to ground surface, the built environment and human beings, caused by an earthquake. This manifests as the quality and quantity of damage based on macroseismic effects. Macroseismic effects of an earthquake are those that can be observed in the field on a large scale without the aid of any instrument.

Damaging effects of an earthquake are broadly classified into three large categorie—ground damage, damage to the built environment, and effect on humans. These effects are incorporated in a descriptive intensity scale. Intensity is denoted by Roman numerals. It ranges from I to X, or I to XII, depending on the scale being used, I being the least and XII being the most damaging. The higher end of the intensity scale describes damage to ground. This is largely dependent on local geology and soil conditions and can manifest in several ways such as surface faulting, liquefaction, landslides, etc. The middle and higher grades, i.e., between VI and X, of most 12-point scales, describes damage to the built environment. The lower grades, i.e. between I and V describe human perception.

Some Common Intensity Scales

Several scales were in use in different parts of the world at different times. Some of the more popular scales with which important earthquakes have been assigned intensity are the Rossi-Forel scale, Oldham scale, Mercalli Intensity Scale, Modified Mercalli Intensity Scale, Medvedev–Sponhover–Karnik (MSK) scale, and European Macroseismic Scale (EMS). Some of these are given in Appendix II.

M.S. Rossi of Italy and Francis Forel of Switzerland developed an intensity scale in the 1880s, which was one of the oldest scales in Europe. This ten-point scale assigned intensity from I to X. It was commonly known as the Rossi Forel Scale, or by its abbreviation, the RF scale. As it was the first intensity scale it achieved wide acceptability at that time. Damage in several important earthquakes in the world are rated according to this scale, including the great Kangra earthquake of 1905 in Himachal Pradesh and the San Francisco earthquake of 1906 in California. However, this scale had limited geographical applicability as it described European houses and at the highest level of intensity, X, too many large effects of an earthquake was lumped together.

The Oldham scale, published in 1899, is the oldest intensity scale that was indigenously devised in India, to map the immense devastation caused by the great Assam earthquake of 1897. Damaging effects of this earthquake are given in Chapter 6. This scale has a seven-point grading, grade I indicating maximum damage and grade VII the least damage. Mercalli, an Italian volcanologist and seismologist, devised another intensity scale in 1902, known as the Mercalli Intensity Scale, or the MI scale. It was more refined than the earlier Rossi Forel scale. Damage in the great Bihar-Nepal earthquake of 1934 is rated according to this ten-point scale. The Mercalli Intensity Scale was modified several times. Cancani linked intensity in this scale with acceleration, a quantitative measure, in 1904. When Sieberg enlarged the text of the Mercalli Cancani scale it became acceptable as an international scale in 1923. Wood and Neumann modified this in 1931 in the USA to account for damage that starts with intensity greater than or equal to VII, the earth's surface deforms at IX, and X indicates near total destruction. It was modified yet again in 1952 by Medvedev who linked intensity with oscillations of a building, which was modelled as a simple pendulum.

Because of so many rapid modifications in the Mercalli scale and the confusion these caused among users, Richter's modification of the Mercalli scale in 1958 came to be known as the Modified Mercalli Intensity Scale or the MMI scale. This was a 12-point scale, with grading from I to XII. As this was more comprehensive than the earlier scales and described earthquake damage more precisely than any other previous scale, it became globally acceptable and popular. It was the principal scale used in India for assessing earthquake damage for a long time.

On the MMI scale, intensity I implied that an event was felt by few only under exceptionally favorable conditions, e.g., by people at rest in upper stories of buildings. Usually, shaking is very feeble and unless the earthquake is very severe, human beings do not experience this shaking. Intensity VI meant that at that particular place one felt the earthquake. Not much damage was associated with it, except may be some cases of fallen plaster, etc. The intervening intensities vary and describe various degrees of shaking.

At intensity IX, general damage to ordinary foundations was noticed. Sand and water fountains, which began on a small scale at VIII, became noticeable. Large and spectacular phenomena of this kind belonged to X. Intensities X - XII were very severe cases of destruction and represented catastrophe, in which nearly all structures collapsed and objects were thrown up in the air.

The more detailed Medvedev–Sponhover–Karnik (MSK) scale was developed from the MMI scale in 1963, and was internationally accepted. This scale defined three types of buildings and three quantities of damage, and classified building damage into five grades. It was the forerunner of the European Macro-seismic Scale, EMS 98, Grünthal (1998), which gave more detailed description of structural damage and vulnerability classes, with relevant photographic support. The EMS – 98 scale defined three quantities of damage, five grades of damage and six vulnerability classes of buildings. As detailed description of structural damage was supported with relevant photographic support the use of this scale became more acceptable, as it became possible to assign intensity with less ambiguity in uncertain cases. Since this scale also included tall buildings, it became even more suitable, in context of the current building scenario. Besides structural effects, this scale also includes earthquake effects on ground and water.

Method of Assigning Intensity

To assign intensity at different places due to an earthquake a team of experienced observers, preferably comprising of a geologist, seismologist, civil engineer, earthquake engineer and an architect, carries out a Postearthquake damage survey of the affected area. A qualitative and quantitative estimate of damage to ground and structures is made at the places investigated. Various types of structures may coexist in the affected area. These may vary in design, quality of building material used and in quality of construction. When an earthquake shakes a building, the seismic response of the building depends on the building material that is used and, more importantly, on the way this material is used. Therefore, details of individual structures and their seismic performance are highly desirable when assigning intensity in the range VI–X. An interview to find out the response of the people who experienced the earthquake is also carried out. A written questionnaire is also distributed, filled, and collected in the earthquake-affected area which helps to assign intensity in the range I–V.

Observations of this damage survey are then compared with descriptions given in the chosen intensity scale, and a value is selected that makes the closest match of damage for a particular place. This is repeated for as many places for which data were collected. This information is plotted on a suitable map. Contour lines are drawn to separate one level of damage, i.e., intensity, from another. For each earthquake using a suitable intensity scale, a set of isoseismals is drawn to give an isoseismal map. These give the spatial variation of intensity, or putting it more simply, geographical spread of the disaster. Isoseismal maps of several earthquakes, such as the Roorkee earthquake of 1975, the Uttarkashi earthquake of 1991, the Latur earthquake of 1993, and the Kashmir earthquake of 2005, are given in Figs. 7.2, 7.3, 7.4 and 7.5, respectively.

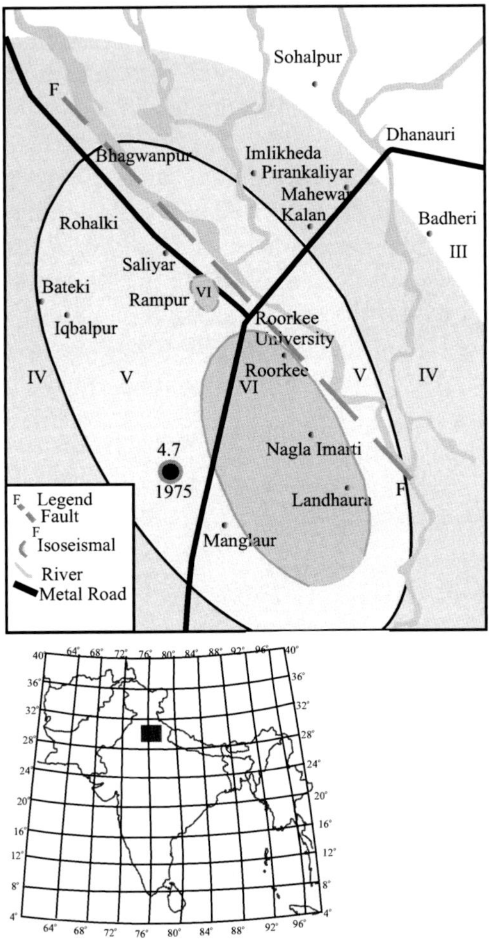

Fig. 7.2 Isoseismal map for the Roorkee earthquake of November 6, 1975. Maximum MMI was VI. The earthquake was felt at Delhi, Dehra Dun, Bijnore and surrounding areas. Damage to older (more than 60 years old) construction in and around Roorkee was minimal. Fissures developed in mud walls, along mortar joints and brick arches of masonry buildings. The main building of the University of Roorkee (more than 120 years old at the time of the earthquake) sustained minor non-structural damage. Loud rumbling noise, no casualties. The epicentral region has thick unconsolidated alluvium (≅3000 m) of the Ganga basin. Isoseismals are elongated along the NW–SE trending Roorkee fault, shown by FF. Origin time 05 h 41 m 30 s IST, Epicentre 29° 48.78′N, 77° 51.8′E, (IMD), 6 km south of Roorkee, on the basis of macro seismic data; magnitude 4.7.

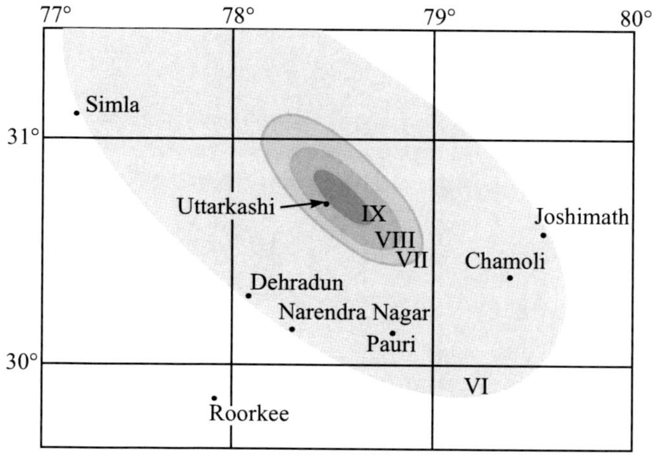

Fig. 7.3 Isoseismal map for the Uttarkashi earthquake of October 20, 1991. MMI (Max) IX indicated destruction of buildings, severe damage to bridges and landslides from steep slopes. MMI VIII indicated general damage to buildings and collapse of stonewalls; MMI VII indicated repairable damage to buildings and fissures in stonewalls; MMI VI indicated that damage to buildings was negligible and the earthquake was frightening; MMI V indicated that all were awakened but no damage occurred to buildings. Epicentre, 30.74°N 78.79°E (PDE), Origin Time 02 h 53 m 16.4 s (IST); Magnitude, m_b 6.5 PDE, M_s 7.1 USGS.

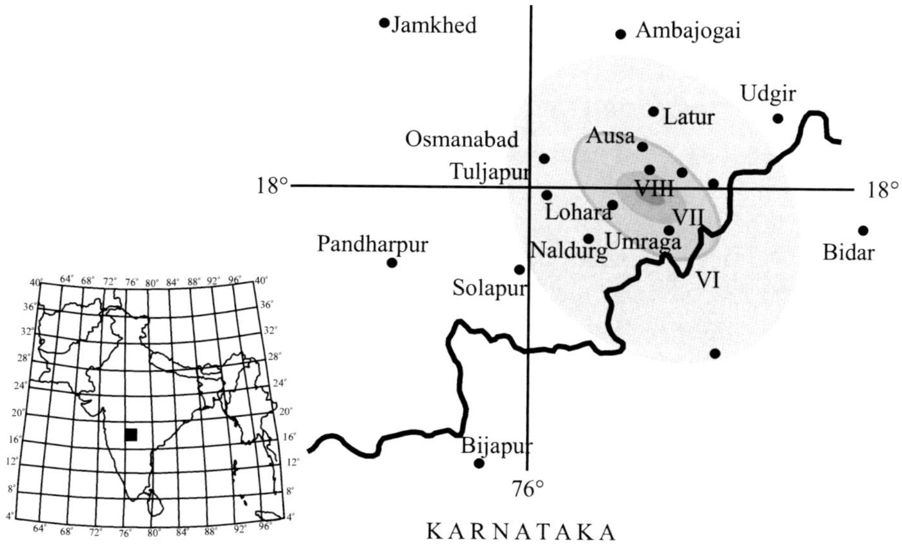

Fig. 7.4 Isoseismal map for the Latur earthquake of September 30, 1993. Epicentre: 76° 35′E, 18° 03′N; based on macro seismic data. Origin Time: 3 h 56 m IST; Magnitude: 6.4; MMI (Max): VIII+

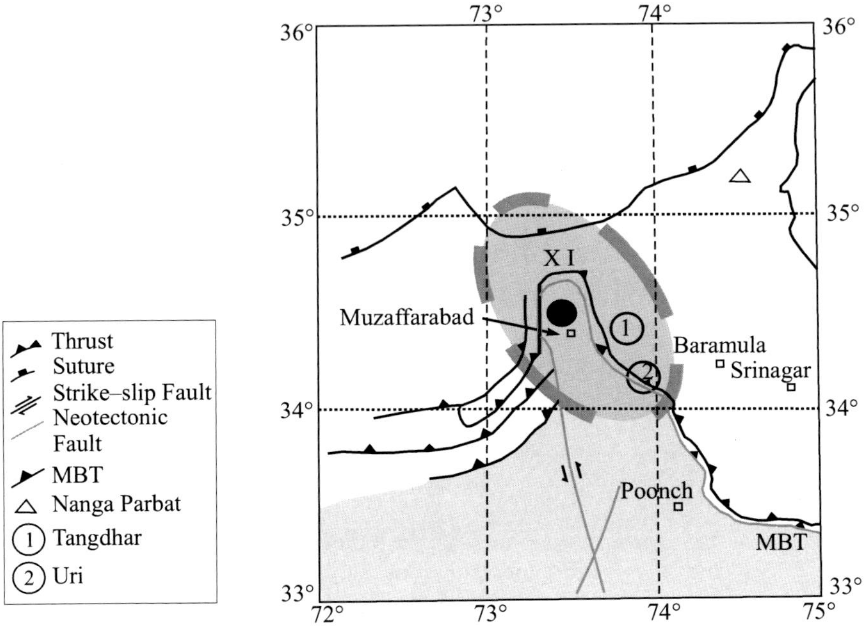

Fig. 7.5 Isoseismal map for the Kashmir earthquake of October 8, 2005. Epicenter is shown by thick dark circle. MSK Intensity XI is within the thick dotted line. It includes Uri, Kamalkote, Salamabad, Panzgam, Naichian, Tangdhar, Chamkote, Tithwal, and Nasta Chun Pass; Intensity was X at Lagama; IX at Mohura, Rajarwani, Kupwara; VIII at Boniyar, Chandanwari, Rampur, Handwara; and VII at Baramulla, Srinagar and Pattan. Epicentral distance of nearest point of Tangdhar and Uri is approximately 25 and 40 km, respectively.

Factors Affecting Intensity

Intensity depends on several inter-related conditions. Some of these are frequency of seismic waves, amplitude of ground shaking, duration of strong ground shaking, epicentral distance, local geology, local soil conditions, fault pattern in the area, magnitude of earthquake, topography, focal depth, quality and type of civil structures, and the materials used, damping in the structure, natural frequency of vibration of structures and population density.

Intensity varies with space. Maximum intensity is usually expected close to the epicenter and it usually reduces as epicentral distance increases. For the Kutch earthquake of 2001 maximum damage occurred in and around the epicenter, which included the following places: Adhoi, Amardi, Bhachau, Chobari, Dudhai, Kadol, Kharoi, Manfara, Rapar, Samakhiali, Trambau and Vondh. In all RC buildings, even those under construction, all columns and joints buckled and failed. Extensive liquefaction resulted in mudflows in Chang nadi for several kilometers between Manfara and Chobari. Earthquake fountains were observed in Bhachau, Samakhiali, Amardi, and Dudhai.

Fissures were numerous in roads within and in roads leading to this region. These places were assigned MMI X (Sinvhal et. al., 2003).

Anjar, a 450-year-old town, is very congested in the old parts. It was destroyed in an earlier earthquake of 1956. Later, new houses were raised on old foundations, and in due course of time, additional storeys were added on top of these. Old portions of Anjar, at an epicentral distance of 40 km, suffered heavy damage, whereas isolated and new four storied modern buildings sustained moderate damage. Anjar, together with Ratnal, Santalpur, and Maliya Miyana were assigned intensity IX because of this kind of damage and because of the numerous ground fissures, earthquake fountains, and several new pools of water and sand craters that developed in these places. Places with lower intensity were further away. Bhuj, at an epicentral distance of about 70 km was assigned MMI VIII. The death toll in congested market areas of Bhuj was heavy and more than 2000 lives were lost, mostly due to collapse of old stone houses and mixed construction. Table 7.1 gives approximate epicentral distance for several places for the Kutch earthquake of 2001 and Table 7.2 gives the intensity assigned to several places in Gujarat. Figure 7.6 gives the isoseismal map for the Kutch earthquake of 2001.

In contrast to this, sometimes, exceptional amount of damage is observed at large epicentral distances. This has more to do with interaction between long-period surface waves and long-period structures. This is given in more detail in Chapter 2, on seismic waves, under the heading Earthquake Damage and seismic waves.

Large magnitude earthquakes have intensities that are higher than those of smaller earthquakes. The great Kutch earthquake of 1819 was assigned intensity XI on MMI scale, whereas a later earthquake in 2001, of magnitude 6.9, in the same district, was assigned maximum intensity X on the same intensity scale (Sinvhal et al., 2003).

Table 7.1 Modified Mercalli Intensity (MMI) VI–to–X assigned to different places for the Kutch earthquake of January 26, 2001 and approximate epicentral distance. Maximum damage occurred in and around Bhachau and decreased away from it, at Bhuj it was less and at Delhi the earthquake was felt without causing any damage.

Place	*Intensity*	*Approximate epicentral distance in km*
Bhachau	X	10
Anjar	IX	40
Bhuj	VIII	70
Mandvi	VII	100
Rajkot	VI	>120
Ahmedabad	VI / VII	>250
Surat	VI	>350

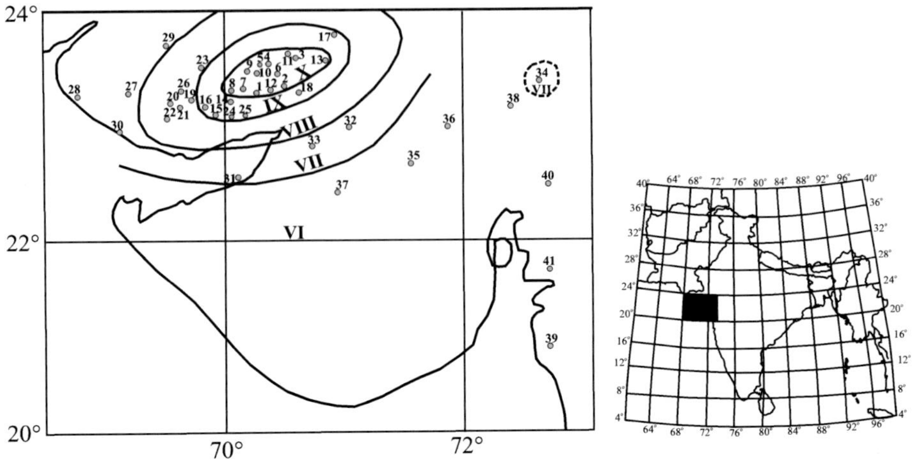

Fig. 7.6 Isoseismal map for the Kutch earthquake of January 26, 2001. Maximum intensity assigned to the meizoseismal area was X on the MMI scale. Intensity reduced as epicentral distance increased. Ahmedabad (34) and Surat (39) posed special problems while assigning intensity. Location of important places of the affected area is numbered and place names are given in Table 7.2.

The human habitat ranges from mud houses, stone houses, adobe houses, brick masonry houses, houses made of hollow concrete blocks, and timber frame structures, reinforced cement and concrete (RCC) structures. The building may be single, double, or multistoried. Sometimes the seismic performance of these structures is dismal. The use of a heavy roof on walls made of large and heavy random rubble stone, together with several other reasons, swelled death lists in several earthquakes. Several earthquakes, e.g., the Uttarkashi earthquake of 1991, Latur earthquake of 1993, Kutch earthquake of 2001 and Kashmir earthquake of October 2005, provide ample examples of this in living memory. More details on seismic performance of stone masonry houses is given in Chapter 11. Criteria adopted for assigning MM intensity for the Kutch earthquake of 2001 is given in Table 7.3. Several ill-designed 4–12 story buildings were vulnerable to damage even in regions that were at large epicentral distances, and were assigned MM Intensity as low as VI. These fared as poorly as rural stone houses in the epicentral region. This is illustrated in Fig. 7.7.

MM intensities assigned during past earthquakes were largely based on observed damage to stone and brick masonry housing of one to two stories. Accordingly, the MMI scale given in IS: 1893–1984 does not consider damage to multistory buildings as a basis for assigning MM Intensities. Damage to a large number of 4-12 story buildings in Seismic Zone III and as far away as 350 km from the epicenter indicated this shortcoming. In contrast, damage to

Table 7.2 Intensity (MMI) of places assigned on the basis of structural damage and ground damage. Serial number refers to numbers in Fig. 7.6.

Serial number	*Place name*	*Intensity*
1.	Bhachau	X
2.	Samakhiali	X
3.	Rapar	X
4.	Manfara	X
5.	Chobari	X
6.	Adhoi	X
7.	Amardi	X
8.	Dhamadka, Dudhai	X
9.	Kadol	X
10.	Kharoi	X
11.	Trambau	X
12.	Vondh	X
13.	Adesar	X
14.	Bhimasar (Rapar)	X
15.	Anjar	IX
16.	Ratnal	IX
17.	Santalpur	IX
18.	Maliya Miyana	IX
19.	Ghadsisa, Kotada, Kukma	VIII
20.	Bhuj	VIII
21.	Madhapar	VIII
22.	Sukhpur	VIII
23.	Lodai	VIII
24.	Gandhidham	VIII
25.	Kandla	VIII
26.	Radhanpur	VIII
27.	Nakhtarana	VII
28.	Naliya, Undot	VII
29.	Khawda	VII
30.	Mandvi	VII
31.	Jamnagar	VII
32.	Halvad	VII
33.	Morvi	VII
34.	Parts of Ahmedabad	VII
35.	Surendranagar	VI
36.	Viramgam	VI
37.	Rajkot	VI
38.	Gandhinagar	VI
39.	Surat	VI
40.	Vadodara	VI
41.	Broach	VI

traditional 1–3 storied buildings was on expected lines and reduced rapidly with epicentral distance.

The effect of local geology and soil conditions when assigning intensity can sometimes be profound. Ground shaking is minimum in stable rock; therefore,

Table 7.3 Criteria adopted for assigning MMI for the Kutch earthquake of 2001. It includes houses made of random rubble stone masonry, houses made of load bearing masonry walls with RC beams and slabs, and tall buildings. Ill-designed 4–12 story buildings were vulnerable to damage even in regions that were otherwise assigned MM Intensity as low as VI.

Building type	*Grade of damage at different MMI intensities*				
	VI	*VII*	*VIII*	*IX*	*X*
Traditional rural houses made of random rubble stone masonry	Moderate	Heavy	Destruction	Collapse	Collapse
Buildings made of load-bearing masonry walls with reinforced concrete beams and slabs for					
• Three stories	Slight	Moderate	Heavy	Destruction	Collapse
• Two stories	Slight	Moderate	Moderate	Heavy	Destruction
• Single story	–	Slight	Slight	Moderate	Heavy
RCC buildings on stilts without earthquake resistant features, 4 to 12 stories	Moderate/ heavy	Destruction	Collapse	Collapse	Collapse
RCC buildings with suitable architectural configuration for earthquake resistance, 4 to 12 stories	Slight	Moderate	Heavy	Destruction	Collapse

(a) (b)

Fig. 7.7 (a) Death toll in densely populated areas of Ratnal was heavy, mostly due to collapse of old stone houses and mixed construction. (b) Performance of several newly built multi-storied buildings in Ahmedabad at an epicentral distance of almost 250 km was a surprise and was worse than

structures founded on such strata are less prone to earthquake damage, and the intensity assigned to such places is lower. On the other hand, geologically recent sediments like alluvium, unconsolidated soil, filled and reclaimed ground are prone to severe shaking and heavy damage, more so if these are thick and the subsurface layers are saturated with water. Such strata absorb a significant amount of seismic energy and amplify long-period surface waves and shake the ground like a bowl of jelly. This condition becomes disastrous at large epicentral distances. Such a situation causes compaction of soft soil, and may sometimes be accompanied with ground damage in the form of subsidence, slumping, fissures, and liquefaction. Ground water may be disturbed, and sometimes earthquake fountains may result. Therefore, structures founded on this kind of soil are prone to heavy damage, and intensity assigned to these places is higher. For this reason, Ahmedabad was assigned higher intensity, VII, while the surrounding regions had intensity VI, as given in the isoseismal map for the Kutch earthquake of 2001 (Sinvhal et al., 2001).

When different soil types are in close contact with each other, intensity may vary by one to two grades in the same place. This difference becomes more prominent when soft soil is in contact with a ridge of hard rock. The latter resists severe shaking, and more damage is observed in structures located in regions of surrounding alluvium. Such effects became spectacular and were observed in the Kutch earthquake of 2001. Figure 7.8 shows isoseismals of the Kutch earthquake overlain over the faults and ridges of the region.

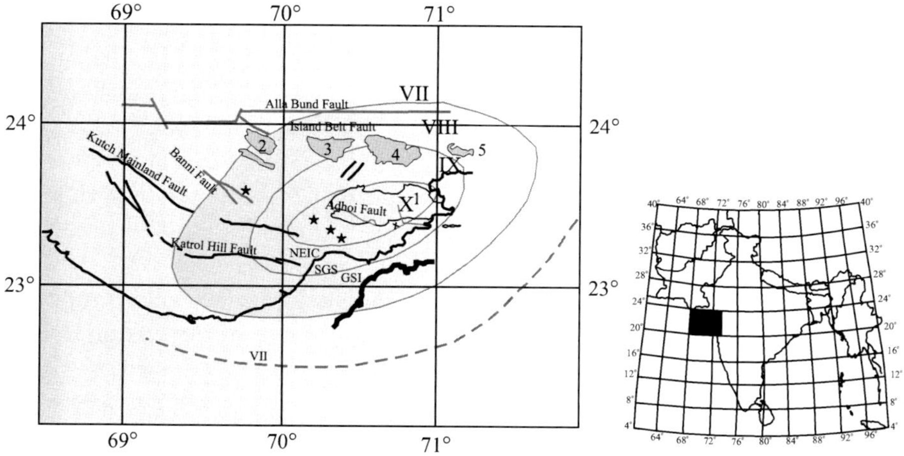

Fig. 7.8 Isoseismal map for the Kutch earthquake of January 26, 2001, shows intensity between X and VII. Map also shows major faults (in red) and ridges (in yellow). Higher isoseismals are elongated in an almost east–west direction. (1) Wagad ridge, (2) Pachcham uplift, (3) Khadir ridge, (4) Bela ridge, and (5) Charor uplift. Epicenters as provided by different agencies (as given in Table 14.2) are shown by star. (See color figure

Relation between Intensity and other Measures of an Earthquake

Repeated attempts have been made to tie intensity with some measurable physical quantity such as magnitude, acceleration, velocity, and displacement. Richter (1958) gave an empirical relation [$\log a = (I/3) - \frac{1}{2}$] to correlate the MMI scale with ground acceleration. In this relation, I is Modified Mercalli Intensity and is treated as a numerical quantity, and a is acceleration in cm/s^2. As intensity is space dependent, it helps in developing an attenuation relation for various regions, which is helpful in estimating seismic hazards and computing seismic forces. For shallow focus earthquakes, epicentral intensity is higher than for deep focus earthquakes. However, felt area for shallow focus earthquakes is smaller when compared to that of deep and intermediate focus earthquakes.

Applications

Postearthquake disaster surveys provide valuable information and have diverse uses. To begin with, these give a qualitative estimate of the geographical extent of the disaster that occurred during an earthquake. From an isoseismal map the area of maximum damage that needs urgent attention is identified and rescue and relief operations are intensified in that area and organized accordingly.

A preliminary estimate of the location of epicentre, especially in the absence of instrumental data, is usually taken to be near the center of the meizo-seismal area. The area with highest damage, i.e. area within the highest isoseismal, is referred to as the meizo-seismal area. The epicenter, when derived from seismograms, may be near one end of the meizo-seismal area, or sometimes it may even fall outside. Spacing between successive isoseismals gives an indication of focal depth, i.e., an isoseismal map may serve as an initial guide to determination of earthquake parameters. Several relationships exist between intensity variation with distance and attenuation and these are widely used in estimation of seismic hazards, and also in assessing size of old earthquakes.

Isoseismals are usually elongated along major subsurface structural trends such as faults, and sometimes these may be indicative of the causative fault. Thus, isoseismals of the Kutch earthquake of January 2001 were elongated along the east-west trending Adhoi Fault, shown in Figure 7.8, and of the Uttarkashi earthquake of 1991 were elongated along the Main Central Thrust (Sinvhal et. al., 1994), and for the Roorkee earthquake of 1975 were elongated along the Roorkee fault, as shown in Figure 7.2.

Isoseismal maps are useful as a preliminary guide for rehabilitation and rebuilding of the damaged area. Since maximum intensity assigned to Kutch earthquake of 2001 was X on the MMI scale, it implies that all future construction within the meizoseismal area should be so designed as to be able to withstand at least this level of intensity, or a grade higher, in a future earthquake. Therefore, information on damaging effects on the built environment of all previous earthquakes in an area is of prime importance to town planners, architects, and engineers, and becomes an important design criterion for rebuilding the damaged area.

Construction activity in an area, which is vulnerable to different earthquake hazards such as faulting, liquefaction, landslides, tsunamis, etc., is best avoided. If absolutely unavoidable, appropriate earthquake-resistant measures become necessary while designing foundations and structures. Some such vulnerable areas are best represented by higher intensities of the seven great earthquakes given in Chapter 6 on seismicity.

All deficiencies in the built environment are exposed at the time of the earthquake and provide a valuable learning experience. At the same time, a study of undamaged structures in the affected area can lead to a better understanding of desirable construction techniques. This has led to an improved understanding of the seismic response of diverse structures and appropriate earthquake resistant structures evolved in several earthquake prone areas. Some of these have a proven safety record to their credit. These structures, indigenously designed and made of locally available light, flexible and strong material, evolved over centuries. This makes bamboo, timber and agricultural residue ideal construction materials. Small houses made of such material just slide about in strong ground shaking without causing serious injury to their inhabitants, and even total collapse is not fatal most of the time.

The commendable seismic response of earthquake-resistant indigenous architecture was amply demonstrated in several recent earthquake disasters. Some of these are illustrated in Figure 15.2. A prime example of this was provided by the exemplary seismic performance of circular huts made of locally available material and known as 'bhoongas' in the Kutch earthquake of January 26, 2001. More than 10,000 people lost their lives in stone and brick masonry structures in the Kutch district alone, and in modern high-rise buildings in Ahmedabad and Surat. In stark contrast to this, life in the bhoongas continued without interruption in the Postearthquake scenario, most of which are located within the Banni depression (Bose et al., 2001). In the distant Andaman and Nicobar Islands, Nicobarese huts in the great Sumatra earthquake of 2004, likewise provided a similar example, as did timber framed three storied houses with walls made of random rubble stone masonry, in "Dhajji Diwari" and "taq" styles in Kashmir earthquake of 2005, (Sinvhal et al., 2005b). Details of these styles are given in Chapter 11. In the aftermath of these destructive earthquakes these instructive lessons acquire a deep meaning.

Lessons learnt from seismic performance of structures are gradually formulated into building codes and existing codes are refined and updated. The Bureau of Indian Standards, (BIS), formulated the first earthquake code in 1962, and updated and expanded this several times. The Koyna earthquake of 1967, which originated in the seismically stable peninsular India, gave a fillip to earthquake studies, as did the disastrous earthquakes of Latur in 1993 and Kutch in 2001. The latest version of the earthquake code, the fifth revision, was published in 2002. The current seismic-zoning map of India was updated after the Kutch earthquake of 2001 and the revised version is given in Bureau of Indian Standards BIS: 1893–2002.

Postearthquake disaster surveys and isoseismal maps have diverse other uses too. Insurance companies use these to decide compensation mode. Also, premium on insurance policies is adjusted according to seismic proneness of the area and type of the built structure. Moreover, data in earthquake catalogues can be made more comprehensive.

Even though intensity has widespread use, it is non-precise in nature. Damage depends on social and construction practice of the afflicted region, and includes different styles, designs, building material and variation in quality of construction, and all this may not always fit into the description given in the intensity scale. Moreover, reports of human perception and eyewitness reports are sometimes subjective and open to discussion. Clearly, instrumental data are desirable. As these instrumental data are not yet adequately available to achieve a better and complete picture of ground shaking, the use of intensity scales continues.

CONCLUSION

Magnitude and intensity are two different and common aspects to describe the size of an earthquake. Magnitude is a fundamental parameter used to quantify and compare the size of large and small earthquakes. It is a definite single number for any given earthquake, is indicative of the energy released at the source, and does not change with change of observation or change of place. It is determined from a seismogram. Intensity, on the other hand, is based on postearthquake damage surveys, is a descriptive scale, is indicative of shaking at that place, is written in Roman numerals, is space-dependent, and varies from point to point in the affected area.

REFERENCES

Arya, A. S., S. Singh, H. Sinvhal, R. Prakash, P. N. Agrawal, K. N. Khattri, B. Prakash and A. Sinvhal, 1977, A macro seismic study of November 6, 1975 Roorkee earthquake, Roorkee, India, in *Proceedings of the Sixth World*

Conference on Earthquake Engineering (6 WCEE), New Delhi, Volume 1, p 255–261.

BIS: 1893—2002, *Indian Standard Criteria for Earthquake Resistant Design of Structures, Part I : General Provisions and Buildings* (Fifth Revision), Bureau of Indian Standards, New Delhi, 40 p.

Bolt, B. A., 2004, *Earthquakes* (Fifth Edition), W. H. Freeman and Company, New York, 378 p.

Bose, P. R., A. Sinvhal and A. Bose, 2001, Traditional construction and its behavior in Kutch earthquake, in *Proceedings of the Workshop on Recent Earthquakes of Chamoli and Bhuj*, May 24–26, 2001, Department of Earthquake Engineering, University of Roorkee, Roorkee, p 151–158.

Cancani, A., 1904, Sur Lemploi double echelle seismique des intensities, empirique et absolue, G. Betir, Erganzungsband 2, p 281–283.

Dambara, T., 1966, Vertical movements of the Earth's crust in relation to the Matsushito earthquake (in Japanese), *Journal of Geodetic Society of Japan*, **12**, p 18.

EQ 87-16, 1987, Report on collection, analysis and interpretation of data (April 1985—March 1987) from seismological laboratories in the Ganga Valley region of Himalayas, Volume IX, Earthquake Engineering Studies, Department of Earthquake Engineering, University of Roorkee, Roorkee, India.

Eremenko, N. A. and B. S. Negi, 1968, Tectonic Map of India, 1 : 2,000,000 scale, and Tectonic guide, Oil and Natural Gas Commission, Dehradun.

Grünthal, G. (Ed.), 1998, European Macroseismic Scale 1998, (EMS-98), Cathiers du Center European de Geodynamique et de Seismologie 15, Center European de Geodynamique et de Seimologie, Luxembourg, 99 p.

Gutenberg, B., 1956, Great earthquakes 1896—1903, *Trans Am Geopys Union*, **37**, p 608—614.Gutenberg, B. and C. F. Richter, 1945, Magnitude determinations for deep focus Earthquakes, *BSSA*, **35**, p 117–130.

Gutenberg, B. and C. F. Richter, 1956a, Earthquake magnitude, intensity, energy and acceleration, *BSSA*, **32**, p 163–191.

Gutenberg, B. and C. F. Richter, 1956b, Magnitude and energy of earthquakes, *Am Geofis*, **9**, 1–15.

Kanamori, H., 1977, The energy release in great earthquakes, *J. Geophys. Res.*, **82**, p 2981–2987.

Kasahara, K., 1981, *Earthquake Mechanics*, Cambridge Univ. Press, Cambridge, 248 p.

Kaila, K. L. and D. Sarkar, 1978, Atlas of Isoseismal Maps of Major Earthquakes in India, *Geophysical Research Bulletin*, **16**(4), p 233–267.

Medvedev, S. V., 1965, *Engineering Seismology*, Israel Progr Sci Transl., Jerusalem, 260 p.

Mercalli, G., 1902, Sulle modificazioni proposte alla scale sismica de Rossi-Forel, *Boll Seismological Italiana*, **8**, p 184–191.

Otsuka, M., 1965, Earthquake magnitude and surface fault formation (in Japanese with English abstract), *Zisin (J. Seismol. Soc. Japan)*, 2nd Series, **18**, 1–18.

PDE, *Preliminary determination of Epicenter*, 1976, US Department of the Interior, Geological Survey, Denver, Colorado.

Richter, C. F., 1935, An instrumental earthquake magnitude scale, *BSSA*, **25**, p 1–32.

Richter, C. F., 1958, *Elementary Seismology*, W. H. Freeman and Co., San Francisco, 768 p.

Sinvhal, A., P. R. Bose and R. N. Dubey, 1994, Damage report for the Latur Osmanabad earthquake of September 30, 1993, *Bull. Ind. Soc. Earthq. Tech.*, **31**(1), p 15–54.

Sinvhal, A., P. R. Bose, V. Prakash, A. Bose, A. K. Saraf and H. Sinvhal, 2001, Damage, seismo-tectonics and isoseismals for the Kutch earthquake of 26th January, 2001, in *Proceedings of the Workshop on Recent Earthquakes of Chamoli and Bhuj*, May 24–26, 2001, Roorkee, Indian Society of Earthquake Technology, p 61–70.

Sinvhal, A., P. R. Bose, V. Prakash, A. Bose, A. K. Saraf and H. Sinvhal, 2003, Isoseismals for the Kutch earthquake of 26th January 2001, *Earth and Planetary Sciences*, **112**(3), p 1–8.

Sinvhal, A., A. D. Pandey and S. M. Pore, 2005, Preliminary report on the 8th October 2005 Kashmir earthquake, Department of Earthquake Engineering, IIT Roorkee, 60 p.

Terashima, T., 1968, Magnitude of micro earthquakes and the spectra for micro earthquake waves, *Bull. Int. Inst. Seism. Earthq. Engg.*, **5**, p 31–108.

Tocher, D., 1958, Earthquake energy and ground breakage, *BSSA*, **48**, p 147.

Tsuboi, C., 1956, Earthquake energy, earthquake volume, aftershock area and strength of the Earth's crust, *J. Phys. Earthq.*, **4**, p 63–67.

Utsu, T. and Seki, A., 1955, A relation between the area of aftershock region and the energy of main shock, Zisin. J. Seismol. Soc. Japan, v 7, p 233–240, (in Japanese)

Wells, D. L. and K. J. Coppersmith, 1994, New empirical relationships among magnitude, rupture length, rupture width, rupture area and surface displacement, *Bull. Seism. Soc. Am.*, **84**(4), p 974–1002.

Wood, H. O. and F. Neumann, 1931, Modified Mercalli Intensity Scale of 1931, *BSSA*, **21**, p 277–283.

8 CHAPTER

Seismic Zoning

INTRODUCTION

Recent history witnessed immense devastation caused by several earthquakes, mostly on margins of plates. Earthquakes can damage ground and the built environment in many ways and claim thousands of human lives. All kinds of earthquake hazards such as ground damage, damage to infrastructure and houses ranging from stone and brick masonry to new multistory buildings were observed in different earthquakes. A systematic study of destruction caused by several earthquakes gave rise to the concept of seismic zoning maps. These maps act as a preliminary guide for construction of important civil structures and in disaster mitigation.

In the chapter on seismicity, Chapter 6, it became obvious that almost half the Indian territory is prone to earthquake damage of intensity MMI VII or higher. As a resurgent India is going through a phase of planned construction activity, it is of paramount importance that development continues unhampered by future seismicity. This may be possible to a large extent, to begin with, by identifying different seismic zones in India. Seismic zoning divides a region into several seismic zones and is best represented by a map. Such maps provide a unified picture of seismicity and seismotectonic framework of the country and are used as a preliminary guide for designing an earthquake-resistant built environment.

BACKGROUND

Several seismic zoning maps were in use in India at different times. The Geological Survey of India (GSI) made the first one, immediately after the Bihar–Nepal earthquake of 1934, Figure 8.1. This map was made on the basis of damage observed in past earthquakes, of intensity VII or higher on the

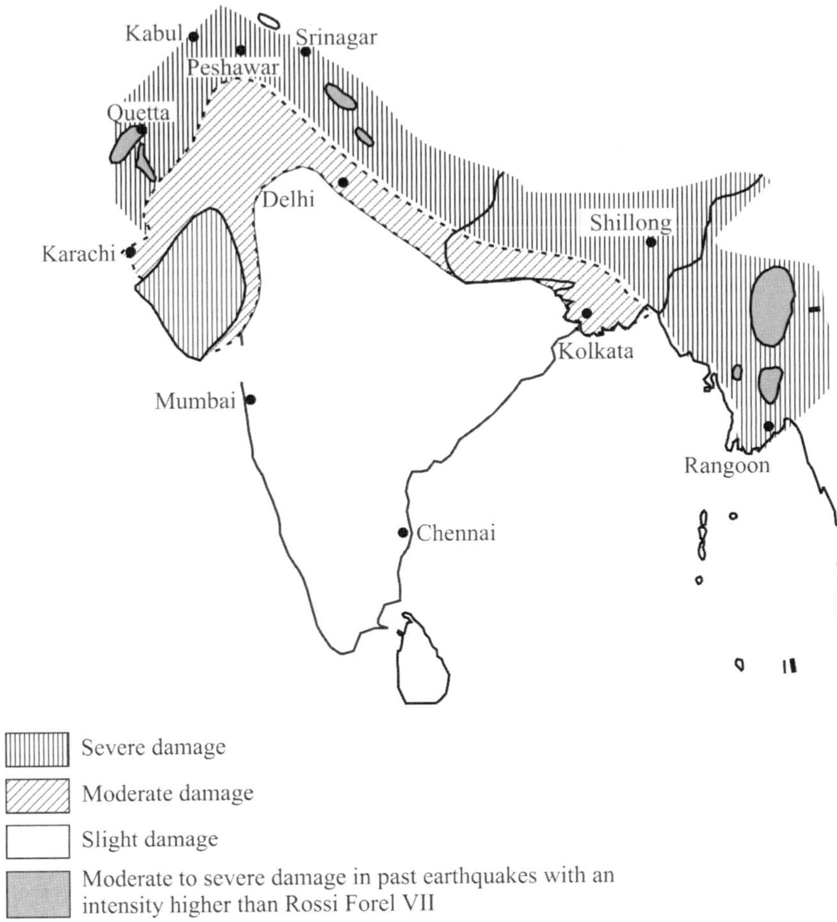

Fig. 8.1 Seismic zones of Indian subcontinent, 1935 (Redrawn after Geological Survey of India).

Rossi-Forel scale. Four different seismic zones were identified and were labeled as zones of severe, moderate, moderate to severe, and slight damage. Quetta, Kabul, Peshawar, Srinagar (Kashmir) and Shillong were placed in zone of severe damage; whereas Karachi, Delhi, and Kolkata were placed in zone of moderate damage, the entire peninsular India was placed in zone of slight damage, and portions of Myanmar were placed in zone of moderate to severe damage.

Professor Jai Krishna's seismic zoning map (1959) was based on a quantitative approach. The distinct spatial pattern of epicentral data for the period 1904–1950 revealed that most epicenters were concentrated north of a line defined by the southern limit of a seismically active zone, where heavy damage occurred in the past. Accelerations were computed with respect to this line, on the basis of Gutenberg–Richter relation (1956). Three of the five identified zones were defined in terms of expected accelerations of more than 30% of acceleration, between 30% and 10% acceleration, and less than 10% acceleration. Two of these zones had their limits 150 and 250 km south and parallel to this line. A zone, north of this line, was identified for very heavy damage and comprised of Ladakh and northeast India. As peninsular India was considered to be a seismically stable region, no special seismic considerations were recommended for the built environment. The line was identified as a region that had the potential to support damaging earthquakes including great earthquakes. It was later approximated as the region defined by the Main Boundary Thrust (MBT) and the Main Central Thrust (MCT).

METHOD OF MAKING A SEISMIC ZONING MAP

A seismic zoning map is made in several stages. It starts with a good earthquake catalogue to assess seismicity. This is followed by estimation of intensity and mapping of isoseismals of the larger damaging earthquakes. Isoseismals of all available large and damaging earthquakes, including great earthquakes, are then drawn on the same map and envelops of different intensities are drawn. The effect of tectonics, plate boundaries, geology, geomorphology, and local soil conditions are then added to the map of envelops. Different seismic zones are then identified (Savarensky, 1967). This constitutes the preliminary seismic zoning map of any region. These are then updated, in terms of accelerations. Seismic zoning is a continuous process and needs to be updated periodically as more data on earthquakes and their association with seismo-tectonic elements become less obscure and method of analysis and preparation is upgraded.

DIFFERENT SEISMIC ZONING MAPS OF INDIA

The Indian Standards Institution (ISI), later renamed as Bureau of Indian Standards (BIS), published the first earthquake code in India in 1962. This included the seismic zoning map of India. This code and map were subsequently revised several times and the fifth revision, published in 2002, has the current seismic zoning map. The first earthquake code has now been expanded to include at least 10 different codes that give detailed design guidelines for different kinds of built environment founded on different kinds of soils and in different tectonic environments.

Seismic Zoning Map of India, ISI: 1893–1962

Geological Survey of India's (GSI) zoning map made in 1935 and 1950, quantitative aspects of Dr. Jai Krishna's map of 1959, and epicentral data provided by India Meteorological Department (IMD) formed the basis of the first seismic zoning map, published by the Indian Standards Institution, ISI:

1893–1962. These included the five great earthquakes: Kutch earthquake of 1819, Assam earthquake of 1897, Kangra earthquake of 1905, Bihar–Nepal earthquake of 1934, and again the Assam earthquake of 1950. Salient features of these great earthquakes are given in Table 6.1. These earthquakes had, and continue to have, a very important bearing on the seismic zoning map of India. Several other damaging earthquakes of magnitude 5 and above were also considered, e.g., the 1960 Delhi earthquake, the 1930 Dhubri earthquake, the 1843 Bellary earthquake, the 1938 Satpura earthquake, and 16 other major destructive earthquakes.

Isoseismals of Modified Mercalli Intensity less than V–to–X and above were drawn on the same map, and when such maps were not available idealized isoseismals were considered. Envelopes were drawn for different intensities. These were modified, where necessary, to accommodate a few smaller intervening earthquakes. Seven seismic zones were identified, as zone 0, I, II, …, and VI. Zone 0 represented areas where probability of earthquake occurrence was least, and it was assumed that if an earthquake occurred it would not damage structures. Most of peninsular India was in zone 0. On the other extreme, zone VI was the severest zone and corresponded to the meizoseismal area of the two great Assam earthquakes of 1897 and 1950. Zone V corresponded to the gap between the two great earthquakes of Assam, and in addition, to meizoseismal areas of 1819 Kutch earthquake, 1905 Kangra earthquake and 1934 Bihar-Nepal earthquake. Other zones were of intermediate severity between zones VI and 0. The map clearly indicated that the Himalayan and Kutch regions were seismically vulnerable, while peninsular India was not, but had isolated centers of activity. Correspondence of each zone with Modified Mercalli Intensity is Zone 0, I, II, III, IV, V, and VI correspond to damage of MMI level less than or equal to V, V, VI, VII, VIII, IX, and greater than or equal to X, in this order.

Seismic Zoning Map of India, ISI: 1893–1966

In view of the tectonic map of India published by the GSI in 1962, it was understood that this additional data also played an important role in seismotectonics. Three tectonic zones were emphasized: the Himalayan tectonic zone, the Kutch region, and the seismically stable peninsular region. Consequently the 1962 version of the seismic zoning map was revised, but general principles followed in the making of the earlier zoning map were retained. The map continued to have seven zones. Smaller islets of zone VI were created within zone V of the earlier map, defined by the meizoseismal area of the great earthquakes, viz. 1819 Kutch earthquake, 1905 Kangra earthquake, and the 1934 Nepal–Bihar earthquake. Boundaries of seismic zones in the Indo Gangetic plains were elongated parallel to the Himalayan arc. Most of peninsular India was still in zone 0, but its size was diminished.

The intensity-magnitude-distance relation was used for computing seismic coefficient for each zone.

Seismic Zoning Map of India IS: 1893–1970

The surprise occurrence of the Koyna earthquake of December 11, 1967, (magnitude 6.5) necessitated a thorough review of seismic status of peninsular India and a consequent revision of the earlier seismic zoning map of India. Additional data on Koyna earthquake of 1967, Coiambatore earthquake of 1900, and North Andaman earthquake of 1941 were included. Magnitude of Bellary earthquake (1843) was revised, and its isoseismals were drawn parallel to zone of minor tremors that extended from Thiruvananthapuram to Chennai, which also corresponded with trend of Dharwar folding (Krishnaswamy, 1977). Isoseismals of Satpura earthquake and Rewa earthquake were upgraded and elongated along the Narmada graben. Isoseismals of Delhi earthquake were upgraded and elongated along the trend of Aravalli folding, and location and isoseismals of Kangra earthquake of 1905 were revised (Srivastava, 1974).

This revised map also retained general principles of the earlier map and incorporated additional data in accordance with the tectonic map prepared by the Oil and Natural Gas Commission, (ONGC) (Eremenko and Negi, 1968). Five principal tectonic units of India were considered. These were: (i) the orogenic unit of Cenozoic era, (ii) fore deep and marginal depression unit, (iii) west coast and Narmada Tapti unit, (iv) Gondwana rift unit, and (v) shield unit.

The Himalayan orogenic belt is highly folded and faulted, and includes the Shillong massif, which is greatly affected by this faulting, Dauki fault, Main Central Thrust, Satlitta thrust and Panjal thrust and others. Some of the major causative faults that supported several disastrous earthquakes of magnitude range 5 and above and also great earthquakes are within this region. The Himalayan foredeep and marginal depression unit contain several active faults in the basement, such as the Patna fault and the Kutch faults. These have supported several disastrous earthquakes including great earthquakes. The west coast and the Narmada Tapti Sone rift zone have faults of Tertiary and Quaternary age. The west coast has supported earthquakes of magnitude range 6.6–7, whereas the Narmada and Tapti rifts have supported earthquakes, the upper range of which was 6.5. The Gondwana rift zone, and adjacent and marginal parts of the peninsular shield has fault movements of Mesozoic age and later. The magnitude range in this region is much less, 5– 6.5. The peninsular shield, of Archean age, has ancient faults with some localized seismo-genic features. Occasional earthquakes that originate in this region have shown maximum magnitude 6–6.5.

Accordingly, parts of the Himalayan orogenic unit were assigned to seismic zone V and IV; the foredeep and marginal depression units to zone IV and III with islets of V, the west coast and Narmada Tapti unit was assigned to zone III with islets of IV, the Gondwana rift unit to zone III and the peninsular shield to I and II with islets of III.

Zone 0 and zone 1 of the previous map were merged together, as were zones V and VI, as earthquake effects in these zones were considered to be similar for purposes of earthquake-resistant design. Therefore, the 1970 version of the seismic zoning map had five seismic zones. The three regions defined by the meizoseismal areas of the 1819, 1905, and 1934 earthquakes, as they appeared as zone V in the 1962 version of the map, reappeared as zone V in the 1970 version.

Zone I was the least active of all zones, followed in severity by zone II, zone III, zone IV and zone V. Zone V was the severest of all zones. Seismic status of several regions was upgraded on the basis of earthquake effects and tectonics. These included Ladakh, Indo Gangetic plains, Moradabad, Delhi, Sohna fault, and the west coast of India. The Narmada Son Damodar graben was assigned to Zone III because of known faults and magnitude of earthquakes that originated there. Zone IV occurred as an eyelet in peninsular India because of the Koyna earthquake.

The earthquakes of Roorkee (1975), Kinnaur (1976), and Great Nicobar (January 20, 1982) occurred after this zoning map was published. Neither of these warranted a revision of the seismic zoning map as the damaging effects observed for these satisfied the conditions laid down in this map. Therefore, the next version of the earthquake code, i.e., IS: 1893–1975 and 1984 adopted this map without any revision. Several damaging earthquakes occurred after the 1984 version of the code was published, e.g., the earthquake in Cachar on December 30, 1984; Dharamsala in 1986; Bihar in August 1988; Uttarkashi on October 20, 1991; Latur on September 30, 1993; Jabalpur in 1997; Chamoli in 1999, and Kutch in 2001. Casualty figures were high in stone houses in the Latur earthquake of 1993. The earthquake code on stone masonry was updated after this earthquake. In the Kutch earthquake of January 26, 2001 the urban landscape with new multi-story buildings was adversely altered, as was the rural environment. Since multistory buildings dotted many cities in India by this time the damage scenario produced by the Kutch earthquake necessitated a thorough and urgent revision of the seismic zoning map, which was published in 2002.

Seismic Zoning Map of India BIS: 1893–2002

Again, like in earlier cases, general principles of the earlier map were retained. In the new map the entire country was divided into 4 seismic zones, II, III, IV and V. Zone I and II of the previous map were merged together to form the upgraded zone II in the current map. Zone II was the least active, and Zone III was of intermediate severity, followed by zone IV. Zone V continued to be the most active zone, and boundaries of zone IV and V were retained from the earlier version of the map without any alteration. The area devastated by the Latur earthquake was upgraded to zone III from zone I. The isolated zone related to the Bellary earthquake was removed. As east coast has a similar hazard potential as the area of Latur earthquake, therefore this area was upgraded from zone II to III and connected with the zone III of the Godavari graben area. Due to neo-tectonic activity the Narmada graben, Mahanadi graben and Godavari graben were assigned to zone III. Some important places within zone II are Ajmer, Allahabad, Bangalore, Bhilai, Bhopal, Hyderabad, Jaipur, Jhansi, Madurai, Nagpur, Pondichery, Raipur, Ranchi, Tiruchirapalli and Vishakhapatnam; and within zone III are Agra, Ahmedabad, Asansol, Bareilly, Bhubaneshwar, Bikaner, Bokaro, Chennai, Durgapur, Goa, Gaya, Jabalpur, Kanchipuram, Kolkata, Lucknow, Mumbai, Nasik, Osmanabad, Pune, Rajkot, Solapur, Surat, Varanasi and Vellore. Zone IV is represented by Almora, Ambala, Amritsar, Chandigarh. Darjeeling, Dehradun, Gangtok, Gorakhpur, Monghyr, Moradabad, Nainital, Patna, Roorkee and Simla. In zone V, damaging earthquakes of severe magnitude were expected to occur frequently with serious consequences to the built environment.

The entire northeastern part of India, consisting of all the seven states in entirety, i.e., Assam, Arunachal Pradesh, Manipur, Meghalaya, Mizoram, Nagaland, and Tripura, was assigned to seismic zone V. This included the populous cities like Agartala, Aizawal, Dibrugarh, Guwahati, Imphal, Jorhat, Kohima, Sadiya, Shillong, Tezpur and Tura. In addition, two elongated eyelets exist in the western Himalayas, one in Kashmir encompassing Srinagar and Baramulla, and the other in Himachal Pradesh. In Himachal Pradesh, zone V included Chamba, Dharamsala, Hamirpur, Jogindernagar, Kangra, Kullu, and Mandi. Most of Uttarakhand was assigned to seismic zone V, and included border districts of Pithoragarh, Chamoli, Champawat, and contiguous parts of Uttarkashi and parts of interior districts of Rudraprayag, Bageshwar, Tehri Garhwal, and Pauri Garhwal. In Bihar Darbhanga, Supaul and Madhubani were assigned to seismic zone V. Andaman and Nicobar chain of islands were included because of the damaging effects of the great Andaman earthquake of 1941. The entire district of Kutch in Gujarat was also within Zone V. In the Himalayan arc islands of zone V almost coincided with the region defined by the three-mega thrusts, the Frontal Foothill Thrust (FFT), Main Boundary Thrust (MBT), and Main Central Thrust (MCT).

In each zone a seismic zone factor was specified for use as a guide in design calculations for ordinary structures. For seismic zones II, III, IV and V it was specified as 0.10, 0.16, 0.24, and 0.36, in this order. Slight damage can be expected in zone II, from earthquakes in the magnitude range 5.0–6.0, MM intensity between VI and VII, with accelerations less than 0.10 g. Zone III can expect moderate damage from earthquakes in the magnitude range 6.0–6.5, MM Intensity between VII and VIII, with accelerations of 0.16 g. Zone IV can expect heavy damage from earthquakes in the magnitude range

6.5–7.0, MM intensity between VIII and IX, with accelerations of 0.24*g* zone V can expect destruction of the built environment from earthquakes with magnitude greater than 7.0, and MM intensity greater than or equal to IX, with accelerations as high as 0.36*g*.

Damage to ground and the built environment in the Sumatra earthquake of December 26, 2004, in Andaman and Nicobar islands conforms to the dictates of this map as given by seismic zone V, but not to coastal areas of Tamil Nadu and Andhra Pradesh. Damage observed in the Kashmir earthquake of October 8, 2005 (Sinvhal et al., 2005) also conforms to the dictates of the seismic zoning map of 2002. Therefore, when this map is modified next, it should take into account damaging effects of the tsunami too, and tsunami genic zones in the Indian Ocean, i.e., subduction zones, as outlined in Chapters 1 and 2.

APPLICATIONS

Seismicity, frequency of earthquake occurrence, damaging effects, and accelerations can be reasonably estimated in each seismic zone. The seismic zoning map is used for designing structures, and the BIS: 1893-2002 earthquake code recommends how seismic forces can be estimated for buildings in different zones. This provides broad guidelines for design and construction of a built environment that is expected to be safer in future earthquakes. All building organizations are obliged to take this map into account and to provide special safety measures in structures. For important and critical structures, such as dams, hydroelectric projects, nuclear power plants, bridges, etc., an additional dynamic analysis is required that deals with synthesizing response spectrum compatible acceleration time history for evaluating design earthquake forces.

Several areas with dense populations, exceeding half a million, are in higher seismic zones. Among these, Guwahati and Srinagar (J & K) are in Zone V, Amritsar, Dehradun, Jalandhar, Jammu, Jamnagar, Meerut, New Delhi, and Patna are in Zone IV, and Agra, Ahmedabad, Asansol, Bareilly, Bhavnagar, Bhiwandi, Bhubaneswar, Chennai, Coimbatore, Cuttack, Dhanbad, Indore, Jabalpur, Kanpur, Kochi, Kolkata, Kozhikode, Lucknow, Mangalore, Mumbai, Nashik, Pune, Rajkot, Surat, Thiruvananthapuram, Vadodara, Varanasi, and Vijayawada are in Zone III. These vulnerable cities require urgent and appropriate earthquake mitigation measures, before the next earthquake takes its toll.

SEISMIC MICRO ZONING

After several revisions of the seismic zoning map of India were made, a need was felt for detailed zoning of smaller regions like major cities with large populations or river valley projects, which have tremendous technoeconomic importance. Seismic micro zoning, an emerging research area, finely subdivides a small area for comprehensive assessment of several earthquake-related characteristics such as identification of source zones, assessment of ground damage, earthquake hazards, vulnerability, ground motion parameters, population at risk, etc.

Micro zoning of Kolkata was given in terms of liquefaction potential. Available bore log data were used, and where this was missing it was interpolated by means of artificial neural networks (ANN). Foundations of new structures need special considerations at places where river channel deposits are saturated by a shallow water table. These are liable to liquefy at a depth of 3 m for ground accelerations as low as 0.17 *g*. Such conditions exist at Salt Lake, Kasba, and Tollyganj (Chakraborty et al., 2005). Liquefaction potential of reclaimed land was estimated to be less than that of river channel deposits.

Micro zoning of Delhi was based on geotechnical parameters such as borehole data and data from soil penetration test (SPT), together with velocity of *P*- and *S*-waves. Four types of micro zones were identified to have liquefaction potential as severe, moderatc, minor, and remote. The severest micro zone corresponded to a region where 150 m deep Holocene deposits consisted of silt and loose sand. This situation exists close to the Yamuna River, e.g., at Yamuna Vihar, Geet Colony, Mayur Vihar, Preet Vihar, Vinod Nagar, and some places in Noida like Udyog Vihar and Sector 62 (Rao and Satyam, 2005). This kind of micro zoning can help in identifying a vulnerable site, and then in selecting a suitable ground improvement technique and a foundation system for a seismically safe built environment.

Micro-earthquake data and detailed tectonics of the Tehri area formed the basis of yet another micro zoning study. As the technoeconomically important Tehri region lies in seismic zone IV of the seismic zoning map of India, as per IS: 1893–2002, it is prone to earthquake shaking with peak accelerations of 0.24 *g*. Also, it is encompassed by isoseismal Rossi-Forel VIII + of the great Kangra earthquake of 1905. This region lies between the MCT and MBT. Two hundred and seventy three micro earthquakes that occurred in the time span between April 1980 and March 1983 were located within co-ordinates 78°–79° E and 30°–31° N.

Seven seismotectonic features were extracted from a circle of fixed radius drawn around each micro earthquake epicenter. These were subjected to a pattern recognition technique based on linear discriminant analysis (Davis, 1973; Khattri et al., 1979; Sinvhal et al., 1979, 1983, 1984, 1986, 1987a, 1987b, 1990, 1991, 1992a,b). These features were magnitude, number of major thrusts, distance from extremity of a major thrust, number of minor lineaments, number of intersections of lineaments, length of major river course/tributary around the epicenter, and number of micro-earthquake epicenters within the circle. Method of extraction of each parameter for a single micro earthquake at the center of the circle is shown in Figure 8.2. These are (1) Magnitude = 1.57, (2) number of major thrusts = 0, (3) distance to closest end of major thrust = 8.9 cm, (4) number of lineaments = 3, (5) number of intersections of lineaments = 1, (6) length of river course/tributary = 3.1, (7) number of Epicenters = 1.

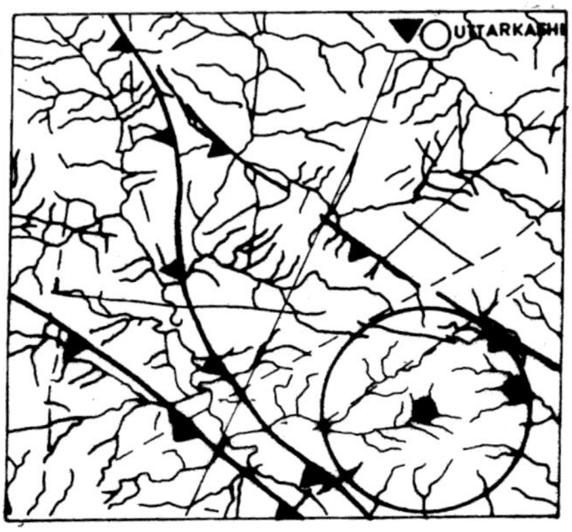

Fig. 8.2 Method of feature extraction for seismic microzonation is illustrated here.

Application of this pattern recognition technique showed an interesting relationship between clustering of micro earthquakes and their proximity to major thrusts. Three distinct types of seismic micro zones, S1, S2, and S3, emerged from this quantitative study. These are shown in Figure 8.3. Seismic micro zone S1 emerged as a highly critical zone and was approximately 100 km^2 in area. It was controlled by its proximity to three major thrusts, the Krol thrust, Garhwal thrust, and Tons thrust. Krol thrust is part of the MBT. Narendra Nagar, Devprayag, Kirtinagar, Chamba, and Jhaknidhar are within this micro zone. The sinusoidal meandering of River Ganga at the southern boundary of this zone indicates neotectonic control in this seismically active region.

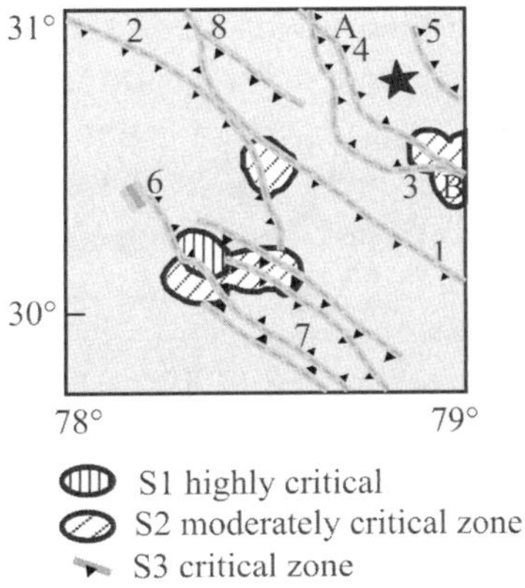

Fig. 8.3 Seismic micro zones and major thrusts: (1) North Almora thrust, (2) Tons–Nayar thrust, (3) Bhatwari Thrust (MCT), (4) Munsiari thrust (MCT), (5) Vaikrita thrust (MCT), (6) Krol thrust, (7) Garhwal thrust, (8) Dunda thrust.

Rishikesh is at its southern extremity. Zone S2 was a moderately critical zone and was characterized by the proximity of two major thrusts. It was found to exist in three separate places. The third seismic micro zone, S3, was characterized by a buffer zone along major thrusts, and was a critical zone.

Ranking of seismic micro zones indicated seismotectonic vulnerability of each zone; it increased with an increase in the number of major thrusts and their intersections. This is because intersections form the locked area of a fault and are locations of stress build up (Talwani, 1989). As movement on one fault is inhibited by an intersecting fault, large stresses can build up at intersections, which may be released as several micro earthquakes. Later studies indicated that each micro zone was liable to a different severity of ground shaking, ground failure, and other related hazards in a future earthquake. Detailed geotechnical investigations and appropriate earthquake engineering interventions are recommended for any large civil structure that is to come up in seismic micro zones S1, S2, and S3.

In a separate study, rupture on the causative fault was computationally modeled for the Uttarkashi earthquake of October 20, 1991 (epicenter 30.75°N 78.86°E; M_s 7.0, m_b 6.5; focal depth 12 km). A detailed study revealed that this earthquake originated on the subsurface manifestation of the Munsiari thrust, which dips at 14° (Joshi et al., 1999a, 1999b). Initially, rupture spread up dip of the Munsiari thrust, then on an almost vertical plane towards the surface, and then propagated from B (30°30′N, 78°55′E) toward A (30°45′N, 78°36′E) on the surface trace of this thrust, as shown in Figures 8.4 and 4.5. Directivity effects maximized toward Jamak, where interlocking of Munsiari and Bhatwari thrusts arrested the propagation effects of seismic waves. This caused massive landslides at Uttarkashi and around Maneri, Bhatwari, Agor and adjoining regions. The trend of isoseismals (Sinvhal et al., 1994) and distribution of aftershocks (Kayal et al., 1992) lend credence to this rupture model. Munsiari Thrust and Bhatwari thrust are part of the MCT.

The most significant observation of this modeling study, when combined with seismic micro zones of Tehri region, was that the hypocenter of the Uttarkashi earthquake was beneath one of the identified seismic micro zones, S2. Moreover, maximum damage observed in the Uttarkashi earthquake was within micro zones S2 and S3, and was explainable on the basis of directivity effects of rupture propagation. Thus, seismic micro zones represent possible source zones where damaging earthquakes may originate in the near future, and also help in explaining the damage these may cause. Therefore, identification of micro zones has the potential to provide basic data indispensable to planning, development, and assessment of earthquake counter measures in local disaster planning.

The Chamoli earthquake of 1999 (epicenter 30° 24′N 79° 25′E) also originated below the postulated extension of seismic micro zone S2, marked

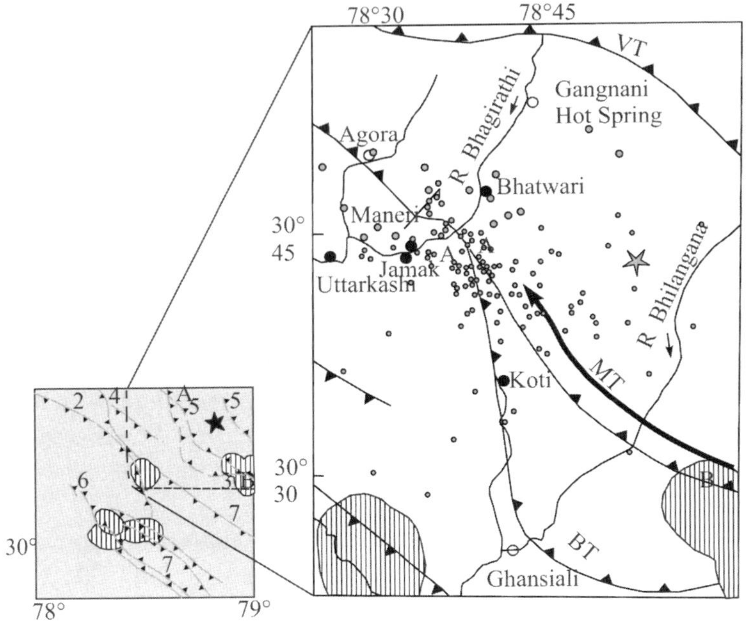

Fig. 8.4 Uttarkashi earthquake originated below point B, according to rupture model shown in Figure 4.5. Hachuring shows micro zone S2. VT—Vaikrita Thrust, MT—Munsiari Thrust, and BT—Bhatwari Thrust. Star denotes epicenter 30.75, 78.86 of Uttarkashi earthquake of October 20, 1991 (USGS). Aftershocks are shown by small dots and most are clustered around point A. Epicenter of Chamoli earthquake is at 30° 24′N, 79° 25′E. Point A corresponds to the Uttarkashi–Bhatwari region.

by B on Figure 8.4. This microzonation technique, therefore, helped in identifying seismic source zones for two damaging earthquakes within a 1° × 1° region in Uttarakhand.

A hypothetical earthquake disaster scenario was developed for the highly critical seismic micro zone, S1, in Uttarakhand for the Narendra Nagar block. In zone S1 River Ganga meanders, and three large thrusts viz. Garhwal thrust, Tons Nayar thrust, and Krol thrust congregate. This indicates that tectonic stresses are building up and could be released in a medium to large-sized earthquake in the future. Destructive earthquakes in the lower Himalayas are in the magnitude range 6–8. Earthquake hazards in any region are best estimated by peak accelerations. These were computed for earthquakes of magnitude 7.0 and 7.5.

Isoacceleration contours plotted for a hypothetical earthquake, with epicenter near Tapowan, at 30° 08′10″N and 78° 20′30″E, were elongated along the MBT. This is shown in Figure 8.5. Almost 59% population of Narendra Nagar block (of Tehri Garhwal district) was found to be vulnerable to damage associated with higher accelerations of 0.41g, as shown in

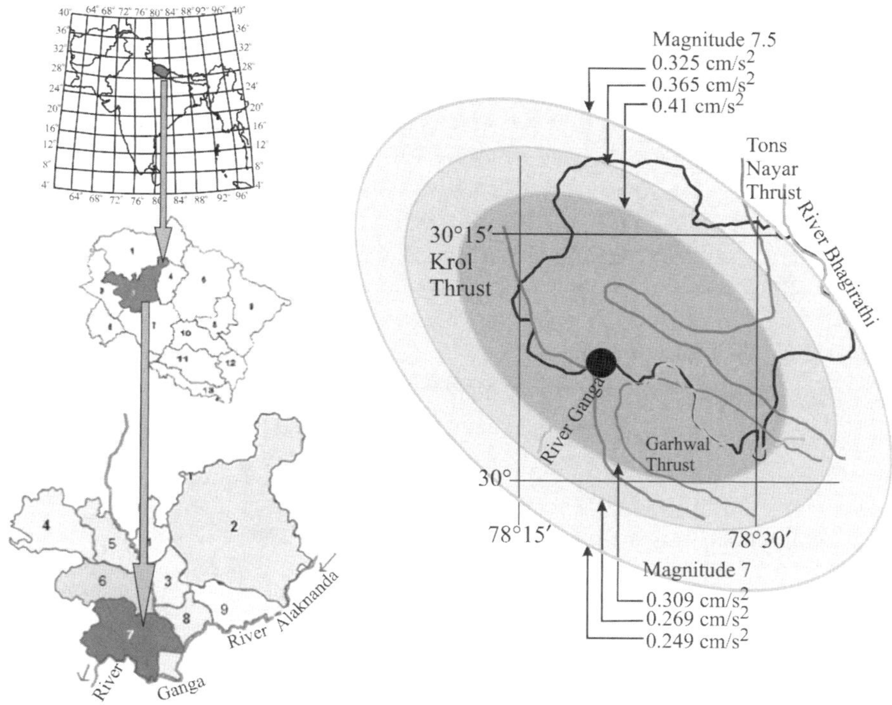

Fig. 8.5 Acceleration contours with epicenter at Tapowan (30° 08′10″N and 78° 20′30″E) for different hypocentral distances elongated parallel to the trend of Main Boundary Thrust. 3 is Tehri Garhwal district, and 7 is Narengra Nagar block. (See color figure also.)

Tables 8.1 and 8.2, whereas in seismic zone IV 0.24 *g* is expected Tables 8.1 and 8.2. This implies that Narendra Nagar block can expect earthquake damage to be much higher than what is expected as per the seismic zoning map of India. This reveals an increased threat perception. Implications of such an earthquake on housing stock, roads, and infrastructure can be profound. Therefore, disaster mitigation strategies, long-term earthquake preparedness, and short-term action plan for emergency management were developed for the Narendra Nagar block (Shankar and Gupta, 2005; Gupta et. al., 2006, 2008).

The risk increases if earthquake magnitude is larger, and may be even higher in the vicinity of faults, riverbeds, confluence of rivers and intersection of fault and river and in areas of higher population. High-altitude villages are expected to be at higher risk due to topographic effects. Forty-seven villages and one urban center, viz. Muni-ki-reti, with the population of 23,695, which is 32.4% of the total block population, are at high risk due to tectonics of the region. Thus threat perceptions and population at risk can be assessed in

Table 8.1 Villages that will be affected by different accelerations in Narendra Nagar block. The last column accounts for both urban and rural population.

Hypocentral distance (km)	*Peak acceleration (cm/s^2)*		*Number of villages*	*Total population*	*Percentage of population*
	Magnitude 7.0	*Magnitude 7.5*			
20	0.309	0.410	93	42,903	59
25	0.269	0.365	34	9,150	12
30	0.249	0.325	81	21,076	29
		Total	208	73,129	100

Table 8.2 Construction material used in houses of Narendra Nagar block.

Material for wall	*Material for roof*	*Type of settlements*		
		Urban areas(%)	*Villages within 2 km of road (%)*	*Villages more than 2 km from road (%)*
Stone	Thatch	2	5	10
	Slate	10	30	70
	RCC	20	35	20
Brick		—	—	—
	Slate	8	—	—
	RCC	60	30	—
	Total	100	100	100

seismic micro zones. This necessitates seismic upgradation of housing stock, structures, roads, and infrastructure, and will vary geographically, with priorities defined by seismic micro zones. This will reduce the uncertainty of potential damage, risk, and postearthquake recovery costs involved. The methodology evolved has the potential to be extended to other vulnerable seismic micro zones.

Since the efficacy of this pattern recognition technique proved useful in identifying seismic micro zones and was established within a limited geographical extent, it was applied to other seismically complex regions, where the required data were available. In Himachal Pradesh, its application led to identification of seismic micro zones around the MBT, MCT, and Chail Thrust. The MBT in this region is known as the Mandi–Sundernagar thrust, and the MCT is known as the Jutogh thrust (Sharma, 1977). Zones S1, S2, and S3 have the same connotations as for the Tehri region, as the kind of data and technique used were the same and both regions were defined by the MBT–MCT seismotectonic environment. These microzones are shown in Figure 8.6.

CONCLUSION

To reduce adverse effects of earthquakes, safe construction of the built environment is of paramount importance and that too at the proper site. There

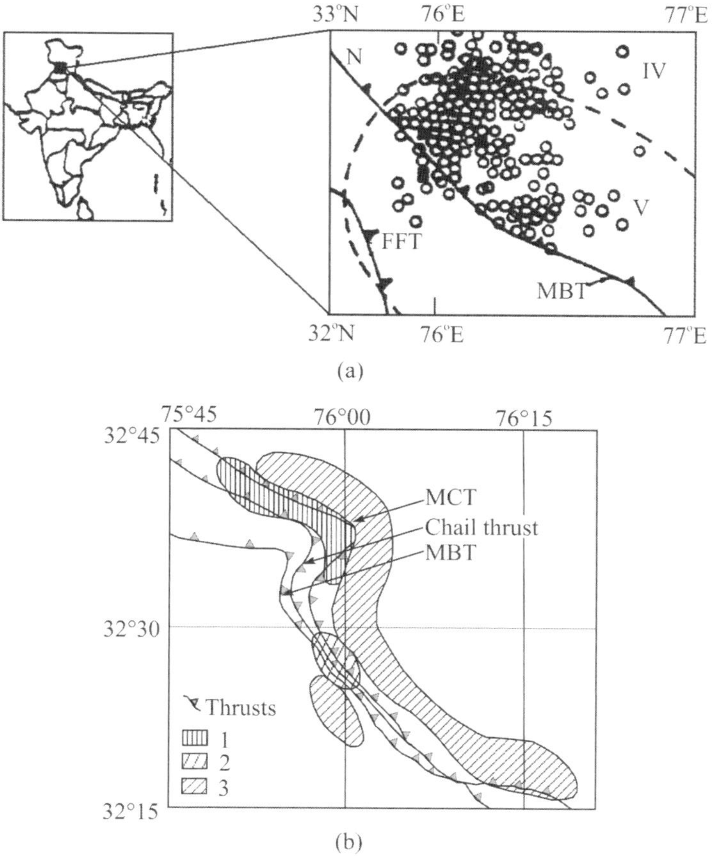

Fig. 8.6 (a) Epicenters of micro earthquakes recorded in the period January 1983 to July 1983 in the area bound by latitudes 32°–33°N and longitude 76–77° E. Seismic zones IV and V are as per BIS 1893–1984. FFT, Frontal Foothill Thrust, MBT, Main Boundary Thrust. (b) Micro zones identified for part of Himachal Pradesh. MCT, Main Central Thrust. 1 Depicts a highly critical micro zone, followed by 2 and 3.

is an urgent necessity to popularize the seismic zoning map of India, BIS: 1893–2002, among earthquake design professionals, builders, contractors, and house owners. This will result in a safer constructed product. Seismic zoning and seismic micro zoning have tremendous potential in mitigating earthquake disasters, both in the long and short term.

REFERENCES

BIS: 1893–2002, *Indian Standard Criteria for Earthquake Resistant Design of Structures, Part I: General Provisions and Buildings* (Fifth Revision), Bureau of Indian Standards, New Delhi, 40 p.

Chakraborty, P., S. Mukerjee and A. D. Pandey, 2005, Application of Neural Network for microzonation of Kolkata city on the basis of liquefaction potential, in *Proceedings of the Symposium on Seismic Hazard Analysis and Microzonation*, September 23–24, 2005, Indian Institute of Technology Roorkee, p 43–60.

Davis, J. C., 1973, *Statistics and Data Analysis in Geology*, John Wiley and Sons Inc., New York.

Eremenko, N. A. and B. S. Negi, 1968, *Tectonic Map of India, 1: 2,000,000 Scale, and Tectonic Guide*, Oil and Natural Gas Commission, Dehradun.

Gupta, I., A. Sinvhal and R. Shankar, 2006, Himalayan population at risk: strategies for preparedness, *Disaster Prevention and Management*, **15**(4), p 608–619.

Gupta, I., R. Shankar and Amita Sinvhal; 2008, Earthquake vulnerability assessment of house construction in Himalayas, *Journal of Design and the Built Environment*, **3**(3), p 1–14.

Gutenberg, B. and C. F. Richter, 1956, Earthquake magnitude, intensity, energy and acceleration, *BSSA*, **32**, p 163–191.

ISI: 1893–1962, *Indian Standard Recommendations for Earthquake Resistant Design of Structures*, Indian Standards Institution, New Delhi.

ISI: 1893–1966, *Indian Standard Criteria for Earthquake Resistant Design of Structures*, (First Revision), Indian Standard Institution, New Delhi.

ISI: 1893–1970, *Indian Standard Criteria for Earthquake Resistant Design of Structures*, (Second Revision), Indian Standard Institution, New Delhi.

ISI: 1983, *Explanatory Handbook on Codes for Earthquake Engineering, (IS: 1893–1975 and IS: 4326–1976), Part I: Explanations on IS: 1893–1975, Criteria for Earthquake Resistant Design of Structures* (Third Revision), Indian Standards Institution, New Delhi, 79 p.

ISI: 1893–1984, *Indian Standard Criteria for Earthquake Resistant Design of Structures* (Fourth Revision), Bureau of Indian Standards, New Delhi, India, 77p.

Joshi, A., A. Sinvhal and H. Sinvhal, 1999a, A strong motion model for the Uttarkashi earthquake of October 20, 1991, in Geodynamics of the NW Himalaya, Eds. A. K. Jain and R. Manickavasagam, Memoir 6, p 329–334, Gondwana Research Group, Japan.(Author: Please check whether the edits in the reference are correct.)

Joshi, A., B. Kumar, A. Sinvhal and H. Sinvhal, 1999b, Generation of synthetic accelerograms by modelling of rupture plane, *ISET Journal of Earthquake Technology*, **36**(1), p 43–60.

Kayal, J. R., V. P. Kamble and B. K. Rastogi, 1992, Aftershock sequence of Uttarkashi earthquake of October 20, 1991 in *Uttarkashi Earthquake October 20, 1991*, Eds. C. P. Vohra and G. D. Sarma, Geological Survey of India Special Publication Number 130, p 203–217.

Khattri, K. N., A. Sinvhal and A. K. Awasthi, 1979, Seismic discrimination of stratigraphy derived from Monte Carlo simulation of sedimentary formations, *Geophysical Prospecting*, **27**, 168–195.

Krishna, J., 1959, Seismic zoning of India, in *Proceedings of the First Seminar on Earthquake Engineering*, p 32–38, University of Roorkee, Roorkee, India.

Krishnaswamy, V. S., 1977, Evolution of the seismic zoning map of India, in *Souvenir of the Sixth World Conference on Earthquake Engineering*, p 77–81, New Delhi.

Rao, K. S. and D. N. Satyam, 2005, Seismic microzonation studies for Delhi region, in *Proceedings of the Symposium on Seismic Hazard Analysis and Microzonation*, Sept. 23–24, 2005, Roorkee, p 213–234.

Savarensky, E. F., 1967, Seismic zoning in *International Dictionary of Geophysics*, Ed. S.K. Runcorn, p 1372–1374, Pergamon Press, Oxford.

Shankar, R. and I. Gupta, 2005, An analytical framework for earthquake preparedness plan: activity, vulnerability and resource potential assessment, *Spatio-Economic Development Record*, **12**, (1), p 30–38.

Sharma, V. P., 1977, Geology of Kulu Rampur belt, H.P., *Memoirs Geological Survey of India*, Volume 106, No. 2, p 235–403.

Sinvhal, A., 1979, Application of seismic reflection data to discriminate subsurface litho-stratigraphy, *PhD thesis*, Department of Earth Sciences, University of Roorkee, Roorkee, India, 218 p.

Sinvhal, A. and Khattri, K. N., 1983, Application of Seismic reflection data to discriminate subsurface litho-stratigraphy, *Geophysics*, **48**(11), p 1498–1513.

Sinvhal, A., K. N. Khattri, H. Sinvhal and A. K. Awasthi, 1984, Seismic indicators of stratigraphy, *Geophysics*, **49**(8), p 1196–1212.

Sinvhal, A., H. Sinvhal and K. N. Khattri, 1986, Mathematical modelling for seismic discrimination, *Journal of Association of Exploration Geophysics*, **7**(2), p 85–93.

Sinvhal, A. and K. N. Khattri, 1987a, Application of seismic reflection data to discriminate subsurface lithostratigraphy, in *Handbook of Geophysical Exploration, Pattern Recognition & Image Processing,* Ed. F. Aminzadeh, Geophysical Press Ltd., London (Reprinted from *Geophysics*, **48**, p 187–224).

Sinvhal, A., K. N. Khattri, H. Sinvhal, A. K. Awasthi, 1987b, Seismic indicators of stratigraphy, in *Handbook of Geophysical Exploration, Pattern Recognition & Image Processing,* Ed. F. Aminzadeh,

Geophysical Press Ltd., London (Reprinted from Geophysics, **49**, p 225–262).

Sinvhal, A., G. Joshi, H. Sinvhal and V. N. Singh, 1990, A pattern recognition technique for microzonation, in *Proceedings of the Ninth Symposium on Earthquake Engineering*, Roorkee, India, p 24–30.

Sinvhal, A., H. Sinvhal and G. Joshi, 1991, A valid pattern of micro zonation, in *Proceedings of the Fourth International Conference on Seismic Zonation*, Stanford University, USA, p 641–648.

Sinvhal, A. and H. Sinvhal, 1992a, *Seismic Modelling and Pattern Recognition in Oil Exploration*, Kluwer Academic Publishers, the Netherlands, 178 p.

Sinvhal, A., H. Sinvhal, A. K. Jain, R. Manickavasagam, A. Joshi and G. Joshi, 1992b, Modelling of Uttarkashi earthquake of October 20, 1991, in terms of seismic microzonation and causative fault, in *Synthesis of the Uttarkashi Earthquake Data, 20th Oct, 1991 and Seismotectonics of Garhwal—Kumaon Himalayas* (Abstract Volume), Ed. G. D. Gupta, Department of Science and Technology, New Delhi.

Sinvhal, H., A. Sinvhal, A. K. Jain and R. Manickavasagam, 1994, Damage pattern in the Uttarkashi earthquake of October 1991, in *Group Meeting on Seismotectonics and Geodynamics of the Himalaya* (Abstract Volume), University of Roorkee, Roorkee, India.

Sinvhal, A., A. D. Pandey and S. M. Pore, 2005, Preliminary report on the 8th October 2005 Kashmir earthquake, Department of Earthquake Engineering, IIT Roorkee, 60 p.

Srivastava, L. S., 1974, Seismic zoning of India (Chapter 2), in *Earthquake Engineering*, Jai Krishna Sixtieth Birth Anniversary Commemoration Volume, Sarita Prakashan, Meerut.

Talwani, P., 1989, Characteristic features of intraplate earthquakes and the models proposed to explain them, in *Earthquakes at North Atlantic Passive Margins: Neotectonics and Postglacial Rebound*, Eds. S. Gregersen and P. W. Basham, p 563–579, Kluwer Academic Publishers, Dordrecht.

9 CHAPTER

Ground Damage

INTRODUCTION

In a large earthquake, due to passage of seismic waves, ground can be damaged in several ways. Faulting is one of them—it can be in the form of either subsurface or surface faults. Ground failure associated with earthquake-induced faults is given in Chapter 4. Some other prominent ground effects include topographic changes, surface distortions, liquefaction, sand boils, mudflows, formation of fissures in ground, and mud volcanoes. Ground water at shallow depth is disturbed due to strong ground shaking and can cause earthquake fountains and sag ponds. Waterfalls, damming and diversion of rivers, change of drainage system, sloshing of water over stream banks, and floods are some other water-related disastrous consequences of earthquakes. Other hazards include landslides in hilly terrains.

TOPOGRAPHIC AND SURFACE DISTORTIONS

Large-scale topographic changes and surface distortions are observed after several large earthquakes. Most of the time, these are associated with the causative fault. Oldham (1899) described surface distortions for the Assam earthquake of 1897 between Tura and Rowamari, which were on river Brahmaputra. This is described in Chapter 6. In the Kangra earthquake of 1905, Dehradun and Siwalik hills showed a rise of 30 cm relative to Mussoorie (Middlemiss, 1910), indicating surface distortion across the Main Boundary Thrust.

Similar large-scale topographic changes can occur along coastlines also. This can happen in two ways, either by submergence and subsidence of coastline, which is accompanied with transgression of sea, or by uplift of coastline, which is accompanied with regression of sea. Both are related to faults and became apparent after the tsunami visited the coastline of Andaman and Nicobar Islands in the Great Sumatra earthquake of December 26, 2004.

Coastlines of southern islands of the Nicobar group of islands showed a large amount of subsidence, which gradually decreased northward and was apparent as uplift in the northern islands of Andaman. This change was obvious from Indira Point to Austen Strait, i.e., a distance of almost 700 km (Shankar et al., 2005; Wason et al., 2006). Going from south to north, Indira Point, the southern most part of India, in the island of Great Nicobar, subsided by a large amount of about 3 m, Car Nicobar by about a meter, and Little Andaman Island and southern part of South Andaman Island by an amount between 94 and 100 cm. The sea transgressed inland in places of subsidence. Low-lying coastal areas were affected the most, e.g., at Car Nicobar a coastal strip almost 3 km wide was inundated. On the other hand, emergence

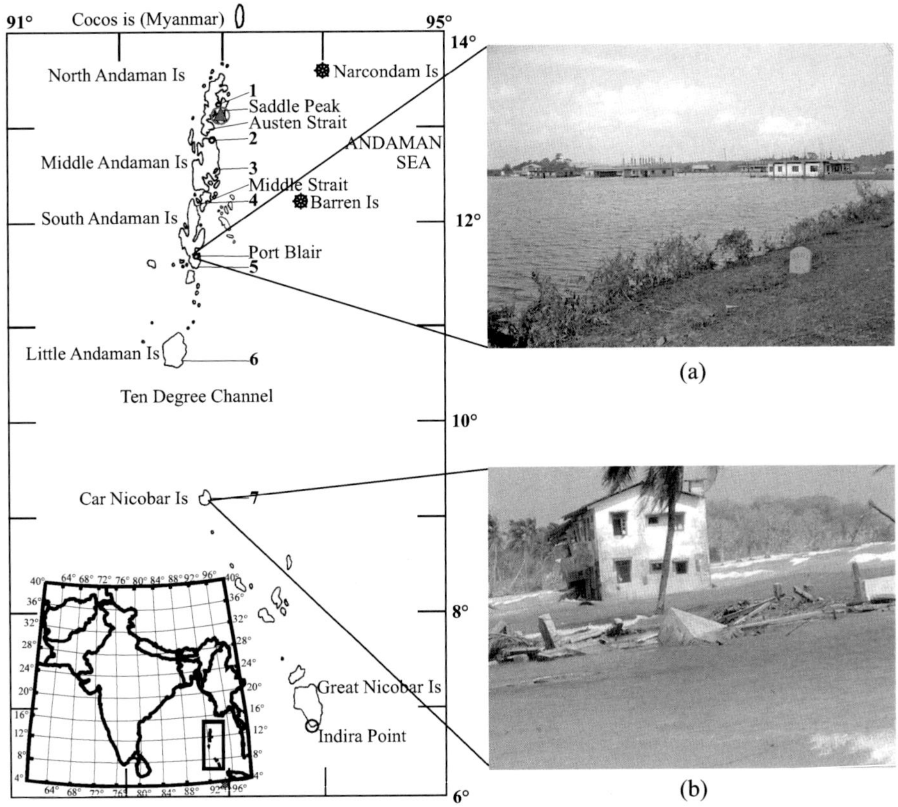

Fig. 9.1 Map shows location of the larger islands in the Andaman and Nicobar archipelago. Inset shows location of Andaman and Nicobar Islands on the map of India, epicenter of the earthquake of December 26, 2004 is shown by star★. (1) Diglipur, (2) Mayabunder, (3) Rangat, (4) Baratung, (5) Chidiya Tapu, (6) Hut Bay, (7) Malacca, ❁ shows volcanic Islands. Inundation and submergence at (a) Port Blair and (b) Car Nicobar.

of new shallow coral beaches and an uplift of about 1–1.2 m were observed below the Austen Bridge, which connects the islands of North Andaman and Middle Andaman. This uplift and submergence is shown in Figure 9.1. Change in elevation at Baratang Island was mainly due to emissions brought about by the mud volcano.

Tidal gauge records taken before and after the earthquake by the Survey of India confirmed these observations. Control points near Port Blair drifted south East by about 1.25 m, while those at Long Island and Vijaygarh situated north of Port Blair, drifted in the opposite direction. This data revealed that the region suffered not only subsidence and uplift of coast at different places but also an anticlockwise twist. The great Sumatra earthquake, of submarine origin, occurred on a convergent plate boundary, on a thrust fault and changed the coastal topography of the islands of Andaman and Nicobar archipelago.

LIQUEFACTION

Liquefaction is a phenomenon in which strength and stiffness of soil is reduced due to strong ground shaking. This takes place in unconsolidated sediments situated at or near the ground surface, or where there is shallow underground water or an aquifer at depths of about 10 m or less. This makes young, unconsolidated sediments, soft soil, river channel deposits, and filled ground on a high water table susceptible to earthquake-induced liquefaction. Repeated shaking by seismic waves often triggers an increase in water pressure in the aquifer. Water-saturated soil rearranges itself in such a way that it essentially becomes a suspension of solids in a liquid. The liquefied sediment not only moves about beneath the surface but may also rise from the pressurized liquefied zone through fissures and 'erupt' as earthquake fountains, mud volcanoes, and sand boils. In addition, liquefaction causes settlement, slumping, and subsidence of ground. It also causes mudflows, which constitute a mixture of water, clay, and silt.

Due to liquefaction, large deformations can occur within the soil and ability of soil to support foundations of structures reduces. This may result in sinking, shift, tilt, fall, or even collapse of structures. Buried objects like pipelines can shift or even float to the surface.

Liquefaction and related phenomena have been responsible for tremendous amounts of damage in several earthquakes around the world. The 1964 earthquake in Japan caused liquefaction at Niigata causing several four-story buildings to tilt by as much as 60°. These were later jacked back into position, underpinned with piles and reused (Wikipedia). During the 1989 Loma Prieta earthquake in California, liquefaction in a lagoon caused major subsidence and horizontal sliding of filled ground in the Marina district of San Francisco. In the Kutch Earthquake of 2001, extensive liquefaction occurred at several places in Kutch, namely in Chang Nadi between Manfara and Chobari for several kilometers, in Kaswali Nadi near Lodai, and in several other places near Rapar, Dandesar (Figure 9.2a and b), Bhuj, Khingarpur, Dharang Godai, Khawda, Samakhiali, Gadsisa, and in marshes below the Surajbadi Bridge (Sinvhal et al., 2003a).

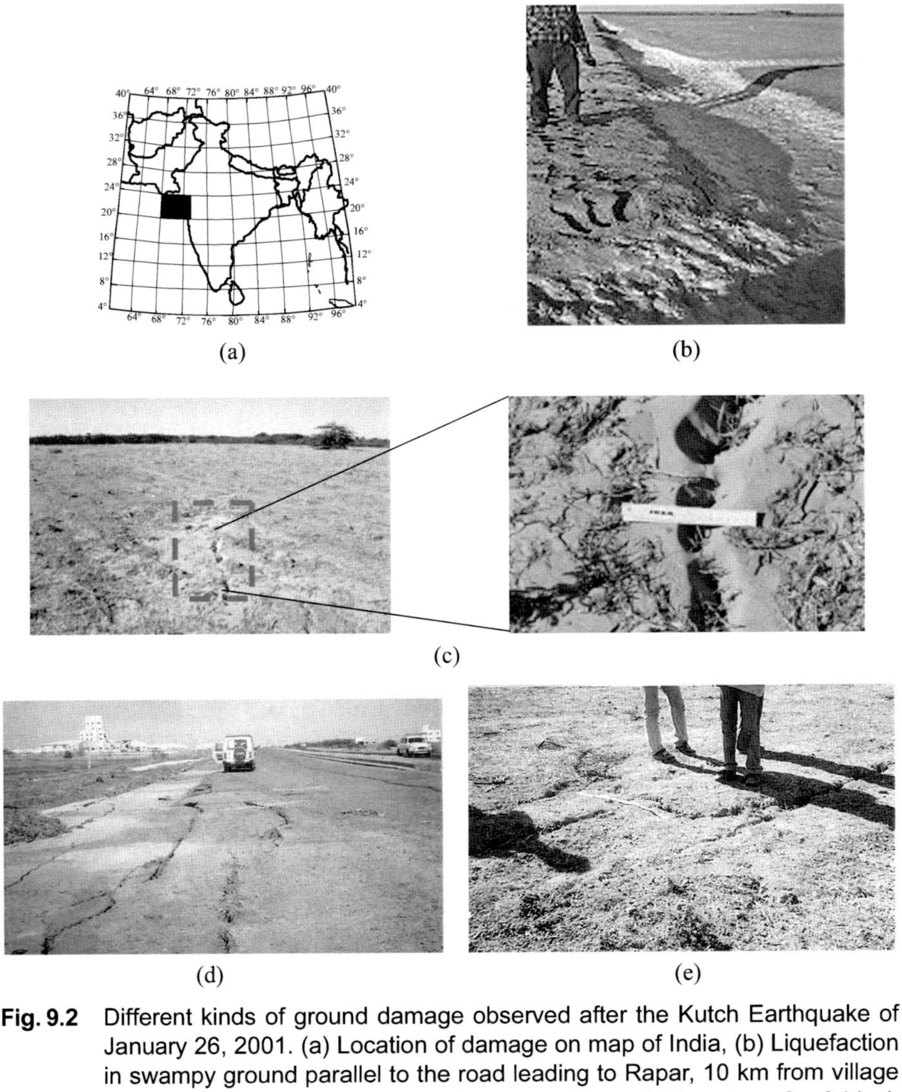

Fig. 9.2 Different kinds of ground damage observed after the Kutch Earthquake of January 26, 2001. (a) Location of damage on map of India, (b) Liquefaction in swampy ground parallel to the road leading to Rapar, 10 km from village Dandesar. The salt layer in the ditch was 6–7 mm thick on top of soft black clay. Shoes sank up to 5 cm in the slippery clay. (c) Evidence of earthquake fountains, which gushed through ground fissures and its detail in the adjoining figure. (d) Longitudinal fissures near Gandhidham on road leading to Bhachau. These spouted saline water, evidence of which is seen as white patches of salt on road. The Kohinoor salt factory, seen in top left hand corner, collapsed partially. (e) Cross-fissures near Mandvi.

The great Bihar–Nepal earthquake of 1934 provided one of the best examples of widespread liquefaction. A 200-km long and 60-km wide belt of liquefaction was formed within Mercalli intensity IX, and was named as the slump belt. Mercalli Intensity X was entirely within this belt. The chief criterion adopted in the demarcation of this slump belt was seismic response of the built environment (Auden et al., 1939). Buildings tilted and sank into the soft alluvium of the Ganga plains, and continued to do so for several days after the earthquake. Floors and walls of sunken buildings were covered by sand up to a depth of 3–4 feet (approximately 1 m). Concentric fissures formed in ground around several buildings. Foundations and floors were completely ruined. All houses were abandoned.

Subsidence of roads and railway embankments was profound within this slump belt. A 2-m high embankment sank and became level with its surroundings. Tanks, lakes, pits, and other depressions became shallower as their bottoms were filled with outpouring sand. Due to strong ground shaking, elevations and depressions approached a common level. This happened in profusion in Champaran, Darbhanga, Muzaffarpur, and Purnea districts. Motihari, Madhubani, Supaul, and Sitamarhi, which were some of the worst affected places, lie within this region.

Parts of many coastal cities and seaports are built on unconsolidated sediments or on filled ground or on land reclaimed from sea. Many cities exist in areas where sand and silt were deposited in geologically recent times, i.e., within the last 10,000 years, and where there is shallow ground water. Some of these may be susceptible to liquefaction in strong ground shaking. It is best to avoid sites that have loose fine sand, soft silt, and expansive clays. If unavoidable, large structures must rest either on a rigid raft foundation or on pile foundation taken to a firm stratum. For light construction, susceptible soil can be suitably improved by compaction, sand piling, or by stabilizing the soil. Also foundations must be sufficiently deep to reach a firm stratum, be wide enough to bear load of the structure, and should use a rich mortar of cement and sand.

FISSURES

Extensive ground fissures are observed in many earthquakes. These can be long, wide, and deep in alluvium. Some fissures spout fountains of water and sand. Water that comes out of fissures as fountains collects in nearby low-lying areas as pools and sag ponds.

In the great Assam earthquake of 1897, long and numerous wide gaping fissures opened in all directions in alluvial plains around the River Brahmaputra. Extreme geographical limits from which fissures were reported were Sibsagar in the east and Bihar in the west, which is an area nearly 600 miles across (960 km) in an ENE–WSW direction. In the north–south direction, these were formed between terai regions of Nepal and Midnapur, a distance of about 300 miles (480 km). This phenomenon was extensive in Goalpara and Kamrup districts, western part of Darrang, Nowgogn, Sylhet, north Cachar, Rangpur, Dinajpur, Rajshahi, Maldah, Purnea, Pabna, Bogra, Maimansingh, and Dacca. The outpouring sand from fissures filled up tanks and wells, and riverbeds were elevated. This disturbed the drainage system and hampered navigation in the Brahmaputra for a long time after the earthquake.

In the great Bihar–Nepal earthquake of 1934, long, wide, and deep fissures were abundant in the entire slump belt. Most large fissures were confined near the epicenter, between Rivers Ganga, Gandak, and Kosi, in Riga, Madhubani, Monghyr, Motihari, Muzaffarpur, Supaul, Purnea, Raxaul, Samastipur, and Darjeeling. A typical fissure at Champaran was 15-feet deep, 30-feet wide, and 300-yards long (approximately 4.57 × 9 × 275 m). At Sitamarhi, a fissure was about 80 yards long and 8 feet wide (70 m × 2 m) and was filled with sand within 1 m of the top. Such fissures were common in the entire affected area.

Deep, wide, and long fissures were formed in topographic highs in the Kutch earthquake of 2001. Wherever fissures were found in abundance, evidence of earthquake fountains, soil liquefaction, and mudflow was often observed nearby, and this is shown in Figures 9.2c and d. At some places, these fissures disappeared, only to reappear a few meters away. Ground fissures also formed in marshes of the Rann of Kutch, below the Surajbadi Bridge and cross-fissures were observed at Moti Undo, near Mandvi, as shown in Figure 9.2(e).

The Andaman Trunk Road (ATR) developed long, deep, wide, and gaping fissures at several places due to the Sumatra earthquake of December 26, 2004. These were observed on the islands of North Andaman, Middle Andaman, South Andaman, and Baratung, at an epicentral distance of almost a thousand kilometers, and the road was difficult to negotiate in large stretches. The fissures were an earthquake effect and not caused by the tsunami.

EARTHQUAKE FOUNTAINS

Where there is plenty of shallow ground water, strong ground shaking often produces earthquake fountains, spouts, and geysers. In this transient phenomenon, there is usually a continuous flow for sometime that gradually falls off. The fountain may contain water, sand, clay, silt, and debris. Water from fountains collects on the surface as pools. These may be produced in the same way and at the same place as artesian wells that exist in many places where earthquake fountains were observed earlier. Strong ground shaking often breaks up local resistance in a shallow and porous aquifer, and builds enough pressure to eject water to the surface in the form of high fountains. Preexisting faults and newly formed fissures may provide a convenient path for this.

Earthquake fountains were reported in several earthquakes. In the Rann of Kutch region 2–2.5 m high fountains of sand and water spouted from ground fissures near the Allah Bund fault in the great Kutch earthquake of 1819 (Oldham, 1928). In the great Assam earthquake of 1897, earthquake fountains occurred in the alluvial plains of the Brahmaputra. In the San Francisco earthquake of 1906, 6-m high earthquake fountains were reported. In the great Bihar–Nepal earthquake of 1934, almost 10-m high, solid columnar fountains spouted from fissures formed along the Ganga River. These continued intermittently for almost 3 h on both sides of the Ganga. Agricultural fields were flooded and standing crops were killed due to the warmth of the water and strong shaking of the roots.

Earthquake fountains were reported in several villages within MMI VIII in the Kutch Earthquake of 2001. In Chang Nadi, and at Gadsisa, sweet water emanated from fountains. Bhachau, Samakhiali, Amardi, and Dudhai also witnessed this transient effect in profusion. An eyewitness, shepherd Murji Khiraj Gaduri, reported 3-m high water fountains emerging from fissures, which continued to spout water during the strong shaking and for about 2 minutes afterward, first muddy then clear. This gave rise to several pools of sweet water in this arid region where drought conditions continued to prevail for 3 years before the earthquake. This indicated that saline seawater of the nearby Arabian Sea did not infiltrate the aquifer at Moti Undo from which this water came. Even 8 days after the earthquake, although the water evaporated due to the hot desert sun yet very deep pugmarks of a dog impressed in soaked clay suggested that a huge quantity of water had collected there earlier, as shown in Figure 9.3c. These fountains were tracked for more than 4 km in a linear stretch (Sinvhal et al., 2003b).

Several earthquake-related water bodies falling within isoseismal X and IX were associated with faults in the Kutch earthquake of 2001. Numerous elongated new pools of water were observed between Samakhiali and Bhuj, along the north edge of the east–west trending National Highway 8A in a stretch of about 40 km. Some of these were about 3 m long and 2 m wide. These and other pools of water were evident between several prominent faults, as interpreted in satellite imageries (Saraf et al., 2001, 2002) and shown in Figure 9.3. Numerous large and small craters were formed during the Kutch earthquake of 2001, which spouted 1-m high sand and water fountains in topsoil in agricultural fields, marshes, and embankments (Figure 9.2c).

(a) (b)

(c)

Fig. 9.3 (a) Pools of water that emerged after the earthquake, as revealed by satellite imageries, are shown by red dots. Isoseismal map for the Kutch earthquake of January 26, 2001 and major faults are also shown. (b) Evidence of pools of water that emerged as earthquake fountains from the fault along the Rukmavati River, near Mandvi in the Kutch earthquake of January 26, 2001. Faulted zone along the dry reservoir of the Wangdi dam, from which water gushed out. (c) Pool of water from fissure at Moti Undo, near Mandvi. (See color figure also.)

SAND BOILS

Sand brought up in an earthquake is sometimes deposited around the spout in a form that resembles a miniature crater, which remains as such for some time till it is eroded. Sand boils can cause local flooding and surface deposit of silt. These were observed in profusion below the Surajbadi Bridge in the Kutch earthquake of 2001.

MUD FLOWS

Mud flows were observed at several places in the meizoseismal area of the Kutch earthquake of 2001 in several newly formed craters, some of which were more than 5 m wide and 2 m deep; this flow was tracked for more than 4 km in Chang Nadi.

MUD VOLCANO

Mud volcanoes are associated with geologically young sedimentary deposits, and are formed at destructive plate margins, mostly due to friction between the subducting plate and the overriding plate, several kilometers beneath the earth's surface. At some point, compressive forces become large enough to squeeze upward and expel gases mixed with mud and water to the surface of the earth. A mud volcano acts like an open pressure valve in the earth's crust.

The Sumatra earthquake of 2004 caused the eruption of several mud volcanoes on the Baratung Island. This tiny island, nestled between the larger South Andaman and Middle Andaman islands, is almost 1000 km north of the epicenter. A big explosion that was heard on the entire island accompanied this rare seismotectonic phenomenon. It marked activation of the volcano immediately after the earthquake. It was reportedly accompanied by fire. When a mud volcano ejects large amounts of gas, which is mostly methane, the gas plume can often catch fire.

The largest and most spectacular mud volcano changed the landscape of the area, as is shown in Figure 9.4. This dome-shaped mound was almost 3 m high and 50 m in diameter. It was composed of a large mass of fine, soft mud, and clay that dried up almost immediately after ejection. A handful of wet cold mud spewed out of several orifices of the volcano into the air even 2 weeks

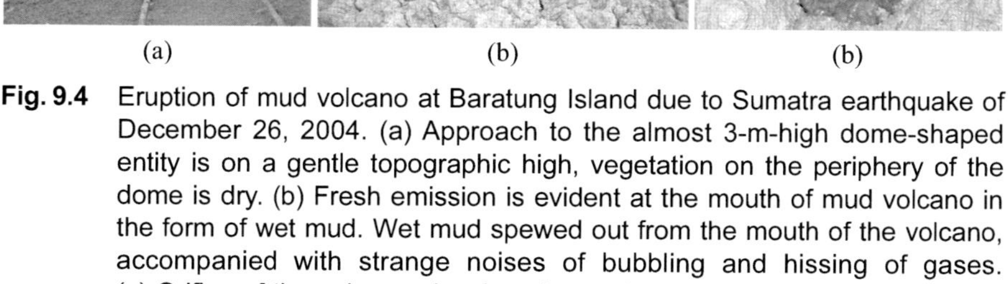

(a) (b) (b)

Fig. 9.4 Eruption of mud volcano at Baratung Island due to Sumatra earthquake of December 26, 2004. (a) Approach to the almost 3-m-high dome-shaped entity is on a gentle topographic high, vegetation on the periphery of the dome is dry. (b) Fresh emission is evident at the mouth of mud volcano in the form of wet mud. Wet mud spewed out from the mouth of the volcano, accompanied with strange noises of bubbling and hissing of gases. (c) Orifice of the volcano showing ejecta of fresh wet mud.

after the earthquake. Belching and escaping volcanic gases accompanied the ejecta. The mouth of the orifice was almost 30 cm high above the dome, with an inner diameter of about 15 cm (Figure 9.4c); (Sinvhal et al., 2005a). Most vents on the unconsolidated dome dried up within 2 weeks of the earthquake. The circular periphery of the mud volcano was surrounded by a dense tropical forest, around which vegetation dried up due to neotectonic activity.

The mud volcano was on a gentle topographic high, indicating that eruptions had occurred earlier also. The Diglipur earthquake of September 14, 2002 (magnitude 6.0, epicenter 13.3°N, 93.3°E) and its aftershock of February 18, 2003, caused a previous eruption of this volcano, which continued to spew mud for several days. Emissions were spread within a diameter of about 70 m, much larger than that caused by the great Sumatra earthquake of December 2004. This was evident in the contact between old and fresh accumulations in the entire uplifted area, as a low dome in the form of dry, eroded, and fissured clay at the base of this reactivated volcano.

The mud volcano at Baratung was caused due to the subduction of the Indian plate beneath the Andaman plate. This produced a large amount of debris, which was pushed up to the surface in the form of soft sediments. The volcano is located on the north–south trending Eastern Boundary Thrust, which extends from Myanmar in the north to Nias Island (off the western coast of Sumatra) in the South, and through most Islands of Andaman and Nicobar archipelago.

GROUND AND SURFACE WATER

Strong ground shaking in an earthquake can sometimes disturb ground water and surface water in a very large area. Damage due to surface water can be in the form of diversion and damming of rivers, sloshing over stream banks, and change in level, color, turbidity, smell, and bubbles of water in springs, rivers, and wells. Several wells get filled with clay and silt that come out with water fountains.

This was evident in the Kutch earthquake of 2001 near Banni and Moti Undo, where dry wells suddenly filled up after the earthquake. In some places, tube-wells yielded sweet water, but salty water was reported in the transient fountain nearby. In other places, the case was reverse, e.g., in Shantli village in Radhanpur Tehsil, 5 km from Lodai, between Manfara and Chobari and at several other places where some wells became unusable. Widespread appearance of earthquake-induced water bodies and channels occurred in the Rann of Kutch, Little Rann of Kutch, between Maliya Miyana and Samakhiali, in the Gulf of Kutch, and in the vicinity of Gandhidham and Kandla seaport (shown in Figure 9.3a).

One year prior to the Latur earthquake on September 30, 1993, ground water conditions in the Meizoseismal area were highly disturbed. Two tube-wells at Killari, about 100 m deep, dried up after a smaller (magnitude 4.3) and earlier earthquake of October 18, 1992. Subsequently, the owner took out his pump sets for fear of damage in the numerous small earthquakes that were expected to follow. After a subsequent shock of October 28, 1992, sound of gushing water was clearly heard even 3 m away from the same well. The pump set was immediately reinstalled, and a very thick discharge of water ensued. The pumps were dry once again, 11 months later, after the main event of September 30, 1993. This was indicative of fluctuations in water level at the same place due to foreshocks and successive aftershocks. Turbid and foul smelling water was reported from nearby wells. (Sinvhal et al., 1994). One-and-a-half kilometer west of the failed water tank at Kawtha, the bore wells showed an increase of output, whereas half a kilometer east of the same water tank the situation was reverse. In Takari village of Paranda taluka, Osmanabad district, bubbles and white smoke emanated from wells and continued for 3–4 hours after the earthquake. These were not isolated instances but were observed in the entire meizoseismal area.

LAND SLIDES

Earthquakes induce landslides in hilly terrains. Large earthquakes induce numerous large landslides that are spread in a wide area. The term landslide describes a wide variety of processes that result in the downward and outward movement of slope-forming materials such as rock, soil, artificial fill, or a combination of these. Strong ground shaking loosens these. A distinct zone of weakness separates the slide material from the more stable underlying material. Major earthquakes in the Himalayan arc have triggered massive landslides. Peninsular and coastal regions of India also have several landslides.

Types of Landslides

Various types of landslides are best differentiated by two factors, the kinds of material involved and the kind of movement of this material. Rock or soil material may move by different modes in an earthquake such as fall, topple, slide, spread; flow, or creep. Classification based on these parameters was given by Varnes in 1978. Other classification systems use additional variables, such as rate of movement and air, water, or ice content of landslide material.

Two major types of slides are rotational slides and translational slides. In a rotational slide, the surface of rupture is curved concavely upward and the slide movement is roughly rotational about an axis that is parallel to the ground surface and transverse across the slide. Figure 9.5 shows a rotational landslide triggered by the Kashmir earthquake of 2005. This curvilinear failure occurred near top of terrace along vertical banks of Jhelum River. It is located between Baramulla and Uri, near Mahura, along National highway NH 1A on left bank

Fig. 9.5 Rotational landslide induced by the Kashmir earthquake of 2005 along River Jhelum near Baramulla. For comparison of scale, note the triple-storey house on top of terrace, of approximate height 10 m.

of Jhelum River. This was the eastern limit of the landslide territory and the incidence and volume of mass wasting increased gradually as epicentral distance decreased.

In a translational landslide, the slide material moves along a roughly planar surface with little rotation or backward tilting. Sometimes slabs of hard sedimentary rock slide down *en masse*. At other times, it may consist of soft debris. One such example, triggered by the Kashmir earthquake of 2005, is shown in Figure 9.6. This gigantic translational type of landslide was triggered between Kupwara and Tangdhar, near Nasta Chun pass, better known as Sadhna pass. The fair weather, unmetalled road, negotiating a steep hill with an almost 40° slope, is the only road connection to Tangdhar and Tithwal. This zigzag road was covered with landslide debris, but due to its strategic importance it was cleared immediately after the earthquake. Roads in this highly thrusted zone were stabilized with protection walls, made of random rubble stone masonry, and could therefore function (after clearing) even at an epicentral distance of about 30 km, in the western syntaxis.

Creep is an imperceptibly slow downward motion of slope-forming soil and weathered rock over bedrock. Movement is caused by shear stress sufficient to produce permanent deformation, but too small to produce shear failure. Continuous creep occurs where shear stress continuously exceeds strength of the material. Progressive creep occurs where slopes and slope-forming material are reaching the point of failure. Tree trunks curved at their base, shown in Figure 9.7, bent fences or retaining walls, tilted poles or fences and small soil ripples or ridges indicate soil creep.

Fig. 9.6 Translational type of landslide at Nasta Chun Pass triggered by the Kashmir earthquake of 2005. The zigzagging fair weather road, negotiating a steep hill with an almost 40° slope, is the only road connection to the Tangdhar bowl. This unmetalled road was covered with landslide debris and cleared immediately after the earthquake. Earlier, it was stabilized with slope protection walls, made of random rubble stone masonry. Note and compare the size of truck, shown within the two rectangular blocks, with that of the landslide. For scale of comparison, the pine trees are 20–30-m high and showed the typical bending at the base, indicative of soil creep.

Falls are abrupt movement of masses of geologic materials, such as rocks and boulders that become detached from steep slopes or cliffs. Separation occurs along discontinuities such as fractures, joints, and bedding planes and movement occurs by free-fall, bouncing, and rolling. Falls are strongly influenced by gravity, mechanical weathering, and presence of interstitial water. Figure 9.8 shows an example of boulders rolling from a heavily jointed face of a steep hill on to a mountain stream, and the bridge over it. The super structure of the single pier Sikh Bridge was damaged due to earthquake-induced rock fall from the adjacent hill. The debris partly blocked the flow of the Qazi nala. More landslides developed from the crest of hills in the background.

Lurching occurs at right angles to a cliff, more commonly, to a stream bank or an artificial embankment and leads to yielding of material in the direction in which it is unsupported. The initial effect is to produce a series of more or less

Fig. 9.7 Pine trees on a hilltop with tree trunks curved at their base, indicating ongoing slow creep in two opposing directions on the MBT, near Uri in Kashmir. Several hilltops also developed large fissures due to slope instability.

Fig. 9.8 Rock fall damages Sikh bridge in Tanghdar in Kashmir earthquake of 2005.

parallel cracks separating the ground into rough blocks. With stronger or longer shaking, the outer of these, adjacent to the bank, slides down, usually holding together and tipping toward the unsupported end. Others may follow in due course of time. Figure 9.9 shows the effect of lurching at Rajarwani near Baramulla.

Fig. 9.9 Lurching at Rajarwani, near Baramulla, induced by the Kashmir earthquake of October 8, 2005. River Jhelum flows in the background.

Landslides in the Himalayan Arc

Because of the northward movement of the Indian plate, the Himalayan ranges in the continent–continent collision zone have been rendered seismotectonically fragile. In these high-altitude areas, topography is rugged, hill slopes are steep, and are sometimes covered with weak, weathered, and unconsolidated material. Moreover, the hills are sometimes formed of faulted, fractured, fissured, jointed, and sheared rock material, with an adverse orientation of bedding planes, unconformities, and contacts. There may be a contrast in permeability and stiffness of materials. River valleys are steep, sometimes nearly vertical. Earthquake-induced landslides are maximized in such conditions. The Main Boundary Thrust (MBT), the Main Central Thrust (MCT), and other faults and thrusts fulfill these conditions and become quintessential landslide territory, waiting to be induced or reactivated by an earthquake or the torrential monsoons. Human intervention can also cause landslides. This can be by excavation of slope or its toe, loading of slope or its crest, draw down of reservoirs, deforestation, irrigation, mining, and artificial vibrations.

Seismic zone V, the most severe zone in the seismic zoning map of India as per BIS: 1893–2002, is particularly vulnerable to earthquake-induced landslides. The great earthquakes of Assam in 1897 and in 1950, Kangra in 1905, and Bihar–Nepal in 1934, repeatedly demonstrated this. More recently, the Uttarkashi earthquake of 1991 and Chamoli earthquake of 1999, both in seismic zone V, and Kashmir earthquake of 2005, induced many landslides in their meizoseismal areas.

In the great Assam Earthquake of 1897, gigantic landslides and rock falls were widespread north of Brahmaputra River and east of the 91° meridians. Tezpur and north Cachar hills marked the eastern limit; while Bhutan, Sikkim, and Darjeeling marked the western limit in the Himalayas. Landslides maximized in and around Goalpara, Sylhet, Cherrapunji, and Tura, and on the southern edge of Garo and Khasi hills. Hillsides facing the valley were stripped bear from crest to base. Oldham (1899) described hillsides so denuded of soil that bedrock stratification was exposed.

Due to the Uttarkashi earthquake of October 20, 1991, landslides maximized along the MBT and the MCT in the valleys of Rivers Bhagirathi and Mandakini (Figure 9.10). Most landslides were located in a belt that was 40-km long, between Ultra in the east and Saura in the west, with an N60°W–S60°E trend, and 2.5 km wide. The Ultra–Saura fault is 4 km north of the epicentral region and is almost parallel to the long axis of isoseismals. The

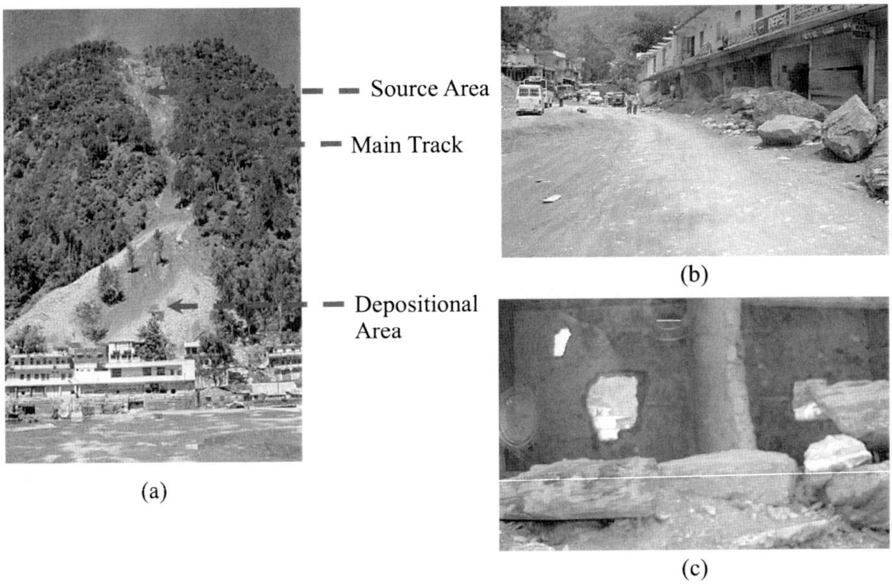

Fig. 9.10 Landslide in Uttarkashi: (a) This continued intermittently for several years from the Varnavrat hill, (b) The effect in Uttarkashi town in 2004, (c) Rolling boulders punched holes through walls in 2004.

areas most affected were the Uttarkashi–Bhatwari–Maneri–Agora region and also Sukhidhar, Dunda, Gangori, Ghansiali, Koti, Sangam Chatti, and Tehri region. Mass wasting was due to rock falls and debris slides and occurred in jointed quartzite. Pre- and postearthquake satellite imageries revealed 47 new landslides and reactivation of 16 old landslides (Vohra et al., 1992; Narula et al., 1995).

River Bhagirathi meanders around the Varnavrat hill in Uttarkashi. The national highway leading to Gangotri gets blocked during successive monsoons due to reactivation of landslide from Varnavrat. Several hotels and homes at the base of this hill nearest to the landslide were buried under the sliding debris. The initiation of this landslide is attributed to the Uttarkashi earthquake of 1991, the disastrous effects of which are continuing even 18 years later, as shown in Figure 9.10.

The Kashmir Earthquake of October 8, 2005, triggered huge landslides of unusual dimensions in the Balakot–Muzaffarabad–Uri region (Pove, 2006; Sinvhal et al., 2005b, 2006). These maximized along the MBT region, and along rivers Jhelum, Kishan Ganga, Kunnar, Neelam, and tributaries of the Indus and Jhelum. Landslides caused tremendous change in topography within the Western syntaxis and in Pir Panjal and Shamshabari mountain ranges. The Jhelum winds through many sharp bends downstream of Baramulla, and has vertical banks and gorges in several stretches. Landslide material from Baramulla to Uri, along the Jhelum, was composed mostly of river-borne material and conglomerates of large rounded pebbles within a soft powdery matrix. The 15–50-m high vertical terraces failed parallel to the river face in fresh vertical knife cut edges in several stretches. Some landslides originated from a height of 100 m above the riverbed.

Effects of Land Slides

Earthquake-induced landslides have many disastrous effects and increase vulnerability and risk of the human habitat and the built environment. Sometimes entire villages, houses, roads, bridges, etc. located on steep slopes are damaged or even completely buried under the debris. Sometimes large stretches of roads slide away. When rocks and debris fall on roads, it causes roadblocks and disrupts the road network. This adds to the difficulties of postearthquake rescue and relief operations. Even if roads are cleared after the earthquake, aftershocks trigger and sometimes reactivate landslides within the same stretch and debris still keeps the roads vulnerable. In Figure 9.6 the winding road was vulnerable to the debris triggered by many aftershocks of the likewise earthquake. The Srinagar–Muzaffarabad road was Kashmir blocked in several large patches by massive landslides; their incidence increased from Baramulla and maximized in and around Muzaffarabad, the epicenter.

Landslide debris obstructs rivers, and alters drainage pattern of the area. It also creates dams on rivers, which after sometime give way and cause extensive floods downstream. Landslides induced by the 1897 Assam earthquake flooded plains in and around Shillong, modified the watercourse in several valleys, and caused large-scale surface distortions in the meizoseismal area. This disastrous scenario was repeated upstream of Brahmaputra River by the great Assam earthquake of 1950. Natural dams and lakes were created in upper reaches of almost every tributary of the Brahmaputra, including Dihang, Dibang, and Subansiri. These swelled after the earthquake and brought down enormous amounts of debris like sand, mud, trees, etc. in landslides. The natural dam across Subansiri burst 4 days after the earthquake and 20 feet (approximately 6 m) high waves claimed 532 human lives downstream. This postearthquake flood scenario was replicated in several tributaries of the Brahmaputra and caused more damage to life and property than the great earthquake. Inundation of rivers swept the countryside for months after the earthquake.

This kind of damage scenario is repeated frequently in the seismically active Himalayan arc. This threat needs to be addressed before the next earthquake and concomitant landslides take further toll. Therefore, earthquake-prone areas in hilly terrains, where large populations are at risk are in need of special earthquake-related attention and protection.

While planning to make a human habitat on precarious hill slopes, considerations of seismotectonic background are paramount (Sinvhal et al., 2006). It is best to avoid construction activity on steep slopes and on preexisting landslides. If this situation is absolutely unavoidable, then mitigation measures involving engineering intervention become necessary. These consider several factors such as ground surface, angle, and material of slope, and faults and drainage of the area. Improvements can be provided to reduce landslide potential. This involves stabilizing and protecting slopes. Slope stability is increased when a retaining structure is placed at the toe of the landslide or when mass is removed from top of the slope. This can be achieved in several ways, by buttressing, providing mechanically stabilized walls, and retaining walls or barriers for holding back debris torrents and rock fall. Soil can be modified or replaced by means of grouting or densification. Stability increases when ground water is prevented from rising in the landslide mass.

The efficacy of these measures was in abundant display in the severely affected areas of Kashmir, when most roads on steep hill slopes continued to function after the Kashmir earthquake due to an elaborate network of slope protection works. Some of these protection measures failed partially, due to strong shaking in a seismotectonically vulnerable area at close epicentral distance, but continued to function as is evident in Figures 9.11.

Fig. 9.11 Slope protection walls kept the winding roads functional in the Kashmir earthquake of 2005. The portion shown is in the region of the Main Boundary Thrust, near Kamalkote in Uri. (See color figure also.)

CONCLUSION

Implication of ground damage to the built environment can be disastrous. If ground is damaged, structures built on it become vulnerable and may be damaged, or collapse partially or totally. Hence competency of ground needs special attention before any structures can be built on it, which will prove to be safe in an earthquake. A submarine earthquake can cause additional destructive effects produced by ocean waves even at very large epicentral distances by tsunamis. This is discussed in the next chapter.

REFERENCES

Auden, J. B., J. A. Dunn, A. M. N. Ghosh, D. N. Wadia and S. C. Roy, 1939, The Bihar-Nepal Earthquake of 1934, *Memoirs of GSI*, Volume 73, 391 p.

BIS: 1893–2002, *Indian Standard Criteria for Earthquake Resistant Design of Structures, Part I : General Provisions and Buildings* (Fifth Revision), Bureau of Indian Standards, New Delhi, 40 p.

Middlemiss, C. S., 1910, The Kangra Earthquake of 4th April 1905, *Memoirs of Geological Survey of India*, Volume 38, 409 p.

Narula, P. L., S. K. Shome, S. Kumar and P. Pande, 1995, Damage patterns and delineation of isoseismals of Uttarkashi earthquake of 20th October 1991, in *Uttarkashi Earthquake*, Eds. H. K. Gupta and G. D. Gupta , Memoir 30, Geological Society of India, 233 p.

Oldham, R. D., 1899, Report on the Great Earthquake of 12th June 1897, in *Memoirs Geological Survey of India*, Volume 29, 379 p.

Oldham, R. D., 1928, The Cutch (Kachh) earthquake of 16th June 1819 with revision of the great earthquake of 12th June 1897, in *Memoirs Geological Survey of India*, Volume 46, p 71–147.

Pore, S. M., A. D. Pandey and A. Sinvhal, 2006 c, Kashmir (Muzaffarabad) earthquake of Oct. 8, 2005: Geotechnical observations, in *Proceedings of Earthquake Disaster: Technology and Management–EARTH 2006,* Volume I, p 1–7, 11–12 Feb. 2006, Motilal Nehru National Institute of Technology, Allahabad.

Saraf, A. K., A. Sinvhal and H. Sinvhal, 2001, The Kutch earthquake of January 26th, 2001: Satellite data reveals earthquake induced ground changes and appearance of water bodies, in *Proceedings of Workshop on Recent Earthquakes of Chamoli and Bhuj*, May 24–26, 2001, Roorkee, Indian Society of Earthquake Technology, p 207–215.

Saraf, A. K., A. Sinvhal, H. Sinvhal, P. Ghosh and B. Sarma, 2002, Satellite data reveals 26 January 2001 Kutch earthquake induced ground changes and appearance of water bodies, *International Journal of Remote Sensing*, **23**(9), p 1749–1756.

Shankar D., H. R. Wason, A. Sinvhal and V. H. Joshi, 2005, Damage due to devastating earthquake (M_W 9) and tsunami of December 26, 2004 in Andaman and Nicobar, India: A perspective, in *Proceedings of the Twenty Second International Tsunami Symposium*, 27–29 June, 2005, Chania, Crete, Greece, p 221–232.

Sinvhal, A. and H. Sinvhal, 1994, Geotechnical aspects of some Indian earthquakes, Indian Geotechnical Profile, in *Proceedings of the Thirteenth International Conference on Soil Mechanics and Foundation Engineering*, New Delhi, p 6–10.

Sinvhal, A., V. Prakash, P. R. Bose, A. Bose, H. R. Wason, H. Sinvhal and A. D. Pandey, 2003a, Ground damage observed in the Kutch earthquake of 26th January, 2001, in *Proceedings of Indian Geotechnical Conference, IGC 2003, Geotechnical Engineering for Infrastructure Development*, Roorkee, India, p 273–276.

Sinvhal, A., P. R. Bose, V. Prakash, A. Bose, A. K. Saraf and H. Sinvhal, 2003b, Isoseismals for the Kutch earthquake of 26th January 2001, *Earth and Planetary Sciences*, **112**(3), p 1–8.

Sinvhal, A., H. R. Wason, D. Shanker, A. K. Mathur and V. H. Joshi, 2005a, Mud Volcano at Baratang, *Sci. Tech.,* 4(1), p 13.

Sinvhal, A., A. D. Pandey and S. M. Pore, 2005b, Preliminary report on the 8th October 2005 Kashmir earthquake, Department of Earthquakc Engineering, IIT Roorkee, 60 p.

Sinvhal, A., A. D. Pandey and S. M. Pore, 2006, The Kashmir earthquake of 8th October 2005, and landslides, in *100th Anniversary 1906 San*

Francisco earthquake Conference (Abstract Volume), 18–22 April, 2006, San Francisco, USA, SSA 836.

Vohra, C. P. and D. Gupta Sarma (Eds.), 1992, *Uttarkashi Earthquake of October 20, 1991*, Special Publication No. 30, Geological Survey of India, Calcutta, 218 p.

Wason, H. R., A. Sinvhal, D. Shanker, A. Kumar and V. H. Joshi, 2006, Ground deformation observed due to the great Sumatra earthquake of December 26, 2004 and tsunami in and around Andaman and Nicobar Islands, in *Proceedings of the Thirteenth Symposium on Earthquake Engineering*, I I T Roorkee, December 18–20, 2006, p 228–237.

wikipedia.org/wiki/landslide

10 CHAPTER

Tsunamis and Earthquakes

INTRODUCTION

Tsunami is a Japanese word that translates as a harbor wave (tsu means harbor and nami means wave). In South America, the name is maremoto. It is a series of gigantic waves triggered in a large body of water by a disturbance that vertically displaces a water column. This phenomenon has catastrophic connotations in low-lying coastal areas, even at very large epicentral distances.

EXAMPLES

Many populated coasts, like those of Chile, Peru, Japan, Indonesia, and Hawaii, have been visited repeatedly by tsunamis. The 1703 earthquake of Awa killed more than 100,000 people in Japan in the tsunami that followed. On June 15, 1896, in the Sanriku earthquake nearly 27,000 were killed on the east coast of Japan. The Lisbon earthquake of November 1, 1755, killed more than 60,000 people in Europe. The Arabian coast of India saw tsunamis due to the great Kutch earthquake of 1819 and again the earthquake of November 28, 1945. The spectacular underwater volcanic explosions that obliterated Krakatoa Island on August 26 and 27 in 1883 created waves as high as 35 m in Indonesia, killing more than 36,000 people in Java and Sumatra. The North Andaman earthquake of January 26, 1941, claimed more than 5000 lives. The tsunami that followed the earthquake of August 23, in 1976, killed 8000 people in SW Philippines. However, the most disastrous of all tsunamis was generated recently, by the Sumatra earthquake of December 26, 2004, which originated in the Indian Ocean.

CAUSE

A tsunami is most often caused by a submarine earthquake, which has a shallow depth of focus, usually less than 50 km, and magnitude usually more than 7.5. It is associated with deep trenches on a destructive plate margin, where the ocean floor is displaced vertically in a dip slip fault by an earthquake. This abrupt vertical displacement in the faulted area displaces a thick column of seawater above it and sets the entire column of water into motion. The result is a sea wave between the top and bottom surface of water. This propagates away from the source of disturbance. This initiates disturbance in the sea and oscillations on the surface of water, with progressively widening wave fronts that propagate to large distances. This process is illustrated in Figure 10.1. For more on subduction zone tectonics, see Chapter 2 on plate tectonics. Submarine landslides, volcanic eruptions, or meteorite impact may also disturb the water and cause a tsunami.

In the deep ocean, a tsunami can have a very large wavelength, of the order of 100–200 km, and very small amplitude in comparison, somewhere between 0.3 and 0.6 m. The period of these long waves can vary from 5 min to almost an hour. Thus, a particle such as a ship on the surface in the open ocean experiences the passage of a tsunami as an imperceptible rise and fall of only 0.3–0.6 m that lasts from any where between 5 min to an hour. In shallow waters, near the coast, the height of the tsunami may build up to several meters. The wavelength of the tsunami and its period depend on the dimensions of the source event and depth of water.

The velocity of a tsunami wave, c, is determined by the formula $\{c = (gD)^{1/2}\}$ (Satake, 2002), where g is the acceleration due to gravity and D is depth of water. In the deep ocean, for an average depth of 4000 m, the wave travels very fast, about 200 m per second, i.e., about 720 km per hour. Thus, depending on the depth of ocean, velocity of these sea waves varies and arrives at different coasts at different travel times. The sharp elevation of the ocean floor near a continental slope, a continental shelf, and a coast, i.e., in the continental margin, slows down the bottom of the sea wave due to friction between ocean waves and land. This considerably reduces velocity of the tsunami, shortens its wavelength, and increases its amplitude substantially. A vast quantity of water then piles up on the coast, into a vertical wall, which can be as high as 15–30 m (50–100 feet) within a short span of 10–15 min. This wall of water crashes on the shore with a tremendous destructive force.

A tsunami often comes in a series of waves, may be three to five major oscillations, separated by small intervals of half an hour or so. The amplitude of the waves gradually decreases and eventually ceases several days after it begins. The retreat of a tsunami from coastal areas can be as disastrous as its approach. Occasionally, on some coasts the first arrival of a tsunami may be a

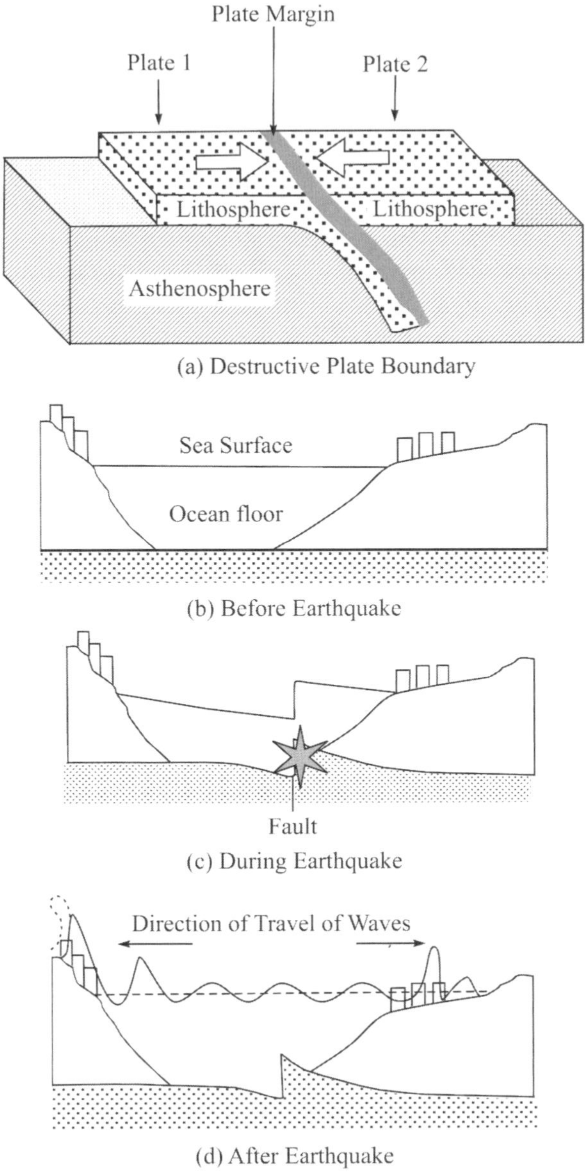

Fig. 10.1 A large submarine earthquake that originates at a destructive plate boundary sometimes causes a tsunami. (a) In a destructive plate boundary plate 1 is the subducting plate, and plate 2 is the overriding plate. (b) The situation of the sea surface and the ocean floor before an earthquake. (c) Hypocenter of a submarine earthquake, shown by the star, displaces the ocean floor vertically in a dip slip fault in the subduction zone. The water column above this is displaced simultaneously. (d) This sets waves in the ocean; and amplitude increases on the seacoast.

trough, the water receding and exposing the shallow sea floor. The succeeding wave crest may arrive a few minutes later.

EFFECTS

The consequence of a tsunami can be catastrophic. Low-lying coastal areas are prone to extensive inundation and run up. Inundation is the horizontal extent of water penetration, i.e., distance between the inundation line and the coast. Run up is the maximum elevation of water on land, i.e., elevation reached by seawater measured relative to some datum. Inundation and run up result in ingress of saline seawater, accompanied with mud and debris, liquefaction, scouring, erosion, flooding, and water logging. Transgression of sea is dependent on local topography. In a bay or river inlet, which narrows rapidly and has a confining effect, the tsunami surges to extreme heights due to continuous decrease in velocity. The built environment is sometimes obliterated in the area of inundation and run up due to impact of sea waves on structures and erosion. This includes coastal structures like jetties, harbours, wharfs, and associated buildings.

Effects of a tsunami vary widely from place to place as tsunamis are reflected and refracted by coastal topography as any other water waves. Coasts that have a landmass between them and the newly faulted sea floor are usually sheltered from the disastrous effects of a tsunami and may be somewhat safe. However, tsunamis can sometimes diffract around such landmasses and may not spare the sheltered area.

THE TSUNAMI GENERATED BY THE SUMATRA EARTHQUAKE OF DECEMBER 26, 2004

The great Sumatra earthquake, with epicenter at 3.27°N, 95.82°E, focal depth 30 km, originated in the Indian Ocean, off the west coast of north Sumatra. Its epicenter was about 250 km south east of Banda Aceh and north of Simeulue Island, on a convergent plate margin. It was about 350 km south-east of Indira Point, which is the southern most point of India and the nearest Indian Territory to the epicentre, as shown in Figure 10.2. Indira point is in Great Nicobar Island, which is the largest of the Nicobar group of islands. The earthquake originated at 00:58:53 UTC (06:29 IST) and was assigned magnitude $M_s = 9.0$, $M_w = 9.3$ (USGS). Parameters of this earthquake are given in Table 10.1.

The aftershock sequence associated with this earthquake continued for several months. These were spread in a region between latitude 0–20° N and longitude 91–98°E and depth ranged from 2 to 110 km (USGS). The magnitude range varied between 3.0 and 7.9 and included several large magnitude aftershocks. The largest aftershock occurred on March 28, 2005,

Table 10.1 Parameters of the great Sumatra earthquake of December 26, 2004, as given by India Meteorological Department (IMD) and United States Geological Survey (USGS).

Agency	*Latitude (North)*	*Longitude (East)*	*Origin time*	*Magnitude*	*Depth of Focus*
IMD	3.34°	96.13°	06:29 (IST)	8.6 M_s	—
USGS	3.27°	95.82°	00:58:49 (GMT),	9.0 M_w	30 km
	3.09°	94.26°	(07:58 local time)	9.3 M_w	

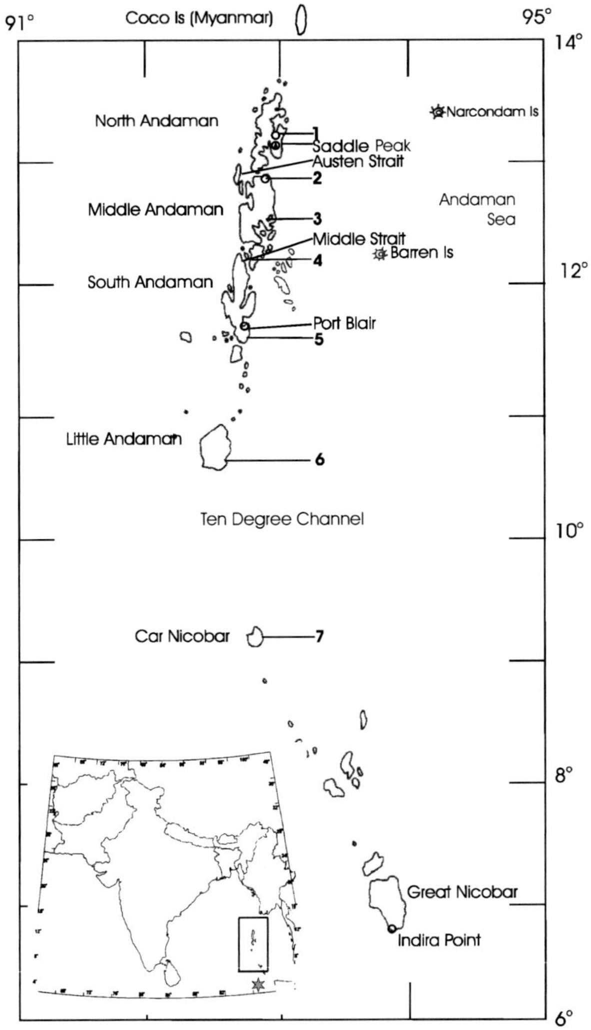

Fig. 10.2 Location map of Andaman and Nicobar Islands. (1) Diglipur, (2) Mayabunder, (3) Rangat, (4) Baratang, (5) Chidiya Tapu, (6) Hut Bay, (7) Malacca, and (8) Port Blair and adjoining areas. ❁ Volcanic Islands. Inset shows location of Andaman and Nicobar Islands on the map of India. Epicenter of the earthquake of December 26, 2004, is shown by

had a magnitude M_S, 8.7, (IMD), was life threatening in a limited area, and did not induce a tsunami. These aftershocks further weakened already damaged structures and continued to spread panic among the affected population.

Geographical Extent of Damage

This great earthquake was followed by a disastrous tsunami in coastal regions of the entire Indian Ocean. It caused extensive damage in an area that was much wider than that directly affected by earthquake shaking. Countries that bore the brunt of devastation included Indonesia, Sri Lanka, India, Thailand, Maldives, Malaysia, Myanmar, and Seychelles. The tsunami spread to the east coast of Africa and affected coastal regions of Somalia, Tanzania, and Kenya. Tsunamis also occurred on the coasts of Coco islands, Mauritius, and Reunion Islands. The tsunami crossed into the Pacific Ocean and was recorded along the west coast of north and South America. Devastation was mostly confined to a narrow coastal belt; about 500–1000 m wide in most places, and extended further inland where the coast was almost flat or where the tsunami went inland due to inlet of a river, or resonated in a bay.

Travel Times

The tsunami traveled away from the epicentral region. Because of the varying distances and ocean depths involved, the tsunami took anywhere from 15 min to 7 h to reach various coastlines. It arrived at Banda Aceh in Sumatra and Car Nicobar (epicentral distance almost 600 km) within minutes of the earthquake. At Port Blair (epicentral distance almost 850 km), the first minor wave started around 7:00 am, i.e., about half an hour after the earthquake originated. Initially five to seven waves were observed every 5 min apart, and then the 3-m high tsunami appeared as a deluge. It arrived in Sri Lanka and on the east coast of India between 90 min and two-and-a-half hours after the earthquake. Coastal beaches of Thailand were struck two hours after the earthquake despite being closer to the epicenter, as the tsunami slowed down in the shallow Andaman Sea. It arrived at Somalia 7 h after the earthquake originated.

Impact of Damage

As coastal areas and beaches on the rim of the Indian Ocean were recently developed, the tsunami that followed the earthquake negated all this development by destroying coastal structures and claimed almost 2,30,000 human lives. In India the Andaman and Nicobar chain of islands were devastated. Coastal states in the Bay of Bengal, namely Tamil Nadu, Andhra Pradesh, Pondicherry, and Orissa, suffered substantial damage. Kerala suffered substantial impact despite being on the west coast, mainly due to diffraction effects of tsunami waves. A glimpse of the damage scenario is given here for Port Blair (Shankar et. al. 2005; Wason et. al. 2006) and for Car Nicobar, which are at an approximate epicentral distance of 850 and 600 km, respectively.

The great earthquake and the tsunami that followed together damaged ground in several ways, devastated civil structures, infrastructure, and the human habitat in all the inhabited islands of Andaman and Nicobar archipelago. The disaster was severe in all the islands of Andaman and profound in all the islands of Nicobar, the latter being closer to the epicenter. Port Blair, was mercifully spared the full fury of the tsunami as the coast was jagged and hilly compared to what was observed at Car Nicobar. The damage scenario was more profound further south, as most of these islands were smaller, (barring Great Nicobar) had low heights, flat beaches, and experienced higher run ups. In some areas of Port Blair, eyewitnesses reported earthquake fountains of clay and sand after the tsunami. These continued for about 3 h after the tsunami, till about 10:30 am IST. This also indicated widespread effects of liquefaction. These effects were later obliterated by sea waves.

Andaman and Nicobar Islands: Location and Seismo Tectonic Features

The Andaman and Nicobar islands, located in the Bay of Bengal, comprise a chain of more than 500 islands distributed in a north–south trending arc spanning about 800 km. The Andaman Islands are in the north and Nicobar Islands are south of the 10° latitude. South Andaman is the longest island, 350 km in length, and 50 km in its widest stretch. Port Blair is the seat of administration for these islands.

These islands have an annual rainfall of 3000 mm/year, and a thick cover of tropical rain forest. Fertile land of these tropical islands produces good timber and crops of coconut, betel nut, cashew, banana, papaya, and various spices like cloves, pepper, and cinnamon. Maximum elevation in the islands is 728 m above mean sea level (MSL) in North Andaman Island. Most hill slopes are gentle, have a soft sedimentary cover, and a thick cover of vegetation. As the tropical sun and silvery sand on palm fringed beaches makes coastal regions a coveted human habitat, large population centers developed recently along the coastline. This included government establishments, some of immense strategic and defence importance, offices, houses, and the vast infrastructure required to support these. All this development was oblivious to the presence and the destructive potential of a large subduction zone in the Bay of Bengal, to seismicity and seismotectonics of the region and disastrous effects of tsunamis generated in these regions by the earlier disastrous earthquakes of 1881 and 1941.

On the destructive plate boundary that exists in the Bay of Bengal, the Indian plate is subducting below the Eurasian plate at an angle of about 30°. Surface manifestation of this subduction zone is the Andaman Sumatra Java Sunda Trench system. The trench axis is about 3000 m deep near North Andaman Island and deeper, 4000 m, near Little Andaman Island and Great Nicobar Island. Significant tectonic units in the epicentral region and around the Andaman Nicobar Islands are the Andaman Trench, the Eastern Boundary Thrust, the Volcanic Arc, Sedimentary Outer Arc Ridge, Andaman Back Arc Spreading Ridge, and the West Andaman Fault. These tectonic features are shown in Figure 10.3.

The N–S trending West Andaman Fault is located east of the sedimentary outer arc ridge. It is accompanied by a complex set of faults. A 60–70-km-wide area, defined by the 200 m isobath, and almost parallel to the Andaman trench, lies between the Andaman trench and the volcanic arc. Part of this is exposed as the Andaman and Nicobar group of islands and is referred to as the sedimentary outer arc ridge. This structural high consists of oceanic crust and sediments scraped off the descending Indian plate. Mergui Terrace defines the Andaman Back Arc Spreading Ridge in the east. The Andaman and Nicobar islands are bound in the east by a spreading ridge and in the west by a subduction zone, so the sedimentary ridge acts as a small tectonic plate. Curray et al. (1982) refer to it as the Burma plate and Dasgupta (1993) as the Andaman Plate. This is a minor plate wedged between the two larger plates: Indian and Eurasian, and the Islands are located on the overriding Andaman plate.

The eastern part of Andaman and Nicobar Islands is occupied by highly deformed rock formations, which are in part volcanic, oceanic, and metamorphic, and occur as a tectonic mélange. In contrast, the western part of these islands is occupied by more coherent and recent formations (sandstone, siltstone, conglomerate). Contact between eastern and western formations is marked by an east dipping thrust zone, the eastern boundary thrust. This regional thrust extends from Myanmar in the north to Nias Island of Indonesia (off Sumatra) in the south. Epicenter of this earthquake is in the vicinity of Nias Island.

Due to the seismotectonic processes, the subduction zone gave rise to a volcanic arc. In the area of interest, it contains the volcanoes of Narcondum and Barren islands. Narcondum Island represents a recently extinct volcano. Barren Island is the only active volcano in this part of the convergent plate boundary. Shallow focus strike-slip earthquakes occur along the West Andaman fault indicating upper-plate seismicity. The Andaman spreading ridge gives rise to many shallow focus earthquakes of moderate magnitude, which display normal fault with strike-slip component.

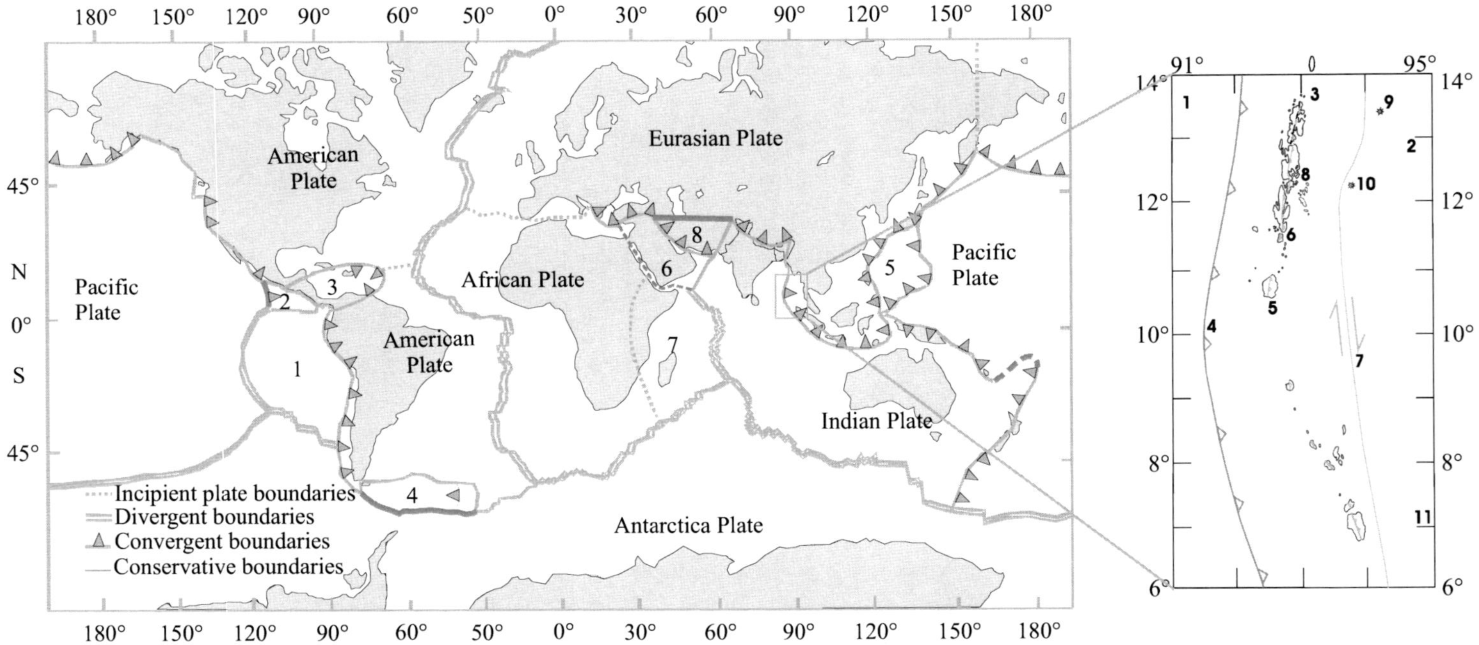

Fig. 10.3 A schematic tectonic map of the area around Andaman and Nicobar region: (1) Indian Plate, (2) Eurasian Plate, (3) Andaman Plate, (4) Andaman trench, (5) Eastern Boundary Thrust, (6) Sedimentary Outer Arc Ridge, (7) West Andaman Fault, (8) Baratung mud volcano, (9) Narcondum Island, (10) Barren Island.

Subduction of the Indian plate beneath the Andaman plate manifests as frequent large magnitude earthquakes in the region. Most of these are concentrated between the Andaman trench and the Back Arc Spreading Ridge, rendering the Andaman and Nicobar region as one of the most seismically active regions in the Bay of Bengal. Three earthquakes, of magnitude M_s 7.2, occurred in this region in the last century. Significant earthquakes occurred in 1881, 1914 (November 16, 09.50°N 94.50°E, E of Car Nicobar), 1929, 1941, 1949 (epicenter 12.00°N, 94.00°E, South West of Barren Island, 23 January) and 1955 (epicenter 7.00°N, 94.00°E off the east coast of great Nicobar on 17 May, Sinvhal et. al., 1978).

The disastrous earthquake of December 31, 1881 generated a tsunami with a run up of 1.2-m on the east coast of mainland India (Oldham, 1884). It was assigned an epicenter between 8.5°N and 10.5°N near Car Nicobar Island, and magnitude M_w 7.9 (Ortiz and Bilham, 2002). The most significant earthquake in recent times occurred on June 26, 1941. This great earthquake was assigned an epicenter 12°50° N 92°50° E, west of Middle Andaman Island, and a depth of focus 60 km. Its magnitude was $M = 8.1$, IMD; $m_b = 8.0$, $M_s = 7.7$, $M_w = 7.7$, $M_o = 4.25 \times 10^{30}$ Nm. It was a large thrust-type convergent margin event. It caused extensive damage to masonry buildings in Middle and South Andaman and Baratung Islands. These places were assigned intensity VIII$^+$ on MMI scale. The tsunami produced by the great earthquake of December 26, 2004, damaged almost the same areas as the tsunami of 1881 and 1941. The tsunami flooded and damaged masonry structures in Port Blair and east coast of mainland India.

The earthquake of January 20, 1982 ($M_s = 6.3$, epicenter 6.94°N, 94.03°E) originated in the sea near Great Nicobar, 50 km south east of Campbell Bay. The approach and berthing jetty were separated by 15 cm, concrete on its piers spalled, a school building collapsed, walls were separated in single-story hollow brick masonry houses, ground developed fissures near a bridge, and rock slid from hill slopes. The highest intensity assigned to this earthquake was MMI VIII (Agrawal, 1983). The North Andaman earthquake of September 14, 2002, sometimes also referred to as the Diglipur earthquake (M_L 6.0, M_w 6.5, epicenter 13.3°N, 93.3°E), was followed by several aftershocks of decreasing magnitude.

As a consequence of this high seismicity, the Andaman and Nicobar Islands have been assigned to seismic zone V, as per the seismic zoning map of India, given by Bureau of Indian Standards BIS: 1893–2002. This is the zone of highest seismicity and is vulnerable to earthquake damage pertaining to intensity MMI IX and above.

About 1200 km of the edge of the overriding plate snapped in the subduction zone, causing the earthquake of December 26, 2004. Moment tensor solution (Harvard) gives strike of the fault as 320° and dip as 11°. The large geographical extent of damage is consistent with the finite fault model, which shows a rupture duration of 200 sec and peak slip of 20 m. Seismic moment released in this plane was estimated as 3.57×10^{29} dyne cm, and rupture propagated northwestward for nearly 400 km with a speed of about 2.00 km/sec (Song et al. 2005). This leads to the conclusion that the tsunami did not start at a point. This great earthquake was accompanied by extensive faulting, and this is to be expected. This forced a massive displacement of water in the Indian Ocean. The sea continued to be rough for several days after this full-moon earthquake. High tides occurred later also, twice daily, and at the time of the new moon, i.e., January 10 and 11, 2005. This continued to hamper rescue and relief operations and continued to cause immense panic among survivors and rescuers.

Car Nicobar

Car Nicobar is the district head quarters of the Nicobar group of islands, and because of its strategic location in the Bay of Bengal it was recently bestowed with rapid development. It lies almost in the center of Andaman and Nicobar archipelago. Coastal regions of this tiny island, of maximum elevation 65 m, were thickly populated. East coast of this island was a thriving and a densely populated area. Malacca had an L-shaped double-story school building, several offices, shops, government residences, jetty, parking spaces, and houses all within a kilometer of the sea front. Coastal areas were cleared of most vegetation, including coconut palms, to make way for the new human habitat.

Inhabitants of these islands are used to earthquakes, but this Sunday-morning earthquake was rather unusual. Rumblings and shaking caused by this great earthquake awakened those, who were still asleep after the Christmas revelry. Every one tried to seek a place of safety, i.e., they came out of their houses. The sea receded immediately after the earthquake, well below the normal low tide. This unusual phenomenon attracted many curious tourists who were savoring the beaches, to venture seaward. The succeeding crest of the sea wave that arrived minutes later proved to be fatal for these and for several thousand others in similar situations in other coastal areas on the rim of the Indian Ocean and claimed a heavy death toll. Only a few survived this ordeal, and that too because they could hold on to a tree trunk while being swept away. Almost 6000 casualties were reported in this island of 23,000 inhabitants, all due to the tsunami alone.

The tsunami washed out the entire Malacca area of its built habitat, jetty, and people. Mountains of debris of uprooted coconut and beetle nut trees, mud, mixed with scattered remains of houses such as tin sheets, timber, RCC hollow blocks etc., was all that was left after the tsunami. Surviving cars and motorcycles, thoroughly battered, were thrown at least a kilometer inland from the sea facing parking space. Some people who held on to the small dome of the seaside temple, which was about 15 m above sea level before the tsunami, were saved when the waves came up to their feet and then receded somewhat by the afternoon. The temple remained partially submerged after the deluge. This indicates a run up in the range between 15 to 20 m. The six surviving double-storied government buildings on the same coastline were all that remained after the tsunami receded, though marooned, indicating large-scale transgression of sea and a concomitant subsidence of coastline, as shown in Figure 10.4(b).

At Katchal, the police station was close to Malacca jetty. Eleven police personnel, including the SHO, were reported missing. The only evidence of the police station after the tsunami was the ground-level RCC signboard. It was a kilometer away from the sea front before the earthquake, and was barely 50 m away after the tsunami, in a bleak surrounding.

The Air Force station at Car Nicobar and residential colony was located on the same sea front, south of Malacca. The double-story sea-facing houses were arranged in neat rows parallel and transverse to the seacoast, the nearest being barely 30 m from the former sea front. These were made of hollow concrete block masonry. The VIP guesthouse was barely 30 m away from the former sea front. The air force hangar was beyond that on the landward side of the residential colony and then came the 2.6-km long RCC airstrip.

The front row of sea-facing houses was completely obliterated by the tsunami that followed the earthquake of December 26, 2004. Hollow concrete blocks at the plinth level were scattered in the back rows. Damage to houses decreased on the landward side due to the shielding provided by the front row of houses. All Roads along the coast in the Air Force Colony were heavily scoured, inundated, or covered with debris and in operational, as shown in Figure 10.4(c). However, the air force operational area, the hangar, and the RCC airstrip survived the effects of the earthquake but were submerged by the debris brought in by the tsunami, and only part of it was useable after water was pumped out and the air strip was cleaned for landing and take off of aircraft, to enable rescue and relief sorties. The air traffic control tower (ATC) was heavily damaged.

Several oil storage tanks were uprooted by the tsunami and floated far inland away from their original place of rest. Five of these large-diameter steel tanks were found entangled within a mountain of debris consisting of cars, building material, trees, etc. The journey of these cylindrical tanks sheared off a coconut forest *en route*, and their passage was stopped only after they got entangled in an upslope coconut grove. Five of these steel tanks were littered in a large area and were scattered amidst debris, at least 50 m away from each other. These cylindrical tanks were found 3 km inland, at a height of

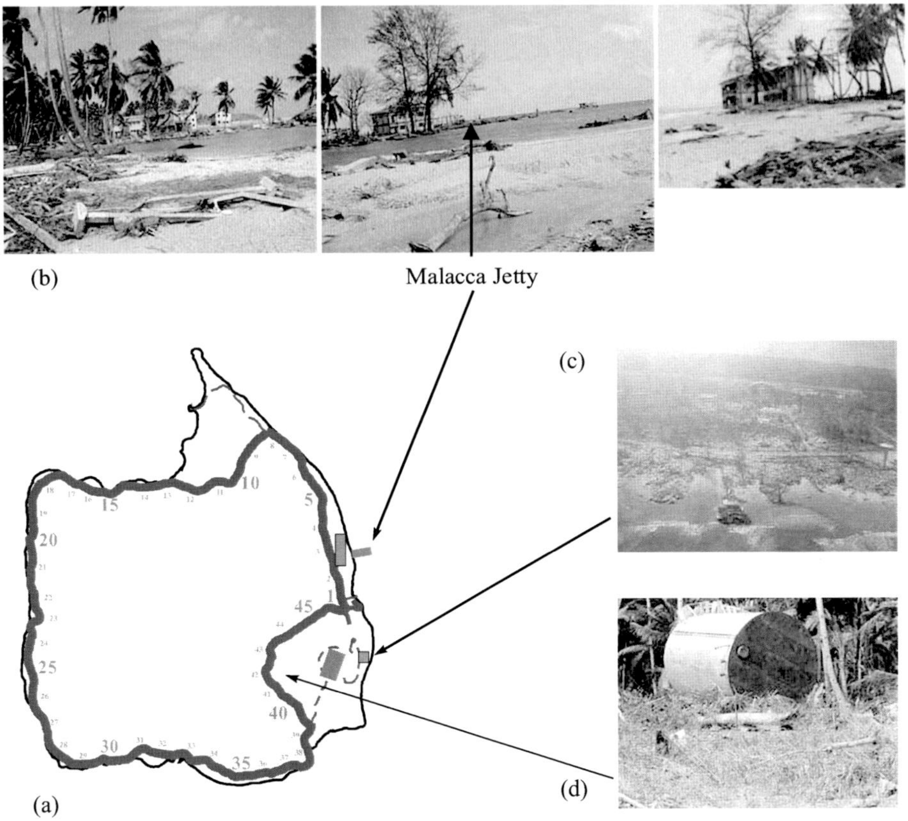

Fig. 10.4 (a) Map of Car Nicobar Island with the main road along the coast shown by kilometer markings. (b) The sea front at the former densely populated Malacca, collapsed and washed out Malacca jetty and a few marooned double-story houses that survived the deluge. (c) Aerial view of sea-facing Air force residential Colony at Malacca. Damage to sea front row of houses was extensive. This decreased on the landward side due to the shielding effect of the front row houses. Airstrip is seen in the background. (d) Oil tanks. (See color figure also.)

more than 30 m above mean sea level when located on Survey of India topographic sheet numbers 87 C/16, C/12, and C/15. This indicates the tremendous kinetic energy and uplift pressure generated by the tsunami that hurled material to such a large horizontal distance inland and to a height of 30 m above mean sea level. This observation led to an estimation of run up of 30 m and inundation 3 km at the air force station and Malacca in Car Nicobar.

Kakana and Kimous are in low-lying areas south of the Air Force Station. Nearly 400–500 human lives were claimed in these two villages alone. Lapathy was a newly developed up market shopping center and all needs of the privileged were fulfilled here. The only buildings that survived the tsunami were the newly built cinema hall, which was a beam and column frame structure, with walls made of RCC hollow blocks, and a two-way sloping roof of asbestos, albeit heavily damaged with all partition walls missing; a school and a few hostel buildings. Everything else either turned into mounds of debris or was swept away by the tsunami.

The coastline moved inland after the tsunami, and landscape changes occurred on the periphery of the entire island. The sea transgressed, in some places, almost 3-km inland after the tsunami, indicating partial submergence of the coastline. The human habitat was completely obliterated in this area. The sea front human habitat turned into eerie ghost places.

The interior of Car Nicobar Island was slightly elevated, maximum elevation of this island is 65 m above MSL, was sparsely populated with neatly laid out government offices and houses made of hollow brick masonry. Consequently, damaging effects were in stark contrast to that witnessed in coastal areas. As the region is subject to frequent large and moderate sized earthquakes, it seemed that the indigenous population was aware of the disastrous effects of earthquakes and tsunamis. This was manifest in their response to the earthquake, both immediate and long term. Indigenous islanders, on hearing the rumbling produced by the earthquake, sought safety in high ground, as they knew by long experience and through folk tales that high-amplitude sea waves sometimes follow an earthquake and cause destruction in coastal areas. These islanders selected safe sites for constructing their houses, in the interior of islands, on high ground and far away from the coastline. Also, they followed a construction practice that seemed to be primitive, but effect of strong ground shaking on these dwellings was minimal, saved their lives in the earthquake, and living conditions continued unhampered.

Traditional Nicobarese huts are made of locally available light building material like timber and bamboo, and are supported on long stilts. Indigenously designed, these are constructed by their occupants. In these huts, even the long stilts were not displaced in most cases and there were no visible signs of structural stress either. One such hut is shown in Figure 15.2a. The use of low-rise RCC hollow brick masonry houses has also caught on with the modern Nicobarese. Both kinds of construction performed surprisingly well on high ground.

Lifelines and Infrastructure

Essential services like electricity, water supply, communications, telephones and mobile services, roads, bridges, seaports, jetties, airports, were adversely affected by the twin onslaught of the earthquake and the tsunami. These were disrupted immediately after the arrival of the tsunami, in varying degrees, in all the inhabited islands of the Andaman and Nicobar archipelago.

Electrical and mechanical equipment was severely damaged in several power plants by the inundation. Debris clogged turbines and rendered them in operational in several hydroelectric power plants, including the 5.25 MW at Kalpong near Diglipur in North Andaman Island and the 20 MW power plant at Bamboo flat near Port Blair and in several others. Transmission and distribution power lines collapsed or were disturbed. Telephone and mobile services were restored in Port Blair within 2 days, and in Car Nicobar via a satellite link on the same day. Several water and sewage pipelines were ruptured, bent, and shifted, including those at Port Blair, Diglipur, and Hut Bay, and a crisis of drinking water ensued. The water treatment plant was totally washed out at Little Andaman Island. The Dhanikari dam, a water supply scheme, provides potable water to Port Blair. Its seismic response is given in Chapter 13 on infrastructure.

Several coastal roads were washed away or were inundated due to the tsunami and the high tides that followed. Others remained unusable due to scouring, piling of debris, and fissuring. The Andaman Trunk Road (ATR), which follows the Eastern Boundary Thrust, developed deep, wide, and transverse fissures in several places, in North Andaman Island, Middle Andaman Island, Baratung Island (Figure 13.2b), and South Andaman Island and became in operational in large stretches. Road communication was hampered and travel within several islands was very difficult in the post-tsunami scenario due to these reasons, and it was further compounded due to scarcity of transport.

Bridges and culverts were severely affected. The Austen Bridge connects North Andaman Island and Middle Andaman Island across the Austen Strait, at an epicentral distance of about 1100 km. This 268 m long and 7.5 m wide bridge, with 16 spans is supported on deep pile foundations. This Bridge survived the earlier Diglipur earthquake of 2002 (M_L 6.0, M_w 6.5) without any damage. However, in the current earthquake, four slabs in the middle of the superstructure shifted and moved upward, almost 50 cm off their bearings, without damaging the concrete slab, giving it the look of backward displaced steps (Wason et al., 2006). Consequently, the bridge was closed even to light vehicular traffic and only pedestrian crossings were permitted on this portion of the ATR. The water level decreased by almost a meter below this bridge indicating an uplift of coast due to the great earthquake.

As most coastal structures like jetties, harbors, wharfs, appurtenant structures, and associated buildings are made on filled ground, these are susceptible to flooding, wave action, liquefaction, scouring, and to differential settlement. The combined effect of the earthquake and the tsunami increases their vulnerability. The dry dock at Phoenix Bay in Port Blair was used for maintenance of boats and ships before the tsunami. At the time of tsunami, two ships were inside the dry dock. One rose with the incoming water and broke the gate, while the smaller ship was damaged. This dock continued to be submerged at subsequent high tides, and hampered maintenance and repair work of damaged boats and ships.

Foundation of the landward portion of a recently made passenger terminal at Haddo Bay in Port Blair was made on filled ground, while that facing the sea was made on stilts. Sinking and collapse of columns and failure of beams observed in the passenger terminal and reception hall, shown in Figure 10.5, was the combined action of the earthquake and the tsunami. Horizontal and vertical shift of canteen that was built on stilts at the wharf was of the order of 50 cm. Fissures 30 cm wide were observed in the road approaching the Haddo Bay. Obviously all this construction was not in accordance with the

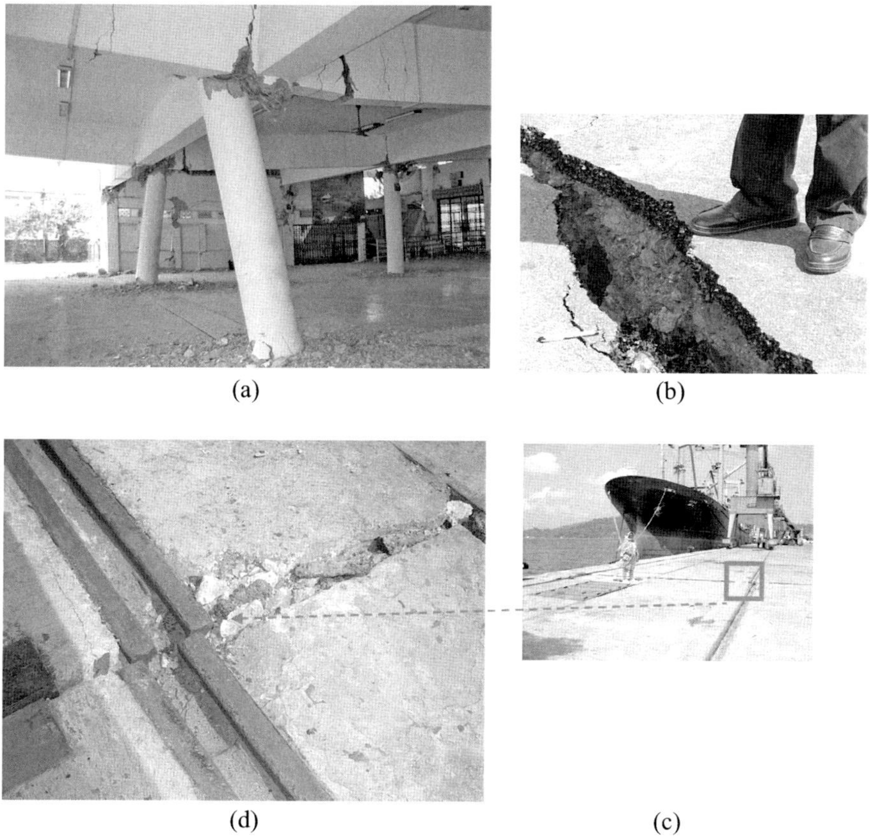

Fig. 10.5 Effect of the tsunami and the earthquake on Haddo Bay at Port Blair. (a) Columns in passenger Hall showed settlement, displacement, shear, and damage at both ends. Beam junction showed exposure of reinforcement and deep vertical fissures. (b) Deep, long, and wide-gaping fissure on road approaching passenger Hall. (c) Haddo wharf with cranes and ship. (d) Rails for movement of cranes displaced due to pounding of adjacent RCC slabs.

earthquake-resistant design as it should be for seismic zone V of the seismic zoning map of India. At Chatham in Port Blair, despite extensive subsidence observed on approach roads and filled ground, the wharf continued to provide shipping services. Harbors and wharfs fared only slightly better than concrete jetties.

Several jetties were bodily washed away, others collapsed, and the distant ones were severely damaged. This adversely affected all sea routes, hampered travel between islands, and slowed down and rendered rescue and relief operations very difficult. A jetty consists of two main portions, the approach jetty and the berthing jetty. The berthing segment of several jetties makes an obtuse angle with the approach jetty. Most of the time, the entire approach and berthing jetty consist of square reinforced concrete piles connected at the top by a neat array of beams, columns, braces, and RCC slabs.

As the tsunami lashed the coast and water level increased, the surviving jetties already weakened by the earthquake were further damaged due to pounding action between adjacent concrete slabs. The approach and berthing jetty separated in many cases, pounding occurred between adjacent RCC slabs, and deep longitudinal fissures and gaping cavities developed in several surviving jetties even at an epicentral distance exceeding a thousand kilometers. Concrete on piers spalled during the earthquake shaking and reinforcement was exposed. Piles on which jetties rested became slender due

Fig. 10.6 Part of the berthing jetty at Junglighat, Port Blair, sheared off its piers, separated from the approach jetty and fell in the water. A barge capsized at this jetty. Piers on which jetty rested earlier are protruding from the sea. (See color figure also.)

to spalling of concrete, shifted sideways, or settled after the earthquake. The berthing jetty at Junglighat in Port Blair collapsed and the approach and berthing jetty separated, as shown in Figure 10.6. The jetty at Aberdeen was heavily fissured and damaged but useable. Malacca jetty in Car Nicobar (Figure 10.4b), the jetty at Hut Bay, Great Nicobar and several other islands and on the east coast of mainland India were washed away by wave action, collapsed or became in operational.

Freestanding walls in school compounds (Figure 13.6b), and Marina park collapsed due to the force of the wave action of the tsunami; in framed structures panel walls showed zigzag cracks between RCC hollow masonry blocks, and at junction of wall with beam and column. The effects described here for Andaman and Nicobar Islands were more profound in the Indonesian Island of Sumatra.

CAUSES OF DISASTER

A combination of several factors makes populations in island arcs, such as the Andaman and Nicobar Islands and coastal areas of mainland India, vulnerable to damage due to earthquakes and tsunamis. High seismicity in a submarine environment on a convergent plate boundary is the largest contributory factor. Lack of appropriate measures, along an extensive and low-lying coastline with a soft and saturated sedimentary cover and bereft of all natural vegetation adds to the vulnerability. It was very surprising to observe that modern and large population centers, including government and defense establishments, offices, and houses, had developed recently all along the coastline, totally oblivious of all these factors. Obviously, the tropical sun and sand makes the coast a coveted building site in all these islands. If the coastal regions were spared the building activity, the loss of human life would have been a very small fraction of what was actually claimed by this tsunami. Damage due to the earthquake alone was not as life threatening as demonstrated by the subsequent aftershock, a great event in its own right, in the same region (M_s = 8.3, March 28, 2005).

WHAT CAN BE DONE?

For mitigating future disasters due to a tsunami, it is pertinent to understand the seismotectonic environment of subduction zones (Sinvhal et. al., 1978) and to avoid vulnerable building activity in low-lying coastal areas surrounding it. If it is unavoidable, then a built environment is safer on high ground, away from the coast. Coastal structures like ports, harbors, and jetties should be made earthquake resistant in the Bay of Bengal and in the Arabian Sea. It seems tribals have lived and learnt from previous earthquakes and tsunamis, and their traditional knowledge is propagated in folk tales. This aspect needs to be respected and harnessed for disaster mitigation. Strategies for disaster preparedness should be made known to those living along the coast so that impact of future disasters can be minimized. Triangulation and leveling studies are required to establish any shift, subsidence, and elevation of islands at the convergent plate boundary between the Andaman Trench and the Back Arc Spreading Ridge. Bathymetric surveys are required to map changes in submarine contours. Installation of a tsunami warning system in the Indian Ocean is inevitable.

CONCLUSION

Submarine earthquakes of large magnitude that originate on a convergent plate margin sometimes become tsunami genic. The coastline on the Indian subcontinent is vulnerable to this disaster not only in the Bay of Bengal but also on the west coast of India, as a similar, albeit smaller subduction zone exists on the Makran coast, near Kutch, as shown in Figure 2.1.

The great Sumatra earthquake of December 26, 2004, and the tsunami that followed claimed almost 2,30,000 precious human lives on the rim of the Indian Ocean, in a wide region stretching from Indonesia in the east to the African continent in the west. This was not the last earthquake on the convergent plate margin in the Indian Ocean. This indicates an urgent need to spare coastal regions from unnecessary building activity in future.

REFERENCES

Agrawal, P. N., 1983, A study of the 20 January 1982 earthquake near Great Nicobar Island, *BSSA*, **73**(4), p 1139–1159.

Curray, J. R., F. J. Emmel, D. G. Moore, and R. W. Raitt, 1982, Structure, tectonics, and geological history of the NE Indian Ocean, p 300 – 450, *in* The Ocean Basin and Margins, edited by E. M. Narrin and F. G. Sehli.

Dasgupta, S. and M. Mukhopadhyay, 1993, Andaman plate, *Tectonophysics*, **225**, p 529–542.

Oldham, R. D., 1884, Note on the earthquake of 31 December 1881, *Records Geological Survey of India*, **XVII**(2), p 47–53.

Ortiz, M. and Bilham, R., 2002, Source area and rupture parameters of the 31 December 1881 M_W = 7.9 Car Nicobar earthquake estimated from tsunamis recorded in the Bay of Bengal, *J. Geophys. Res.*, **108**(B4), p 1–16.

Satake, K., 2002, Tsunamis, in *International Handbook of Earthquake and Engineering Seismology*, Part B, Eds. W. H. K. Lee, H. Kanamori, P. C. Jennings and C. Kisslinger, p 437–451, Academic press, San Diego.

Shankar, D., H. R. Wason, A. Sinvhal and V. H. Joshi, 2005, Damage due to devastating earthquake (M_W 9) and tsunami of December 26, 2004 in Andaman and Nicobar, India: A perspective, in *Proceedings of the Twenty Second International Tsunami Symposium*, 27–29 June, 2005, Chania, Crete, Greece, p 221–232.

Sinvhal, H., K. N. Khattri, K. Rai and V. K. Gaur, 1978, Neotectonics and time–space seismicity of the Andaman–Nicobar region, *BSSA*, **28**, p 399–409.

Song, Y. T., C. Ji, L. L. Fu, V. Zlotnicki, C. K. Shum, Y. C. Yi and V. Hjorleifsdottir, 2005, The 26 December, 2004, tsunami source estimated from satellite radar altimetry and seismic waves, *Geophysical Research Letters*, Article No. L20601.

Wason, H. R, A. Sinvhal, D. Shanker, A. Kumar and V. H. Joshi, 2006, Ground deformation observed due to the great Sumatra earthquake of December 26, 2004 and tsunami in and around Andaman and Nicobar Islands, in *Proceedings of the Thirteenth Symposium on Earthquake Engineering*, I I T Roorkee, December 18–20 2006, p 228–237.

11 CHAPTER

Stone and Brick Masonry Houses

INTRODUCTION

Stone masonry houses have proved to be the biggest killers in several earthquakes. More than 10 lakh people were killed worldwide in the twentieth century alone, mostly due to collapse of stone houses. Description of destructive effects of the great earthquake of 1905 in Kangra region of Himachal Pradesh, which killed more than 19,000 people, is still valid more than a hundred years later, when more than 90% casualties among the 86,000 killed were in stone houses due to the Kashmir earthquake of October 8, 2005. Brick masonry houses fare slightly better.

When more than 10,000 people were killed in the Latur earthquake of 1993, the tragedy focused attention on the disproportionately large human losses, all in stone houses, confined to a very small area and that too due to an earthquake of moderate size, magnitude 6.4, (Sinvhal et. al., 1994, 1995). This calamity occurred in the seismically stable peninsular region, within the erstwhile seismic zone I, where seismic risk was least. Because of the heavy human losses in this earthquake, Latur and other seismic zones I were subsequently upgraded to seismic zone II in the seismic zoning map of India, BIS: 1893–2002, given by the Bureau of Indian Standards. Some other recent earthquakes like the Uttarkashi earthquake of October 20, 1991, Chamoli earthquake of 1999, Kutch earthquake of January 26, 2001, Bose et. al., 2001, 2004, and Kashmir earthquake of October 8, 2005, (Pandey et. al., 2006, Pore et. al., 2006, Sinvhal et. al., 2005) all in seismic zone V, the severest seismic zone, witnessed the tragic performance of stone masonry houses. Let us discuss the reasons behind this scenario.

STONE WALLS

In a stone house, walls are made of heavy, large, and uneven shaped or round stones. Most of the time a stonewall is load bearing and very thick. In reality, it consists of two closely spaced walls, each with a thickness less than half that of the wall. These are called laminations, or wythes. During construction of stonewalls, one large stone is placed from the inside and another stone is placed from the outside. This process continues till the desired height of the wall is achieved. To give a tidy exterior look, the larger face of stone is placed on the outer surface of the wall and the angular face is placed on the inner side of the wall, i.e., all angular faces are placed on one side and larger faces on the other side. This makes an unstable vertical configuration of stone in each lamination. This is random rubble stone masonry.

Space between laminations and between stones is filled with mortar and smaller stones. Mortar consists of mud or clay, is brittle, serves as a filler material only, and does not provide any bonding between stones. Moreover, it wears off after a few seasons of rains and strong winds. In more prosperous areas, the outer wall is often plastered with mud, lime, or cement to give it a smooth appearance and to seal it from outside air.

Seismic Response of a Stone Masonry Wall

Such a primitive form of stonewall provides satisfactory living conditions and adequate behavior in normal situations but is found to be inadequate when shaken by an earthquake. These are vulnerable to strong ground shaking produced by an earthquake and are extensively damaged in meizoseismal areas. The nature of damage to such walls showed a significant similarity in various earthquakes, irrespective of time and space.

As stonewalls are exceptionally brittle, they have low strength in bending and tension, and are unstable under reversal of seismic load. During strong ground shaking, loose, uneven shaped stones slide out of each lamination. This aspect is shown in Figure 11.1(a). Dry mud mortar and small stones that are filled between laminations and between gaps in angular stones are also shaken out of place. This further destabilizes and worsens the unstable vertical configuration of random rubble, often leading to their total collapse. Sometimes the two laminations split vertically, as shown in Figure 11.1(b), separate out, or bulge. In addition, long walls fail, corners collapse, and extensive fissures develop near openings, as shown in Figure 11.2. This makes random rubble stone the worst construction material in earthquake prone regions. More than 75% of such construction collapsed totally within meizoseismal areas of several disastrous earthquakes.

The roof is made of different materials and designs in different places. In Kutch, it is made mostly of clay (Mangalore) tiles, which is supported on an inclined bamboo grid. In Latur region, it is flat and heavy, and is sometimes

(a)

(b)

Fig. 11.1 (a) Collapse of stone houses revealed the use of large uneven shaped stones and dry mud mortar, at Gubal, and (b) failure of a stone wall showing wythe failure at Killari, in the Latur earthquake of 1993.

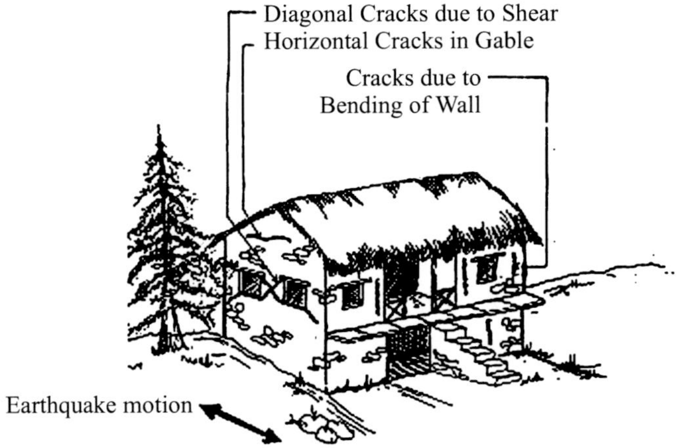

Fig. 11.2 A stone house is prone to different kinds of damage that can be induced by an earthquake.

made of timber on which a thick plaster of mud is laid. This roof is sometimes replaced with RCC roof. In Kashmir, the roof is usually inclined and light, and is made of a timber frame and metal sheeting.

A roof that rests on such load bearing stone walls collapses to the ground as soon as walls collapse from underneath. A heavy roof compounds the catastrophe. When big, heavy stones and roof material start falling inside small rooms the occupants hardly have any chance or time to escape. Thus, houses in which walls are made of heavy uneven shaped stone and a heavy roof soon turns into a heap of rubble, or worse, a grave. All this happens in a very short span of time, may be within less than half a minute or so during which the strong ground shaking lasts. Most earthquake engineers would prohibit construction of such stone masonry houses in seismically prone regions as these have several inherent deficiencies and some of the most undesirable characteristics as far as their seismic performance is concerned.

Despite their known deficiencies and dismal seismic performance, stone masonry houses continue to be popular throughout the world. This is due to easy and abundant availability of stone, simplicity and speed of construction, and minimal need of technical know how and manpower. Walls in many rural houses are made of other weak materials such as sun burnt clay bricks, known as adobe, or of mud. These too have contributed heavily to earthquake death lists.

Most traditional rural houses are made of random rubble stone masonry (Type A structures as given in several intensity scales, such as Modified Mercalli Intensity Scale). Stone is laid in mud or lime mortar and walls are thick. Sometimes, the central wall is very high, almost 5 m, and this is considered as a status symbol in villages of Kutch. The roof is made of clay (Mangalore) tiles, which is supported on an inclined bamboo grid. This roof is sometimes replaced with RCC roof. Mostly Type A structures were heavily damaged within the meizoseismal area of Latur, Kutch, Uttarkashi, and Kashmir earthquakes and were responsible for bulk casualties. The seismic response of stone houses in these earthquakes is shown in Figure 11.3.

Stone houses that have no earthquake-resistant features cannot resist high inertia forces generated by even a moderate-sized earthquake. But contrary to common belief, it is surprisingly easy and practical to rectify deficiencies in stonewalls. This involves some modification in design of the house and introduction of a few simple earthquake-resistant features. The necessity of these earthquake-resistant features increases as severity of seismic zone increases. With the same locally available material as are commonly used and with a little extra and judicious use of other materials such as timber, cement and steel, stonewalls can be strengthened to withstand earthquake shaking. The desirable seismic response of this too has been observed in several earthquakes and is illustrated in this chapter.

Earthquake-resistant Features in Stone Masonry Houses

Based on observations in several earthquakes, and the need for a safer stone house, the Indian Society of Earthquake Technology (ISET) at Roorkee published *A Manual of Earthquake Resistant Non-Engineered Construction* in 1981. The Bureau of Indian Standards brought out several earthquake codes, with appropriate illustrations, like IS: 4326–1993,

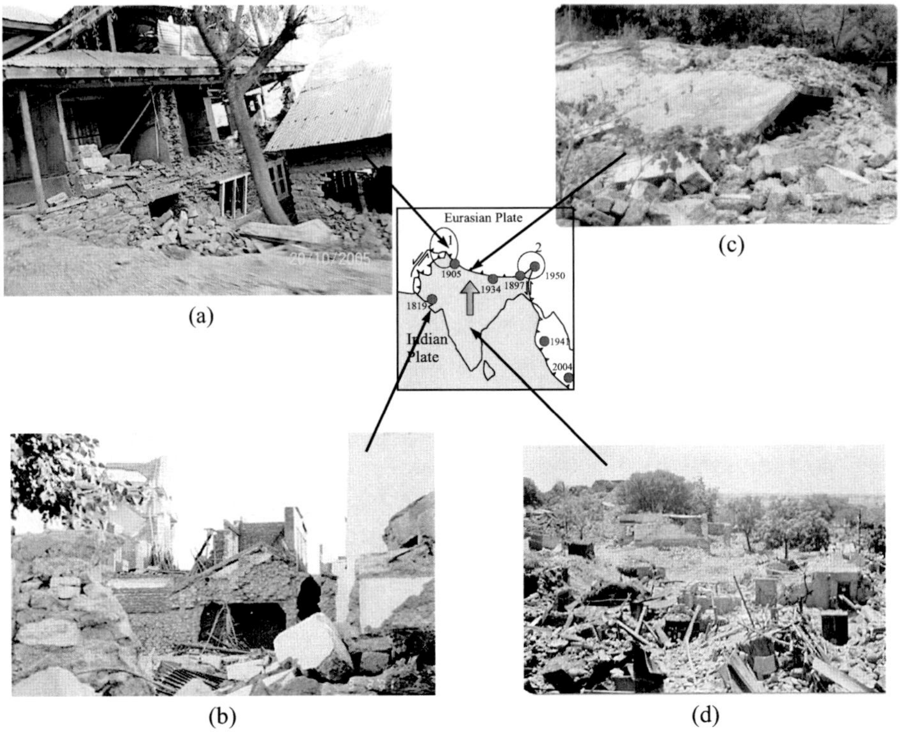

Fig. 11.3 Stonewalls and earthquake disasters are almost synonymous with human tragedy, whether in the Himalayan Arc, or in peninsular India. All houses shown here were made of random rubble stone masonry. (a) In Tangdhar, after the Kashmir earthquake of October 8, 2005, (b) In Bhuj, after the Kutch earthquake of January 26, 2001, (c) A collapsed house in Bhatwari after the Uttarkashi earthquake October 20, 1991. The heavy concrete roof was supported on walls made of a mixture of random rubble stone masonry and concrete blocks. (d) The devastated village of Killari after the Latur earthquake of September 30, 1993. (See color figure also.)

Earthquake Resistant Design and Construction of Buildings—Code of Practice; IS: 13828-1993, Improving Earthquake Resistance of Low Strength Masonry Buildings—Guidelines; IS: 13827–1993, Improving Earthquake Resistance of Earthen Buildings—Guidelines. Illustrated and useable literature on this is available in other places also like Thakkar, et. al., (1994) and Paul, et. al., (2002).

An improvement in the construction method of stonewalls can cut down the death toll dramatically in an earthquake. If stones that are flat at the upper and lower face are used, it provides a more compact and stable vertical configuration of stone, which is a desirable feature. Performance of this building material further improves if side faces are also flat. This is known as dressed stone. Use of dressed stone has several advantages. It provides a more compact and stable vertical stack of stones, reduces gaps between adjacent stones, reduces the amount of mortar required, and above all is more difficult to dislodge in an earthquake. This is shown in Figure 11.4.

Fig. 11.4 The use of dressed stone makes its dislodgement more difficult in an earthquake. This figure shows the seismic response of a column made in dressed stone at Killari, after the Latur earthquake of 1993. The column twisted and opened vertically from the centre, but continued to perform its intended function, viz., holding up the roof.

Use of rich mortar can further improve the seismic performance of stone masonry. Mortar that uses sand, lime, and cement has bonding properties that are better than that of clay or mud. For example, a mixture of lime and sand in a proportion of 1:3, or cement and sand in a proportion of 1:6, is adequate in seismically stable regions. In higher seismic zones, a richer mortar is required, i.e., the proportion of cement is increased. In that case, proper curing is necessary to increase bonding.

A stonewall can be further strengthened if the two laminations are somehow forced to behave as a single wall unit. This objective can be achieved in several ways. The simplest way is to stitch the two laminations together at regular horizontal and vertical intervals throughout the wall. The use of a long stone spanning the two laminations accomplishes this objective. This is shown in Figure 11.5(b). These long stones are also known as through stones or bond stones. If such long stones are not available, then the same objective can be achieved by other available means. Two smaller stones of three-fourth width of the wall can be used in conjunction, or a concrete block or a steel dowel can be equally effective. Wooden blocks, well treated to withstand weathering and insect action, can also be used in regions where rainfall is scanty; this is shown in Figure 11.5(a). This binds the two laminations together. Moreover, space between two laminations acts as an insulator from extreme temperature conditions, whether hot, like in peninsular India, or cold, like in the Himalayan arc, so this modification continues to provide thermal comfort in diverse climatic conditions.

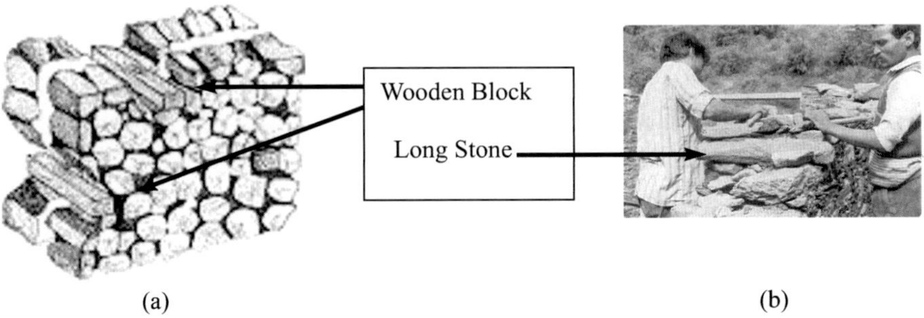

Fig. 11.5 A stonewall can be strengthened by binding together the two laminations of a wall. This can be achieved in several ways like using a sturdy material that is as long as the thick wall. This is shown here for: (a) a timber block and (b) a long stone.

Tying all walls together, and that too at several levels, is an immensely effective earthquake resistant measure. This ensures that all walls act in unison (as far as possible) and together counter the earthquake force. This is achieved by placing continuous bands around the house at several convenient horizontal levels in the wall. These can be at the bottom of the wall, i.e., the plinth band, at bottom of the window frame, i.e., the sill band, at the top of window and doorframes, i.e., the lintel band, and at top of wall, i.e., the roof band. These are shown in Figure 11.6.

The sill and lintel bands evenly divide the wall in three vertical portions. The lintel band is the most important band and incorporates in itself all door and

Fig. 11.6 The walls of a house can be made earthquake resistant if several bands are introduced around the house. These can be horizontal bands at plinth, sill, lintel, and roof level. These can be tied into vertical bands at corners, at junctions of rooms, and can be part of door and window frames.

window lintels. The roof band is needed if the roof is made of thatch, tin, timber, or asbestos sheets. A roof made of reinforced cement and concrete (RCC) or reinforced brick and concrete (RBC) also performs the function of the roof band, and then a separate roof band is no longer needed. The roof band has the added advantage that it prevents rainwater from seeping into walls, and reduces the effect of dampness, which in turn lowers maintenance cost. A gable band, enclosing the triangular portion of masonry at the gable end, is required in gable walls.

These bands are made either of timber or of steel, are as wide as the wall, and are continuous at corners and at junctions of walls. If timber is used, then it should be well seasoned, to avoid shrinkage in dry weather and expansion in wet weather. If steel bars are used, then these should be long and continuous, should be bound together at regular intervals of 15 cm by thin steel stirrups, and should be covered with a rich mix of cement and concrete during masonry construction.

A stonewall can be further strengthened if several vertical bands are also introduced in to it. These restrain horizontal slip of stone and connect horizontal bands by providing a lateral load resisting system. Vertical bands placed at corners and junctions of walls strengthen two adjacent walls simultaneously. These bands may also be placed along vertical sides of openings in which case these can act as door or window frame. Vertical bars must be firmly anchored into the plinth band and continue from the foundation to the roof band or roof slab at the top. With these added lateral and vertical bands, stone masonry walls become better equipped to resist an earthquake. These bands provide a framework that helps in arresting the propagation of earthquake-induced cracks. The story of breaking a single stick versus an entire bundle is apt here too. Besides providing seismic resistance, these bands increase resistance to wind and blast loading also.

Too many and very large openings in a wall for doors and windows are best avoided, as are openings close to cross-walls and at edges of walls, as these weaken the stone wall. Length of all openings in a wall must be less than half the length of the wall for a single storey house, and less than this for a double storey house. Openings should be well spaced out in the wall, and should maintain a stipulated minimum horizontal and vertical distance between any two openings.

Very thick walls give a false sense of strength, which is belied in an earthquake, sometimes with tragic consequences. In Latur, it varied between 40 and 80 cm. Such thick walls are best avoided as they increase earthquake forces and cause more damage or collapse. A reduced thickness of 35–45 cm is adequate to ensure seismic safety and for thermal comfort of residents. Very long stonewalls are also best avoided as these are prone to out of plane collapse. The span of a wall between cross-walls should be less than 5.0 m, and longer walls require buttresses. For exceptionally long walls, i.e., longer than 7 m, a thick horizontal band with steel bars and a rich mortar is required. Exceptionally tall walls are also undesirable and a maximum height of 3.0 m is adequate in most cases.

Thus, the addition of several simple features in just the construction of stonewalls improves the seismic response of a stone house manifold and makes it a worthy dwelling. Several stone masonry houses, which incorporated principles of earthquake resistant design withstood strong ground shaking in several major earthquakes.

TIMBER FRAMED CONSTRUCTION

Many stone houses use a timber frame. Strong ground shaking racks and distorts the timber frame. Sometimes stonewalls collapse but the roof, supported on the timber frame, resists the earthquake shaking and does not collapse, sometimes even when it is heavy. Thus, the use of timber frame prevented complete collapse of stone houses and saved precious lives in several earthquakes. This desirable aspect was observed in the older houses in the meizoseismal area of several earthquakes, and an example is shown in Figures 11.7–11.9. Seismic performance of more sophisticated variations of these earthquake-resistant measures were indigenously developed in the western syntaxis, and were observed in the Kashmir earthquake of 2005 (Sinvhal et. al., 2005).

Fig. 11.7 A heavy roof supported on a timber framework did not collapse on its residents and saved them. In comparison neighboring houses, which did not have such a timber roof, killed their residents. This was a typical scene in and around Killari, after the Latur earthquake of 1993.

Fig. 11.8 The walls of this house were made of random rubble stone masonry. The roof was light and was made of timber and corrugated galvanized iron (CGI) sheets, and rested on a timber frame. These two earthquake resistant measures proved to be a desirable aspect in the Kashmir earthquake of 2005. The residents of this house survived the earthquake (in Uri after the Kashmir earthquake of 2005).

(a) (b)

Fig. 11.9 (a) The desirable use of timber bands in a brick masonry structure meant for storing apples in Baramulla, and (b) the use of horizontal and vertical bands saved this house of composite construction in Tangdhar. Walls in ground floor were made of random rubble and show failure of vertical lamination, while those on the upper floor were made of brick masonry, i.e., construction of mixed masonry.

Taq

Taq is a traditional form of local construction in Kashmir. It consists of load-bearing walls made of random rubble stone masonry and timber bands placed on top of these walls. These bands were tied together at each floor level. This constituted a timber framework in the horizontal plain. Vertical members were mostly nonexistent, except at locations of openings for doors and windows, so no framework, in the conventional sense existed in the vertical plane. This feature, i.e., a timber frame in the horizontal plain, proved to be remarkably resilient even when support from underlying stonewalls was partially withdrawn, as shown in Figure 11.9b. This was observed in many houses in Baramulla District of Jammu and Kashmir.

TIMBER FRAME WITH MASONRY INFILL

In a more sophisticated timber framework, both horizontal and vertical timber bands are used. The two together reduce the possibility of out of plane failure of walls, and wherever this provision existed total collapse of wall was absent even in strong ground shaking. Use of timber leads to enhanced damping and thereby better shock absorbing capacity. This was traditionally and extensively used in regions of rugged mountain terrain, with locally available material and showed an exemplary seismic performance even when the house was located close to the epicentre and on the Main Boundary Thrust, in Baramulla district of Kashmir.

Dhajji-diwari

This desirable feature was further refined by the introduction of additional timber members within the larger framework, in view of frequent large magnitude earthquakes in this region. These additional timber members were distributed in different directions in the wall like horizontal, vertical, and sometimes even diagonal and formed a very elaborate timber framework. Vertical timber members were spaced 75–100 cm apart and created a patchwork of in-filled masonry. Panelled walls were filled with burnt clay bricks or by sun-dried mud bricks. In its various forms, this is commonly known as brick nogged wooden frame construction, or, *dhajji-diwari*. In Persian, it means a 'patch-quilt wall'. In urban and semiurban areas interior walls were in *dhajji-diwari* and peripheral walls were in random rubble stone masonry. Consequently the outer stone masonry walls collapsed but the inner walls showed only moderate damage, which in most cases was life saving. In rural areas, inner walls in *dhajji-diwari* were filled with stone masonry laid in mud mortar. Even though these suffered extensive damage yet they saved the lives of all their occupants. An example of this is shown in Figure 11.10. Walls in the *dhajji-diwari* system were observed to be thinner than those in the taq system.

(a) (b)

Fig. 11.10 (a) A severely damaged three-storey timber framed stone masonry house at Komalkote, in Uri, after the Kashmir Earthquake of October 8, 2006. Situated on the Main Boundary Thrust, this house showed exemplary seismic performance as it used earthquake resistant features such as (b) in filled panel walls, locally known as "*dhajji diwari*".

Thus taq and *dhajji-diwari* modes of construction showed the efficacy of judicious use of timber in a vulnerable region like seismic zone V. The presence of these features implied that there was local awareness of earthquake resistant design and it was ably implemented in the older houses of the region. Seismic performance of these worthy human dwellings was found to be superior when compared to other forms of construction in the same locality, especially newer houses. Long dormant periods between major damaging earthquakes probably led to abandoning robust construction practices, with tragic consequences.

Several other earthquake-resistant, life-saving variations of this traditional mode of construction were also observed in the Kashmir region. Generally the balcony beams, of timber, were supported at tips at all floor levels. The super structure was sometimes made entirely of timber and placed on top of stone masonry walls. The roof in most cases was light and in some cases when it fell to the ground it was found to be intact and was capable of being re-used.

BRICK MASONRY

Seismic performance of rectangular blocks of dressed stone is, most of the time, comparable to that of other building materials of similar shape such as burnt brick or hollow concrete blocks of adequate strength. The extent of damage in brick masonry is, as observed in several earthquakes, much less than in random rubble stone masonry. In meizoseismal areas of several earthquakes, where more than 75% stone masonry houses collapsed, only about 50% brick buildings were damaged. One to three storey houses are short period structures, and are vulnerable to damage in the epicentral region because of the high-frequency content of body waves.

Most well-built one to four storey brick masonry houses consist of load bearing brick walls. These support reinforced concrete beams and floor slabs. Columns may be made with or without reinforced concrete. Usually, different elements of a building such as floor, wall, and roof are not tied together. In an earthquake, these vibrate independently, hit each other and get damaged or collapse. Many of these Type B and C structures were heavily damaged within MM Intensity IX and above in the Kutch earthquake of 2001. At several places such new houses could be seen rising above the debris of random rubble stone masonry houses like in Figure 11.11. Luxury holiday homes built in Kutch (in Manfara, Anjar, Samakhiali, and Gandhidham) were patronized by rich non-resident Gujaratis settled in Mumbai and abroad, who had constructed these for family get-togethers, old folks, and for retirement. These houses were exceptionally well finished and furnished with all possible modern amenities. Most suffered structural and non-structural damage of Grade 4.

Fig. 11.11 A modern reinforced concrete-framed building rising above the debris of stone buildings in a thickly populated region of Anjar (MMI IX), after the Kutch earthquake of January 26, 2001.

Commonly observed deficiencies in brick masonry are attributed to several factors. These are the absence of connection between perpendicular walls, absence of connection between walls and roof, and absence of horizontal and vertical bands. This causes damage in the form of separation or collapse of walls and corners and diagonal cross-fissures between openings, as shown in Figure 11.12. Individual buildings should be adequately separated to prevent pounding. Considerable damage was observed due to pounding between closely spaced adjacent buildings, as shown in Figure 11.13. Remedial measures for brick masonry are simpler than those for random rubble stone

Fig. 11.12 Diagonal cracks in brick masonry wall, in Civil Hospital at Killari, due to the Latur earthquake of 1993.

Fig. 11.13 Pounding between adjacent buildings at Gandhidham. Both buildings were new. RCC roof top water tank on top of building on left toppled on to the building on right and pierced through its roof and fell inside the rooms. This was due to the Kutch earthquake of 2001, at an epicentral distance of 40 km.

masonry. Plinth, lintel, roof, and gable bands and vertical bands are required and the mortar gets richer in cement content for higher seismic zones.

COMPOSITE CONSTRUCTION

Sometimes an undesirable combination of rubble and brick masonry is used in which case conventional defects of each type of masonry were found to coexist. Sometimes brick masonry walls are raised on old stone masonry walls to add new floors to an existing building. Lack of any bonding or interlocking between the two kinds of masonry or the two floors of the same structure become disastrous in an earthquake. This is shown in Figure 11.14. Sometimes the thick outer wall is in stone masonry and the inner wall is in brick masonry. The two masonries, dissimilar in many aspects lack any inter-connection, and result in failure of outer walls in many cases.

(a)

(b)

Fig. 11.14 (a) In this house, the ground floor was in random rubble stone masonry, and upper floor was added later and was in brick masonry. Defects of both kinds of masonry coexisted. There was no structural connection between the two kinds of masonry. When one of the brick walls on the upper floor overturned, the stone masonry on the lower floor collapsed. This was at Killari due to the Latur earthquake of 1993, (b) A double storey house with mixed construction, thick stone walls on the ground floor and timber frame and half brick wall on the upper floor in Anjar (MMI IX), after the Kutch earthquake of 2001.

SITE EFFECTS

Local surface geology plays an important role in the seismic performance of a stone and a brick masonry house, just like for any other civil structure. Incompetent soil such as land fill, loose and fine sand, soft silt, expansive clay, or compressive soil is liable to subside or liquefy in strong ground shaking. When a structure is founded on such soil, it can result in a large differential settlement of the structure and other kinds of damage. This is shown in (Figure 11.15). When foundation soil is poor, the plinth band becomes necessary.

(a) (b)

Fig. 11.15 Four blocks in Paldi area of Ahmedabad (MMI VI–VII) after the Kutch earthquake of January 26, 2001. The ground floor formed a soft and weak story and was used as a car park. Due to the earthquake, the interior columns of the ground floor were totally crushed and collapsed due to several reasons like a deficient foundation, inadequate capacity of columns, and filled site. The staircase was located at the center of the back edge of the building, which created eccentricity and additional shear leading to shear failure of front columns. Columns near the staircase were not completely crushed. The ground floor column was excavated for rescue in one of the blocks, revealing an inadequate base area of footing. Almost every column in the building failed. Two blocks toppled and tilted towards each other at an angle of 30° due to similar shear failure of front columns, crushing all cars and vehicles parked there. Mercifully nobody was killed in these apartments.

Its use has the added advantage that it can also reduce dampness that seeps in from the foundation. For rocky, hard, or firm soils plinth band is not too critical. The necessity of vertical reinforcement increases for weak foundation soils as severity of the seismic zone increases. For soft soil in seismic zone III, it is required at joints and corners. In seismic zone IV, vertical reinforcement is required at corners for all types of soils, and in seismic zone V, it is required at joints, corners, jambs, and openings also. The number of steel bars to be used increases with the severity of the seismic zone, and so does the diameter of the steel bars. A rich mortar of cement and sand in the ratio 1:3 is required in such vertical bands.

CONCLUSION

Because of the heavy loss of human life in stone houses, a prejudice has developed against their use—it is not a suitable building material for earthquake prone regions. Although partially true, yet it is not economically feasible to wish away the use of stone as a building material. Damage observed in stone houses suggests that the fault lies less in the material and more in the way it is used. The same applies to brick masonry houses also. In fact, any structure if not adequately designed and any building material, if not properly used, has immense potential to inflict damage in an earthquake. The dismal seismic performance of multi-storey buildings in the Kutch earthquake of 2001 is a case in point. This aspect has been discussed in the next chapter.

REFERENCES

BIS: 1893–2002, *Indian Standard Criteria for Earthquake Resistant Design of Structures, Part I: General Provisions and Buildings* (Fifth Revision), Bureau of Indian Standards, New Delhi, 40 p.

Bose, P. R., A. Sinvhal and A. Bose, 2001, Traditional construction and its behavior in Kutch earthquake, in *Proceedings of the Workshop on Recent Earthquakes of Chamoli and Bhuj*, May 24–26, 2001, Department of Earthquake Engineering, University of Roorkee, Roorkee, p 151–158.

Bose, P. R., A. Sinvhal, A. Verma and A. Bose, 2004, Implications of design and construction decisions on earthquake damage of masonry buildings, in *Proceedings of the 13th World Conference in Earthquake Engineering*, Vancouver, Canada, p 9.

Indian Society of Earthquake Technology, 1989, *A Manual of Earthquake Resistant Non-Engineered Construction*, University of Roorkee, Roorkee, India, 158 p.

ISI: 4326-1993, *Indian Standard Code of Practice for Earthquake Resistant Design and Construction of Buildings*, Bureau of Indian Standards, New Delhi, India, 34 p.

ISI: 13827–1993, *Improving earthquake resistance of earthen buildings—Guidelines*, Bureau of Indian Standards, New Delhi, India, 14p.

ISI: 13828–1993, *Improving earthquake resistance of low strength masonry buildings—Guidelines*, Bureau of Indian Standards, New Delhi, India, 11p.

ISI: 13935–1993, *Repair and Seismic Strengthening of Buildings—Guidelines*, Bureau of Indian Standards, New Delhi, India, 22 p.

Pandey, A. D., S. M. Pore and A. Sinvhal, 2006, Kashmir (Muzaffarabad) earthquake of October 8, 2005: Damages to non-engineered constructions, in *100th Anniversary 1906 San Francisco earthquake Conference* (Abstract Volume), 18–22 April 2006, San Francisco, USA, SSA 874.

Paul, D. K., Y. Singh and M. K. Ruhela, 2002, *Guidelines for Earthquake Resistant Buildings*, Department of Earthquake Engineering, IIT Roorkee, Sponsored by TISCO, Roorkee, 87 p.

Pore, S. M., A. D. Pandey and A. Sinvhal, 2006, Kashmir (Muzaffarabad) Earthquake of Oct. 8, 2005: Observations on buildings, in *Proceedings of the Seminar on Impact of Earthquake and Tsunami on Architecture,*

Structural Design and Coastal Protection Works, Military Engineer Services, 3–4 March 2006, Port Blair, India, p 35–47.

Pore, S. M., A. D. Pandey and A. Sinvhal, 2006 a, Response of ancient monuments and traditional constructions to Kashmir (Muzaffarabad) earthquake of October 8, 2005, in *Proceedings of the National Seminar on Bharatiya Heritage in Engineering and Technology*, May 11–13, 2006, Indian Institute of Science, Bangalore, India, p 1–24.

Sinvhal, A., R. N. Dubey and P. R. Bose, 1994, Damage to the built environment in the Latur- Osmanabad earthquake of September 30, 1993, in *Proceedings of the Tenth Symposium on Earthquake Engineering*, Roorkee, India, p 19–27.

Sinvhal, A. and P. R. Bose, 1995, Damage to stone houses in the Latur–Osmanabad earthquake of September 1993, in *Proceedings of the Fifth International Conference on Seismic Zonation*, Volume 1, p 623–630, Nice, France.

Sinvhal A., A. D. Pandey and S. M. Pore, 2005, Preliminary report on the 8th October 2005 Kashmir earthquake, Department of Earthquake Engineering, IIT Roorkee, 60 p.

Thakkar, S. K., B. Chandra and P. R. Bose, 1994, *Earthquake Resistant Houses* (in Hindi), Department of Earthquake Engineering, University of Roorkee, Roorkee, Sponsored by Rajiv Gandhi Foundation, New Delhi, India, 26 p.

12 CHAPTER

Multistorey Buildings

INTRODUCTION

Earthquakes in Mexico (1985), Philippines (1990), Japan (1994), Taiwan (1999), and Turkey (1999) offer ample examples of collapse and damage of multistory buildings, and those too at large epicentral distances. More than a hundred multistory buildings were ruined for the first time in India by the Kutch earthquake of 2001. Some of their occupants were rendered homeless, injured or worse, were killed. In Surat, located at an approximate epicentral distance of about 350 km of this 6.9 magnitude earthquake, several 4–12-story buildings having reinforced concrete frames with plain masonry infill were destroyed. In Ahmedabad, located at an epicentral distance of 250 km, more than a hundred four-story buildings and several 10–12-storey buildings were damaged beyond repair. Most of these multistory buildings were located in seismic zone III and IV, and Kutch is in the severest seismic zone V. The seismic response of Krishna Complex in Surat is given here as an example of several kinds of damage observed to tall buildings (Sinvhal et. al., 2001, 2004a).

KRISHNA COMPLEX

Krishna complex was constructed between the years 1989 and 1991 and was founded on soft alluvium of Tapti River in Surat. This building complex had four interconnected towers in a row. Each tower consisted of a basement, parking on ground floor, and ten additional stories. Three of these towers were on one side of the lift shaft. Two common service cores, one for staircase and another one for lift, serviced these. The staircase well was in plain masonry and the lift well was in RCC. Neither of these was well connected to the floor diaphragm. The ground floor had abnormally slender rectangular columns.

Large balconies on all floors were heavily cantilevered and some were later converted into rooms. A reinforced concrete overhead water tank, of 40,000-L capacity, rested on top of the lift shaft.

This tall building was at an epicentral distance exceeding 350 km. Because of the strong ground shaking produced by surface waves due to the Kutch earthquake of 2001, of moderate size, magnitude 6.9, the lift shaft, with three towers on one side, created eccentricity and torsion. The parking on ground floor acted like a soft story and created a vertical irregularity of stiffness, mass, and geometry in the building. Due to these and other factors, the rooftop water tank toppled over and fell on to the adjacent tower, at the free end of Krishna Complex. Due to the impact of this fall, the entire tower collapsed. This tank is visible on top of the debris in Figure 12.1(b). The debris precariously supported the adjacent surviving tower. Casualties and injuries were high in this building.

The same building is shown after clearance of debris in Figure 12.1(c) Excavation of foundation of the column, staircase, and lift shaft revealed that

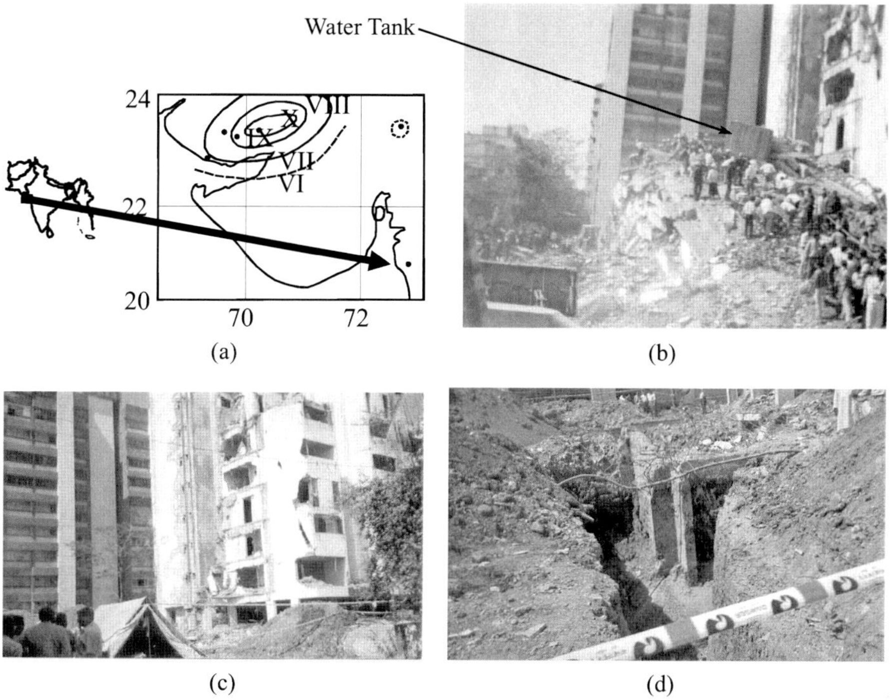

Fig. 12.1 Affect of long-period seismic waves on multistorey buildings at large epicentral distances is shown for Krishna complex in Surat, which was at an epicentral distance of approximately 350 km. (a) Isoseismal map for the Kutch earthquake of 2001. Arrow points to Surat. (b) Concrete water tank on top of debris of collapsed tower shown by arrow. (c) The same tower after the debris was cleared. (d) Excavated foundation.

it was 135 cm deep below the basement floor, and 2 m × 2 m in plan. These isolated shallow footings, in the absence of tie beams, were insufficient to resist earthquake forces that developed in this tall building even 350 km away from the epicenter. In an adjacent building seen in the background of Figures 12.1(b) and (c), residents hacked down a similar tank for fear that it would be similarly disastrous in aftershocks. Krishna Complex was demolished in March 2001.

Fig. 12.2 This roof top concrete tank toppled onto the ground, from a five-storey hotel building in Bhuj and trapped and killed three workers who were fleeing from this building.

The seismic response described for Krishna complex in Surat was not an isolated instance but was repeated in several tall buildings. For example, in Panchratna apartments also in Surat, the exterior seemed to be unharmed, decorative tiles stuck on the exterior were intact, and even the lift was functional. A closer inspection showed deep structural cracks in staircase; lift well, beams and columns and the exposed reinforcement showed rusting and were in a poor state of maintenance. Let us see the reasons for this kind of damage.

Multistorey buildings serve as residential, commercial, educational, administrative, office, and hospital buildings. In shaking a multistory building, an earthquake will relentlessly seek out every possible weakness. Most of these weaknesses deal with characteristics of the site at which the building is located, its foundation, planning and architectural configuration, structural details, nonstructural elements, construction materials, and supervision of construction at the site. These are briefly discussed in this chapter. Once the seismically induced defects are known, their causes can be better understood and consequently appropriate solutions can be formulated and adopted in new buildings and vulnerable older buildings can be strengthened.

SITE SELECTION

Choice of a suitable site plays a very important role in seismic performance of any structure, more so for a multistory building. If the site is not properly selected, then even an earthquake-resistant building can be ruined. Some building sites can be prone to seismically induced ground failures such as large permanent ground deformations associated with currently active fault zones. High topographic relief makes the site susceptible to landslides, and low-lying coastal sites are vulnerable to tsunamis even at very large epicentral distances. That multistory buildings are vulnerable not only to near earthquakes but also to distant earthquakes has been known for a long time. The reasons for this are twofold, local geological and soil conditions, and frequency content of seismic waves.

Local geological and soil conditions play a very important role in seismic performance of tall buildings. Soft sediments usually have low damping values. Seismic waves arriving from the basement rock and traveling through these soft sediments amplify the ground motion at the surface and increase duration of strong shaking. Sediments of previous lakes, thick alluvium, river deposits, filled ground, or marine sediments are susceptible to such a scenario—a situation usually found in sedimentary basins and in coastal areas. Therefore, buildings in such areas show more damage compared to similar buildings founded in areas of hard rock.

Long-period surface waves are dominant at large epicentral distances compared to body waves and can be greatly amplified in sites that have soft soil. Multistory buildings have long fundamental periods of vibration and when this closely matches with the frequency content of long-period surface waves, it leads to near-resonance conditions. Such conditions make tall buildings vulnerable to even moderate-sized distant earthquakes. This aspect is dealt in detail in Chapter 3 under the heading Earthquake Damage and Seismic Waves.

This phenomenon, coupled with several other aspects, contributed to the partial collapse of Krishna complex and other tall buildings in Ahmedabad and Surat. Ahmedabad and Surat have an abundance of soft alluvium of Sabarmati River and Tapti River, respectively. Also, in some places low-lying areas were filled and tall buildings were founded on these. Damage in Gandhidham is partly attributed to soft marine sediments.

Damaging earthquakes frequently keep revisiting the same seismotectonic environments in which great earthquakes occurred earlier. The great Kutch earthquake of 1819 and the Anjar earthquake of 1956 damaged almost the same areas as the Kutch earthquake of 2001, even though the latter was of moderate size. It is pertinent to visualize the seismic response of tall buildings in the area of influence of great earthquakes of India.

Urban areas are now dotted with tall buildings and the deleterious effect of a moderate-sized earthquake, magnitude 6.9, on such buildings has been illustrated for Krishna complex, at an approximate epicentral distance of 350 km. If an earthquake of greater magnitude were to originate now in the Himalayan arc, the kind of which are frequent in the Himalayas, it will have a larger geographical spread of destructive influence. Four great earthquakes occurred in the Himalayan arc within a span of 53 years, in the years 1897, 1905, 1934, and 1950. There has been no earthquake of comparable size in the same arc after 1950, and such an earthquake is imminent. The Kashmir earthquake, of magnitude 7.6, was considerably smaller than a great earthquake, yet it caused the collapse of one wing of the posh Margalla towers in Islamabad, at an epicentral distance of almost 100 km.

Such an earthquake will cover an area that will be defined by an arc parallel to the Himalayan arc, of at least a width of 400 km, and would include a large part of the Indo Gangetic and Brahmaputra basins, i.e., sedimentary basins with soft sediments, on which several major cities are founded. Jammu and Kashmir, Himachal Pradesh, Punjab, Haryana, Delhi, Uttar Pradesh, Uttarakhand, Bihar, Sikkim, Bhutan, West Bengal, all the seven states of North East India and large portions of Pakistan, Nepal, and Bangladesh lie within this arc. This region supports more than half the population of the country and some of these are densely populated. Population has trebled since the last great Himalayan earthquake occurred in 1950. The stock of multistory buildings is rapidly increasing in this region and if some of these too are being built with the same motivations, considerations, and designs, as was evident in a progressive state like Gujarat, then there is cause for immense worry as this indeed indicates a credible chilling scenario. The disaster would be magnified manifold, compared to that brought about by the earlier great earthquakes, as given in Chapter 6, or by the moderate-sized Kutch earthquake of 2001.

Thus, tall buildings are susceptible to local geological and soil conditions and long-period effects of seismic waves even at large epicentral distances, and this factor can be overlooked in design only at great peril. Best building sites are provided by hard and competent rock. Compact sediments and stiff-soil with a large bearing capacity are the next best sites.

FOUNDATION

Thick alluvium, loose and compressive soil like fine and soft sand and silt, expansive clays, uncompacted, filled and reclaimed ground are liable to loose strength and liquefy during strong shaking in an earthquake. Such conditions may prove to be incapable of holding load of a tall and heavy structure, which could tilt, sink, or collapse partially or completely. These conditions are best avoided. If unavoidable, then design of foundation and superstructure need special considerations.

The Kutch earthquake of 2001 revealed several deficiencies in design of foundation of multistorey buildings. Isolated footings were provided for columns and that too at a shallow depth, as shown for Krishna complex in Figure 12.1(d). For elevator shafts and staircase wells foundations were again undesirably shallow, generally 135 cm below basement floor level and 2 m × 2 m in plan, even in soft soil, i.e., the foundation was inadequate. Raft foundation or pile foundations would have been more suitable for 10–12-story buildings in such conditions. Tie beams were absent at foundation level, where provided in rare cases, these were too far apart. When the desirable aspect of tying individual footings with beams at the plinth level was followed it was again at a shallow depth of about 75 cm below floor of the basement. Pile foundations were unknown except in Gandhidham and Kandla port where the soil was prone to liquefaction. In many buildings, some parts were on pile foundations and others on shallow foundations, and damage was concentrated at junction of the two parts.

For an appropriate foundation of a tall building, there is a need to evaluate the bearing capacity of soil. If it is found to be incompetent, then soil must be improved, compacted, and stabilized and the foundation must be sufficiently wide and deep to reach a firm stratum, and plinth bands must have closely spaced ties. If firm strata cannot be reached, i.e., it is too deep then a tall structure must rest either on a rigid raft foundation or on deep pile foundations.

PLANNING AND ARCHITECTURAL CONFIGURATION

In addition to deficiencies of site and foundation, most reinforced concrete multistory buildings that were ruined in the earthquake had several critical deficiencies in planning and architectural configuration. Some of these, observed in Ahmedabad and Surat, are listed here. The problems of architecturally ill-planned buildings are very difficult to remedy after these are built.

In the area affected by the Kutch earthquake, parking space was provided on the ground floor for cars and scooters, as in Krishna complex. It was a ground storey with an assembly of columns and absent walls. This situation was akin to a tall and heavy box supported on inadequate stilts, as shown in Figure 12.3. This is a common and popular aspect in modern tall buildings. Such buildings were commonly referred to as 'buildings on stilts' or buildings with a 'flexible ground floor' or a 'soft story'. In such buildings, because of this and other weaknesses, failure invariably occurred at the soft storey. This

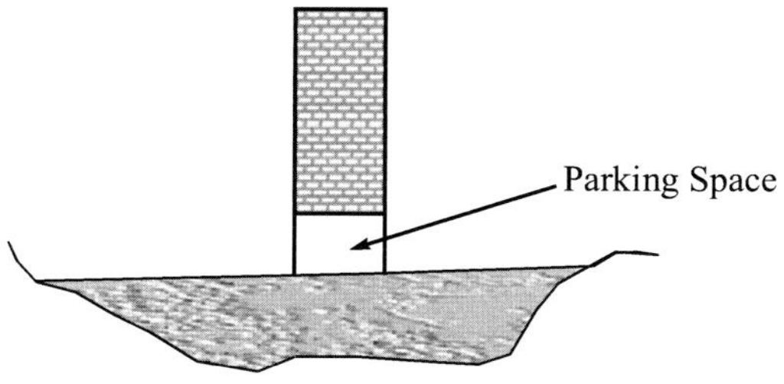

Fig. 12.3 A tall building founded on soft soil, with ground floor for parking, and isolated footings, is seismically prone to damage.

was the major contributor to collapse and damage of multistory buildings throughout the damaged area. Disastrous effect of this vertical irregularity was further compounded by several other shortcomings in columns that supported the upper floors.

Sometimes, columns sank in soft soil, buckled or collapsed, and ruined the entire superstructure. Shallow individual footings for columns and elevator shafts without plinth beams in soft soil were common in the affected area. Some columns had slender and rectangular sections, with an undesirable width to thickness ratio of three or more, and most of the time all rectangular columns were aligned in the same direction, as in Krishna complex, Figure 12.1(c). Normally, column dimensions should be in the ratio 1:1 or 1:1.5 or 1:2 and the minimum dimension should be 300 mm.

Sometimes columns were not continuous throughout the height of the building. At times these were placed above the soft storey on beams that protruded outward and were heavily cantilevered. These were referred to as loaded cantilevers. This feature helped to increase floor area and also provided large spacious rooms and balconies in every apartment. Sometimes brick masonry walls were constructed on top of these heavily cantilevered balconies to convert the balcony into a room. This practice was rampant in the affected area, and was also evident in Krishna complex. Sometimes columns were discontinued on top floors, or terminated in the beam at the first floor without any anchoring. Further, when tall columns were not tied to each other in upper floors and were unsupported each column vibrated independently during the strong shaking and behaved like a "floating" column. It was observed that vertical steel bars separated out and concrete within them crumbled, i.e., brittle failure occurred in critical regions due to shear. This is shown in Figure 12.4. Most of the time quality of concrete in columns was poorer than that in floors. This was largely because a large quantity of concrete was required when the floors were cast, so, it was economical to hire a large mechanical mixer. On the other hand, casting of columns required a smaller quantity of concrete, and in this case use of such a machine was avoided. This resulted in weak columns.

To save on floor space, sometimes columns were either flush with wall thickness or were embedded along masonry partition walls of upper floors that were barely half brick thick, irrespective of structural consequences. To save on the height factor, the beam was sometimes as thick as the floor slab, or, sometimes the beam was thicker than the slender column. Moreover, columns and beams, in general, did not form a consistent grid pattern.

Shape, size, and geometry of a building play a very important role in seismic performance of a building. Buildings that have asymmetric plans with shapes like L, C, T, E, U, and Y, and symmetric plans like X or H, etc., have an undesirable seismic response as distribution of lateral loads is uneven and

Fig. 12.4 Failure of ground floor column in Anjar in a new six-storey building, waiting to be occupied at the time of the Kutch earthquake of 2001.

stresses are concentrated at junction of wings. This leads to torsion, twisting, and large interstorey drift at the free end and sometimes leads to brittle collapse of the building. Seismic response of buildings with such shapes can be improved if it is separated into several smaller symmetric and rectangular parts, and gaps are provided throughout the structure height, which are wide enough to rule out pounding between adjacent parts during earthquake shaking. For example, an L-shaped plan can be changed into two rectangular plans and a separation joint can be provided at the junction.

Multistory buildings require a simple architectural configuration. Horizontal and vertical symmetry is preferable, compact plans close to a circle or a square are ideal. These have a better seismic response compared to a complex one, all other parameters being similar. Moreover, simple architectural details are easier to formulate in drawings, and to implement rather than complicated ones.

Openings for doors and windows were excessively large, large spans were un-reinforced, and masonry walls were unsupported, all of which are undesirable features as far as seismic response is concerned. Windows break due to distortion of frame; glass windows are fractured along planes of weakness. It is also necessary that no major changes, like alterations, change of occupancy, addition, deletion of all inside partition walls in a storey, conversion of balconies into rooms, etc. are made during the service life of the structure. If this is unavoidable, then structural and seismic implications should be considered.

The shaft for the staircase was often there in plain masonry (rather than in reinforced concrete), and the elevator shaft was in reinforced concrete. Neither of these was adequately connected to the reinforced concrete floors. This deprived multistory buildings of a potential lateral load-resisting path and proved to be a failure at the time of the earthquake. In tall buildings with undesirable plan shapes like C, X, H, etc., the staircase and elevator shaft were often at the junctions of wings and adjacent towers. Thus, the service core between the adjacent wings attracted the seismic force initially, and because of inadequate connection with floor slab at each storey, it separated, failed, and in several instances, one wing of the building complex was ripped off. While escaping via such a vulnerable staircase, many residents were killed. One such example is shown in Figure 12.5. A separate staircase is a desirable feature.

Another tall building in Gandhidham had five stories, and part of the ground floor was a soft story, which was used for parking. The middle arm of this C-shaped building had shops at ground level, which extended to the upper floor in the form of a plaza. This building had two lifts and four staircases. The

(a) (b)

Fig. 12.5 Collapse of one tower and seismic performance of staircase in a tall building in: (a) Ahmedabad and (b) Gandhidham. (See color figure also.)

staircase was raised spirally along the sides of the lift well. Connection between the floor slab and lift core was missing. RCC water tanks were placed on top of the staircase and the lift core in all wings. Balconies, like in many other places, were converted into rooms. Due to change in floor area of flats, placement of filler walls changed from third floor upward, which introduced a change in vertical stiffness. People connected with the salt trade lived in these 300 apartments. Torsion occurred due to horizontal and vertical irregularities. Only one stiff core of lift shaft remained erect after the earthquake, and remains of the staircase spiraling around the lift shaft, and the detached and remaining failed storey is clearly visible in Figure 12.5(b).

STRUCTURAL DETAILS

Structural elements of a multistory building deal mainly with ductility aspects of column, beam, and frame, and also with slab, wall, staircase, and lift shafts. Irregular distribution of mass and stiffness causes horizontal and vertical eccentricities in a tall building and makes it vulnerable to seismic forces. Adequate provisions of ductile design and detailing were absent almost everywhere, as observed in several collapsed columns and beams (Bose et al., 2002). Before the Kutch earthquake, seismic vulnerability of multistory buildings was not addressed by earthquake codes in India. The earthquake code, BIS: 13920–1993, dealt with details for achieving ductility in reinforced concrete buildings with five stories or less, subjected to seismic forces in seismic zone III. These were not provided anywhere even for buildings with more stories, mainly because the existence of this earthquake code was unknown to designers of these buildings.

For a desirable seismic response, all structural components in a tall building should be strong, stiff, and ductile, and there should be a balance of strength and stiffness between members, connections, and supports. These must be tied together so that they act in unison to resist dynamic forces produced by an earthquake; otherwise these are prone to fail one after the other. It is easy to break any number of individual sticks, but when the same sticks are tied together into a bunch then it is very difficult to break them. Closely spaced ties increase ductility and confines steel at ends of beams and columns. For a ductile frame, partial side sway of structure can be minimized if ductility provisions are so detailed that inelastic deformation develops in beams before it develops in columns.

NONSTRUCTURAL ELEMENTS

Nonstructural elements of a building are those components that are, as the name indicates, not part of the structural system like vertical support components (columns, piers, walls, etc.), horizontal components (beams, slabs, etc.), or any other structural element used for bearing the load of the building. A wide variety of elements constitute non-structural elements of a building. Some of these are heavy and unanchored into the structure and some of these are provided after the structure has been completed. Architectural components of non-structural elements consist of nonload-bearing walls, stairways, rooftop water storage tanks, chimneys, balconies, doors and windows, false ceiling, exterior facing, plaster, glass panels, glazing, ornamentation, and several kinds of additions and alterations in the building. Mechanical and electrical components consist of equipments, elevators, escalators, fire-fighting systems, pumps, boards for electrical panels, air conditioning system, etc. services for gas, water, sewerage requires pipelines and plumbing (Bose et. al., 2004).

Seismic response of non-structural elements in a tall building can lead to an adverse behaviour of structural elements. When a multistorey building begins to shake, anything that is supported by or attached to it will be affected by seismic vibrations and is liable to deform. Damage to and by non-structural elements, and their replacement can amount to almost half the value of the building, and can at times become injurious and fatal.

Heavy appendages like sunshades, weighing more than the wall on which they are anchored, when not tied to the main structural system of the building, behave independently and not in consonance with the structure of the building when shaken by an earthquake. This can damage the building beyond repair. Nonload-bearing walls are liable to overturn if not properly tied to the main structural system.

Seismic forces are amplified with height and are maximized at the top of a building due to an inverted pendulum effect. This makes water tanks, chimneys, unbraced parapet walls and ornamentation situated at the roof level of tall buildings particularly vulnerable. Ignoring all this, heavy concrete water tanks were provided on rooftop of many tall buildings. These usually rested on plain masonry pedestals, either on mumpty or on lift well, and their inclusion seemed to be more of an afterthought. Due to strong shaking, the water tank developed torsion, rotation, and rocking motion, and at times toppled over with a menacing impact, sometimes puncturing the roof slab and falling on the floor below, or overturning onto a neighboring building, as shown in Figure 11.13, or falling on to the ground, as shown in Figure 12.2, or as shown in Figure 12.1(b), pan caked an entire tower of Krishna complex. This kind of seismic performance of water tanks caused havoc in the Kutch earthquake, in the meizoseismal area and even at large epicentral distances, and was the cause of many casualties.

Ideally, the water tank should be placed at a location where it does not cause mass eccentricity in a vertical plane, and the tank and its supports should be tied to and integrated with the main structural system of the building, to take care of the dynamic forces introduced by the earthquake. This desirable aspect can be part of architectural design. Water stored in underground water tanks is more likely to be available after an earthquake than water stored in tanks on the roof of a building. For a desirable seismic response all non-structural components must either be properly integrated into the main structural system or be effectively isolated from it.

LACK OF COHERENT CONSTRUCTION

Collapse and severe damage often results when a complex structure does not behave as an integral unit. This can happen in many ways. Damage of buildings due to lack of good connections is common. Sometimes an annex is added in contact with a building. Earthquake motion affects the two parts differently, and strains may develop at their junction. If separate units are individually well consolidated, strains may act to distort both, subjecting them to effects which would not have occurred if they were separated, or had been so connected as to respond coherently. Floors must be well connected to walls to ensure integral action during the earthquake. Otherwise, walls in upper floors are likely to slide away.

CONSTRUCTION MATERIAL AND SITE SUPERVISION

Dynamic forces produced in an earthquake are proportional to mass of the building. Therefore, building materials that are light and also have a high strength to weight ratio are preferable. The best construction material for high-rise buildings is good quality structural steel. RCC is the strongest and most earthquake resistant type of construction, when well designed and well executed.

Rerolled steel bars, with a high carbon content originating from shipwrecking yards, were commonly used as reinforcement. A large quantity of water was used to increase the workability of concrete admixture. Water–cement ratio was as high as 0.7 (by weight) against a desirable ratio of 0.45. This resulted in porous concrete that led to a rapid corrosion of the reinforcing bars. An additional drawback with this was that concrete was poured in sections and when reinforced concrete structures were damaged significant movement was noticed at these construction joints. The junction between these pours is called construction joints. To ensure proper bonding, these joints should be kept clean during construction, but this was often overlooked. Construction joints were located in columns at critical regions, i.e., at top and bottom of columns for convenience. Locally, this was referred to as Topi construction. Ideally construction joints in columns should be located at mid-height. All longitudinal reinforcing bars should not be spliced at the same section, as was observed at several places, but splices in longitudinal reinforcing bars should be staggered.

Moreover, the best quality of plaster falls and cracks when there is other structural damage. On the other hand, a mediocre quality of plaster, which uses a large quantity of water to increase the workability, rather than a suitable admixture and water–cement ratio, does not provide adequate bonding, and cracks due to shrinkage. This falls copiously during strong shaking. Plaster cracks open and close regularly with changing seasons, but are first noted and reported after an earthquake. Moreover, deficiencies in structural formwork (shuttering) were common. In particular, reinforcement cage was not properly tied and spacers were avoided. Twenty-one days required for curing were reduced to speed up the construction activity. This led either to a larger concrete cover of reinforcement thereby reducing strength of the member, or smaller concrete cover thereby increasing corrosion of the reinforcing bars. Thus, quality of construction and workmanship play an important role in the desirable seismic response of structures.

WHAT CAN BE DONE

Before considering design and construction of multistorey buildings, several important aspects need thorough evaluation. Most of these are briefly cited above, and include seismicity of site, long period effects of surface waves, local geology and soil conditions at the site, design of an adequate foundation, and only after that architectural and structural aspects should be adequately catered to design either a stiff or a flexible tall building. The height of building and its time period, seismic effect of material to be used, frequency content of ground motion, seismic coefficient, dynamic forces introduced by a design earthquake at the base of the tall building, and its modes of vibration and displacement need to be estimated.

Reinforced concrete multistory buildings, sometimes in a deteriorated state, (Sinvhal et. al., 2004b) are in use all over the world. New ones are coming up rapidly everywhere, sometimes lacking appropriate seismic considerations. Agencies that are involved in design, construction, repair, restoration, strengthening, and retrofitting work of a building are not always aware of disastrous consequence of an earthquake. Therefore, there is an urgent need to incorporate earthquake-resistant measures in all existing and in new multistory buildings. The funding agency should include this as a prerequisite for giving any financial support. This action will mitigate to a large extent the disastrous effects of an earthquake on multistorey building.

CONCLUSION

It is high time to recognize that multistory buildings are vulnerable in an

greater life and economic losses and that too concentrated within a small area. Seismic response of Krishna Complex in Surat has been taken as an illustrative example as it showed almost all kinds of possible earthquake-induced damage in a tall building. New design concepts often originate as a result of damage observations made in previous earthquakes. Earthquake-resistant design of large and tall structures is still vigorously debated in the earthquake engineering profession.

REFERENCES

BIS: 13920-1993, Indian Standard Code of Practice for Ductile Detailing of Reinforced Concrete Structures Subjected to Seismic Forces, Bureau of Indian Standards, New Delhi, India.

Bose, P. R., A. Sinvhal, A. Bose, A. Verma, Pranab, Saurabh, 2002, Implications of planning and design decisions on damage during earthquakes, in *Proceedings of the 12th Symposium on Earthquake Engineering,* Roorkee, India, p 561–568.

Bose, P. R., A. Sinvhal, A. Bose and A. Verma, 2004, Impact of Kutch earthquake on non structural elements and appendages of buildings, in *Proceedings of the 13th World Conference in Earthquake Engineering*, Vancouver, Canada.

Sinvhal, A., P. R. Bose, A. Bose and V. Prakash, 2001, Destruction of multistory buildings in Kutch earthquake of 26th January 2001, in *Proceedings of Workshop on Recent Earthquakes of Chamoli and Bhuj*, Roorkee, May 24–26, 2001, Indian Society of Earthquake Technology, p 451–460.

Sinvhal, A., P. R. Bose, A. Bose and V. Prakash, 2004a, Multi storied buildings and Kutch earthquake of 26th January 2001, in *Seismic Hazard—Proceedings of International Conference on Seismic Hazards*, October 2001, IMD & DST, New Delhi, India, p 8–14.

Sinvhal, A., P. R. Bose, A. Bose and A. Verma, 2004b, Retrofitting of a deteriorated building and its seismic resistance—a case study, in *Proceedings of the 13th World Conference in Earthquake Engineering,* Vancouver, Canada.

13 CHAPTER

Lifelines and Infrastructure

INTRODUCTION

Every community is dependent on a network of lifeline services and infrastructure facilities. When these get damaged, the earthquake-related tragedy is compounded manifold. Their immediate restoration is usually very difficult and alternative arrangements become necessary, so that at least immediate rescue and relief operations can be speeded up. Thus, damage to lifelines and infrastructure amounts to valuable time lost in the postdisaster scenario. Failure of lifelines and infrastructure not only severely strains quality of life after the earthquake, but also the economy of the afflicted community in the long run (Prakash et. al. 2004a). Water, electricity, hospitals, and medical facilities are lifelines of any community. Infrastructure deals with transport systems, communication facilities, industry, and educational facilities.

Immediately after the Kutch earthquake of January 26, 2001, all lifeline services like water and electricity supply snapped, and several hospital buildings collapsed, that too at a time when these were needed the most. Telephones were put out of order. Loss of transport systems hampered emergency response, as several roads, bridges, railway lines, railway station at Bhachau, airport at Bhuj, and Kandla seaport were adversely affected. Several schools and industrial structures collapsed partially or completely. A similar dismal scenario was repeated in coastal areas of the Indian Ocean by the Sumatra earthquake of December 2004, and again in the rugged Himalayan terrain by the Kashmir earthquake of October 2005.

WATER SUPPLY

In a postearthquake scenario, a crisis of water, especially drinking water, may ensue. In the earthquake-affected community, it may be disrupted, or contaminated or, in a more severe situation, it may not be available at all. Water supply schemes, reservoirs, canals, pipelines, overhead municipal tanks and rooftop domestic storage tanks may all be affected. Canals may be damaged due to slumping or emergence of ground water and sand, or may be offset by a fault. Water pipelines, especially old and weak ones, are damaged even in small earthquakes by slumping and subsidence of soft ground. Initial failure may only be a small crack in the pipe. In the 1906 San Francisco earthquake as the pipeline carrying water to the city followed and crossed the fault line repeatedly, large pipes were completely ruined by rending or compression.

Overhead municipal tanks are also affected by strong ground shaking produced by an earthquake (Prakash et al., 2001a; Pore et al., 2005). The RCC water tank at Kawtha collapsed in the 6.4 magnitude Latur earthquake (Sinvhal et al., 1994). Overhead concrete water storage tanks in several multistory buildings were a big disaster in the Kutch earthquake of 2001 (Bose et al., 2004). An example for Surat, at an epicentral distance exceeding 350 km, is shown in Figure 12.1b.

ELECTRICITY SUPPLY

Earthquakes often cause electrical power failures. This may be due to damage to transmission and distribution lines, transmission towers, snapped wires, loosely hanging and damaged electrical equipment, electric substations, and power plants, whether hydroelectric, thermal, or super thermal. In the 20 MW power plant at Bomboo Flat near Port Blair, electrical and mechanical equipment was severely damaged by the effects of inundation and silting by the tsunami. In the 5.25 MW hydroelectric power plant at Kalpong in North Andaman Island, the turbines were similarly damaged. Collapse of transmission towers at Middle Strait made distribution impossible.

The Gujarat Electricity Board had a standard design for 66 KVA substations. The control, monitoring, and relay panels were housed in a single-story rectangular building. The RC roof of this building was supported on load-bearing unreinforced stone masonry walls with reinforced concrete bands at lintel and plinth levels in all four walls. The lintel band was above windows in sidewalls, and just below the roof in end-walls, to accommodate a big rolling shutter door for transit of large equipment. This made the lintel band discontinuous and its optimum benefit was lost (Prakash et al., 2001b). As a consequence of this, a large number of these substations were severely damaged within MM Intensity X, on the Wagad Ridge, at Chobari, Bhachau, Amardi, Bhimasar (Rapar), Dudhai, Adhoi, Balasar, Barudia, Kharoi, Sapar, Trambau, and Vajepar in the Kutch earthquake of January 26, 2001, and electricity supply was hampered throughout the affected area.

Rarely, a large earthquake may cause a dam failure. Due to the California earthquake of 1925, moderate shaking of intensity VIII in soft soil damaged an earth fill dam in Santa Barbara. This was possibly due to forces that developed in the Sheffield reservoir, which acted like a soft hammer on the dam (Richter, 1958). However, dams, which are designed when seismotectonics of the area and earthquake parameters are accounted for, have proved their mettle in several earthquakes. Seismic performance of Maneri dam, within intensity X of the Uttarkashi earthquake, Makni dam within intensity VIII of the Latur earthquake (Sinvhal et al., 1994) and the underground Uri hydroelectric project, within intensity IX of the Kashmir earthquake was exemplary. The Dhanikari dam, a water supply scheme in Port Blair, developed several vertical hairline fissures through which water leaked into the inspection gallery. This concrete dam, 182 m log and 32.2 m high, was at an epicentral distance of 850 km and repairable and minor damage was caused by the Sumatra earthquake of December 26, 2004.

MEDICAL FACILITIES

In the emergency created by an earthquake, when medical facilities are ruined the survivors and the injured are left without any medical help at a critical time. Several hospitals and structures housing medical facilities showed partial or complete collapse in several recent earthquakes (Sinvhal and Bose, 1996; Sinvhal et al., 2001a). For this reason, it is very important that hospitals and all other medical facilities be adequately designed to resist earthquakes. Spread of infectious diseases and epidemics in a postearthquake scenario can be arrested only if medical facilities are available when they are needed the most.

The Latur earthquake of September 30, 1993, ruined the civil hospital at Killari. Latur was located within the safest seismic zone, Zone I, of the seismic zoning map of India, in Maharashtra. This single-storey brick masonry building, in cement sand mortar, with RCC slab for roof was an L-shaped building. It was situated close to a seasonal nala, and probably on artificially filled ground. This government facility was well equipped with medical instruments and staff and catered to the needs of the surrounding region. The moderate-sized Latur earthquake, of magnitude 6.4, in which maximum damage was of intensity MMI VIII+, gave it a near field vertical jolt. This was manifest as deep cracks and severely crushed mortar between loosened and displaced bricks over the (iron) collapsible side entrance to the hospital. Several doors located near the intersection of the two wings of the L-shaped building were jammed, due to the torsion introduced by the asymmetric configuration. Nonstructural elements were also extensively damaged. Wall tiles, washbasins, plaster, and notice boards fell off their supports from walls. Large glass windows were broken due to distortion of frames. Lighting fixtures snapped and tube lights dropped to the floor. Cribs and beds were strewn with debris. Large X-shaped fissures developed between large openings in walls, as shown in Figure 13.1.

(a) (b)

Fig. 13.1 Seismic response of civil hospital: (a) at Killari, due to the Latur earthquake, and (b) at Bhuj, due to the Kutch earthquake.

Similarly, the Republic day earthquake of 2001 in Kutch wiped out many hospital buildings and clinics. The civil hospital at Bhuj was constructed in 1952. The load-bearing stone masonry walls were almost 45 cm thick, as shown in Figure 13.1b. Several additions and alterations were made later to the original building. At the time of the earthquake, 150 of the 180 admitted patients, 3 head nurses, 4 staff nurses, and 4 supporting staff were reportedly killed within this building. The ruins showed a total absence of any earthquake-resistant measures like through stones, earthquake bands, vertical steel, etc., not that these would have helped much for a double-storied, random rubble stone masonry building in seismic zone V. Incidentally, this hospital earlier provided succor to victims of the Anjar earthquake of 1956, but it was probably a single-story building then.

TRANSPORT SYSTEMS

Roads, bridges, railways, airports, seaports, jetties, wharfs, and harbors constitute some important elements of transport systems. Loss of transport lines hampers emergency response and rescue and relief operations after the earthquake and makes recovery much more difficult.

Roads

Roads, whether with asphalt topping or unpaved, can be damaged in several ways by an earthquake. In regions of undulating topography or in mountainous terrain, roads are prone to blockage by landslides and undercutting from below. This effect is shown in Figures 9.6, 9.8, 9.10 and 9.11. Roads can get washed away or be inundated by postearthquake floods. This scenario gets more severe for coastal roads in a tsunami as these may sometimes get heavily scoured, or debris may be deposited on them, as seen in Figure 10.4(c).

Strong ground shaking may cause settlement or liquefaction of the underlying soil layers, which in turn may cause deep fissures in roads. Many culverts settle down.

National Highway 8A was damaged in several places due to the Kutch earthquake of 2001, as seen in Figure 13.2(a). All roads leading to Rapar and Bhachau and those between them were fissured extensively, as was the road between Gadsisa and Ganga Rampar (west of Bhuj). Similarly, the Andaman Trunk Road (ATR) developed long, deep, and wide fissures at several places in North, Middle, and South Andaman Islands and at Baratung, as shown in Figure 13.2(b), and these were difficult to negotiate, as numerous fissures were transverse to the road. This was the effect of the Sumatra earthquake of 2004 and not of the tsunami that followed.

(a) (b)

(c)

Fig. 13.2 (a) Longitudinal fissures on national Highway NH 8A, between Gandhidham and Bhachau. The Kohinoor salt factory, seen in top left corner, collapsed partially, (b) Transverse fissures across Andaman Trunk Road in Baratung due to the Sumatra earthquake of 2004, (c) Clogged streets in Gubal, due to the Latur earthquake of 1993 (left). (See color figure also.)

In congested areas, debris from collapsing houses and buildings from both sides fall on to narrow streets and clog them. This is shown in Figure 13.2(c). This further hampers and delays postearthquake rescue and relief operations, and critical time is lost in cleaning streets. In an unusual situation, the debris may trap those on the streets. Four hundred school children of class seven, together with 40 teachers from various Government schools, were taking out a Republic day procession in Anjar, through Khatriwadi. This was a thickly populated area with three- to four-story houses on either side of narrow streets. Due to the Kutch earthquake, these mid-rise buildings collapsed on to the streets and trapped, buried, and killed school children and their teachers in 6-m high mounds of debris. Streets were blocked and heavy earth moving equipment, trucks, and dumpers used for removing debris could reach the site only after clearing the debris *en route*.

Bridges

Bridges meant for railways, for highways, or for small roads sometimes get seriously damaged due to strong ground shaking. A bridge may be affected by faults in its vicinity. Bridge piers may be displaced or sheared. The bridge deck may be subjected to pounding, unseating, a bridge span may collapse, or the superstructure may develop fissures or may fail completely. Abutments may develop instability, and road embankments and approach roads may settle down or be fissured.

The Surajbari Road Bridge, situated between Kandla Port at Gandhidham and Saurashtra, on National Highway NH 8A, is an important road link between Delhi and Mumbai. This balanced cantilever bridge with 36 piers trends NW–SE, and crossed the marshes of the Little Rann of Kutch at its narrowest portion. It is situated across the North Kathiawar fault, which trends NE–SW. Damage in this bridge was observed due to several reasons. One of the main reasons was strong ground shaking of the marshy soil, which led to settlement and separation of soil all around the circular wells supporting the piers. Pounding between supported span and cantilever span, and displacement of deck, led to misalignment of bearings on piers. The superstructure of the bridge shifted toward the north-end abutment causing the bridge deck to separate from the south-end abutment by about 200–275 mm. This is shown in Figure 4.6. The north abutment cracked and fissured due to pounding by the bridge deck, resulting in long (larger than 30 m), wide (2–15 cm), and deep fissures on the approach road at the north-end.

Lateral shift of superstructure with respect to pier 6 was clearly visible in shifting, twisting, detachment, and crushing of steel plate bearings. Fall of steel cover exposed steel rods in the deck. In some cases, horizontal displacement exceeded the length of the bearing and the deck impacted the pier vertically as the deck settled by an amount equal to the thickness of the bearing plate.

During the reverse motion, the bearing plates collided horizontally with each other and became detached. Since the approach road at north-end of the Surajbari Bridge, on NH 8A, was severely fissured and the bridge had other kinds of damage too (Sinvhal et al., 2001b), in the initial stages immediately after the earthquake only light traffic was allowed to cross the damaged bridge.

In the Kashmir earthquake of 2005, in general, most bridges did not suffer any significant damage, and the failures and damage that were observed were not due to failure of design of steel bridges but were attributed to other causes. Landslides led to failure of abutments and wing walls of the Aman Setu, rock fall claimed the Sikh bridge over Qazi Nalla, as shown in Figure 13.3, and the bridge at Sarai Bandi succumbed due to the presence of a fault, the Main Boundary Thrust (MBT).

Sarai Bandi in Uri is situated in the vicinity of the Main Boundary Thrust. This village has a single span Bailey bridge. This steel bridge is approachable in a straight stretch from the south, while from the north it has a curved alignment. Landslides completely blocked the road head at the south-end of the bridge. At this end, 300-mm wide fissures developed parallel to the slope surface of embankment in approach road, which also showed an equal amount

Fig. 13.3 Damage to Sikh bridge due to Rock fall in the Kashmir earthquake of 2005, in Tangdhar.

of vertical settlement. Due to horizontal and differential displacements at both ends, the bridge rotated in plan. This was evident from outward displacement of bearing plates. These were displaced diagonally opposite to each other, such that displacement occurred on right side of SW end bearing plate and left side of NE end bearing plate. Displacement at both ends was almost equal; about 450 mm. Cross-girders and the deck showed signs of twisting and in-plan bending. There was no visible damage to main trusses, or masonry in abutments (Sinvhal et al., 2005, 2006, Pandey et al., 2006). Slope failure including that of stone pitching occurred at northeast end of bridge.

Railways

The railways include railway lines, stations, tunnels, and bridges etc. Change of level in soft ground, usually due to slumping and subsidence, can put a railway line out of service. Rails were bent in the great Assam earthquake of 1897 (Oldham, 1899), and rail lines snapped and broke due to strike slip faulting in the Baluchistan earthquake of 1892. The Kutch earthquake of January 26, 2001, damaged the railway station at Bhachau and several railway bridges. Therefore, rescue, relief, and rehabilitation material could not be sent by rail to the affected area. The 60-km long railway track (Western Railway), linking Bhuj with Gandhidham, was in the process of gauge conversion from meter gauge to broad gauge at the time of the earthquake and was to be inaugurated on January 29, 2001, on the auspicious day of Basant Panchami. But the earthquake that originated 3 days earlier changed all this. The railway bridge at Maliya Miyana, parallel to the Surajbari Bridge, was nearing completion at the time of the earthquake and suffered minor repairable damage.

Railway Bridge number 48, at Dholawa, east of Bhuj, is an arch bridge, with four spans of 9.15 m each. It was made of unreinforced dressed stone masonry. It was being widened with reinforced concrete jackets on both sides. Due to the 6.9 magnitude Kutch earthquake, the crown of all four arches developed several wide, deep, and zigzag fissures, impairing the arch action of the bridge. It was no longer safe to carry the weight of the train until this bridge was strengthened (Sinvhal et al., 2001c). The newly made RCC jacketing was unharmed. Rail traffic on the new broad gauge rail track between Bhuj and Gandhidham resumed 5 months after the earthquake, after necessary remedial corrections.

The stone masonry railway station at Bhachau, in a semiurban setting, collapsed completely. The ground floor entrance lobby, offices, and rooms on the railway platform, cabin from where signals were given for change of railway track, lever frame, and internal equipment were reduced to rubble. However, it was remarkable that the meter gauge and broad gauge lines were restored and made functional within 4 days of the earthquake. Meanwhile, the affected people used stationery trains on the platform at Bhachau and at Ratnal as temporary shelters.

Sea Ports and Coastal Structures

All coastal structures like seaports, harbors, jetties, wharfs, and associated buildings are made on or near filled ground. These are susceptible to differential settlement and other effects due to the combined effect of the earthquake and the tsunami. Damage to these can be tremendous. Earthquake effects on coastal structures were witnessed in the Kutch earthquake of 2001 and in abundance in the Sumatra earthquake of 2004.

Gujarat is the westernmost coastal state of India, closest to the oil-exporting nations of west Asia. With the loss of Karachi port to Pakistan after partition of the country in 1947, pressure on the Mumbai port intensified. Therefore, an alternate port was developed at Kandla, along with several smaller ports along the Kutch and Saurashtra coastline. Kandla lies in seismic zone V as per seismic zoning map of India, BIS: 1893–2002, and it is also prone to cyclones. Kandla port now handles 17% of India's total cargo. This is the port of choice for crude oil imports from west Asia as the landed cost is the most favorable here. It serves the hinterland of Jammu and Kashmir, Punjab, Haryana, Himachal Pradesh, Delhi, and Rajasthan. These states are well connected by national highways and railways to Kutch. Agricultural and other exports from these states are preferably routed through Kandla port. Oil refineries, petrochemical, and fertilizer industries and other related industries developed in Gujarat as a result. Reliance Industries developed one of the world's largest refineries at Jamnagar. Crude oil imported at Kandla port is pumped via a submarine pipeline to the Jamnagar refinery. Extensive damage due to soil liquefaction occurred at Kandla Port, and several buildings and jetties were later demolished and reconstructed.

After the tsunami generated by the Sumatra earthquake of 2004, the list of affected transport systems was extended to include coastal structures like harbors, wharfs, jetties, and lighthouses. Their response is given in Chapter 10 on tsunami.

INDUSTRY

The industrial scenario is adversely affected due to earthquakes. This is also true of the tourist industry, (Sinvhal et. al., 2001e, 2002, 2003; Pore 2006; Prakash et al., 2001c, 2002). Losses can accumulate due to damaged industrial structures and installations. But good engineering design and its execution shows worthy performance in an earthquake. The oil industry is endowed with several desirable engineering practices and therefore the seismic performance of structures most of the time is exemplary. An example is given here to illustrate this point as observed in the Kutch earthquake.

The Kadol oil field in Cambay basin, at an epicentral distance of 240 km, is the type area for oil wells in the Cambay basin. It was discovered in 1960 and has a productive area of 250 km^2. Oil is produced from a depth of 1400 m from the Kalol formation. Crude oil from several oil wells is collected at a group gathering station (GGS) through pipelines. This oil then goes to the refinery at Vadodara, through pipelines. Another GGS collects oil from 24 wells, which are located within a 6-km radius. Three of these wells are for water injection. A self-flow well, which produces oil 24 hours a day and has been doing so for the last 2 years and from which oil is expected to flow without interruption for the next 20 years, was undisturbed by the Kutch earthquake. Oil spillage occurred at several GGS. Twenty tanks filled with a combination of oil and water sloshed from side to side during the Kutch earthquake of 2001. The steel structures above the oil well, known as Christmas trees, were also unharmed by the earthquake in the entire region. Other oil wells and GGS at Kalol showed exemplary seismic behavior. Production from oil wells continued without interruption even after the earthquake.

Crude oil is supplied from Kandla port to oil refineries through steel pipelines that are situated along the highway leading to the Kandla Port. To mitigate temperature stresses, and in deference to the dictates of seismic zone V, expansion joints in the form of loops were provided at intervals, either in the horizontal or vertical plane. The 50–60-cm diameter pipelines were supported on steel frames, concrete pedestals, or on top of steel bars having about 8 cm diameter at regular intervals. To reduce contact stresses at supports, a steel plate (0.8–1.0 cm thick) was welded to the base of the pipe at support locations. A similar plate was provided at supports on top of the steel frame. A steel bar of 2–3 cm diameter was provided perpendicular to the length of the pipe at supports on top of concrete pedestals. Long stretches of straight lengths of pipes did not have any restraints against transverse movements, except that provided by steel friction at supports. During the Kutch earthquake, some pipelines moved transversely and were dislodged from their supports. At some joint locations, pools of oil collected on the ground due to leakage at joints, and the joints were repaired subsequently (Prakash et al., 2001d, 2004b). Fires can sometimes result from these and also from broken gas lines.

COMMUNICATIONS

An earthquake can cause failure of communication facilities and communication-related buildings in a very wide area. These are related to telephones, television, radio, fax, postal services, e-mail, mobile, satellite communication, etc. Due to the Kutch earthquake, telephone services were disrupted in Bhachau, Rapar, Anjar, Gandhidham, and Bhuj. Damage to unreinforced masonry walls occurred in a residential building of the post and telegraph department at Anjar. Emergency services via a satellite link were restored within a week. A similar scenario developed in the tsunami-genic earthquake of December 2004, in Andaman and Nicobar regions.

SCHOOLS

Schools house our children in large numbers and are, therefore, important buildings. A large number of these are destroyed and severely damaged in earthquakes (Sinvhal et. al., 2001d). This ruins educational infrastructure in the earthquake-affected region. Most schools are constructed in increments according to availability of funds. Whenever some funding is received, a few rooms are added to the existing building. This construction practice is adopted at most places. This can lead to construction weaknesses between old and new parts of a building, and discontinuity in longitudinal beams. However, from the observed damage pattern, it appeared that schools are designed and constructed no better than ordinary buildings. Most educational buildings have two or three stories with plain brick masonry infill walls and reinforced concrete floors. Damage observed was due to crushing of columns, fall of nonstructural masonry pillars, fall of beams, diagonal cracks in walls, damage to appendages, shape of building plan, use of random rubble stone masonry, etc.

Haritpawan Gurukul of Swaminarayan is a residential high school at Ganga Rampar, west of Bhuj. The three wings of this school building were in a C shape, as shown in Figure 13.4. The east wing had three stories. The middle wing was a combination of two and three stories. The west wing had three stories and a basement, was built later, and finishing work was in progress at the time of the earthquake. All wings had long verandahs on all floors.

Fig. 13.4 Haritpawan Gurukul of Swaminarayan High school at Ganga Rampar west of Bhuj. This place was assigned damage intensity VIII on the MMI scale and was at an epicentral distance of 70 km. In this school building, failure of ground floor columns caused failure of upper stories. The decorative dome on the terrace fell off in all the three wings. (a) West wing of school, (b) Detail of shear failure of short columns in the west wing.

Decorative nonstructural items like fictitious columns, arches, and railings provided in these verandahs created a pleasing elevation. However, this converted the front columns into short columns, which attracted larger earthquake forces because of increased stiffness.

Damage to the west wing was concentrated in the ground floor. All peripheral columns toward the open end of the wing failed. Lack of ductile detailing in the form of larger spacing of ties and splicing of all bars at the same level in columns added to further damage. More damage to this newer wing as compared to the old wing was due to change in quality of construction. This resulted in extensive damage to the building, and it was later partly demolished. A meager 3-cm spacing separated the three wings, which is insufficient according to the IS code for earthquake-resistant design in seismic zone V. Therefore, the west and central wing together behaved like an L-shaped unit. This behavior created an eccentricity in plan and sheared the columns at free-end of this wing. This effect was compounded by the torsion component of ground motion, which was observed in the twisting of freely kept ornamental objects on the boundary wall of the school. Decorative domes were constructed on columns above all wings on the terrace. These appendages collapsed. The dome above the central wing showed a twisting type of failure. The east wing moved in the N–S direction and behaved independently. This was borne out by the fact that a wall built on a projection and also a stair case lattice (jali) in this wing fell toward the south. Hairline cracks appeared between the frame and filler walls.

Shri Ramji Ravji Lalan College in Bhuj, better known as the Lalan College, has several building blocks. The two story front block had a long continuous beam supported on thick masonry piers at lintel level in both the stories. Masonry walls were raised on top of this beam to support the transverse beams of T-beam floor. This building had too many and very large openings for doors and windows. This college suffered extensive damage to its various blocks. All structural elements built separately and at different times, without any structural connection, behaved independently. The masonry pier collapsed, which caused the collapse of longitudinal beams, which in turn led to collapse of the entire upper floor and roof. The use of a small amount of vertical steel in the masonry piers could have saved these blocks. The vast open field of this college served as the supply depot for relief material in the postearthquake scenario.

The L-shaped RC-framed building of Modern School in Gandhidham withstood the earthquake well. Separation of infill walls from the frame was observed at several places in this double-storey brick masonry building. A typical feature of this school was the verandah in front of the ground floor classes. It was supported on masonry piers of approximate dimension 16″ × 16″. According to one of the teachers of this school, the children used to jump from the verandah to the open ground and created disturbing noises during school time. To circumvent this daily problem, the open verandah was closed by erecting several closely spaced nonstructural brick masonry pillars of size 8″ × 8″. These had no structural connection to the main system, and were anchored neither to the floor nor to the ceiling. A steel door was provided in the center of the verandah, which was locked during school hours. This served the intended purpose and children stopped making noise near classrooms. During the earthquake of January 26, all nonstructural masonry pillars fell inside the verandah, as shown in Figure 13.5. One shudders to think what could have happened if school children were crowded there and the steel door was locked at the time of the earthquake. Several supporting masonry piers developed cracks at the top end. This practice to support the verandah slab on a longitudinal beam supported on masonry piers needs to be curbed or at least vertical steel should be provided in such pillars that should be anchored in either the beam or the roof slab.

Fig. 13.5 L-shaped Modern School in Gandhidham. The nonstructural masonry pillars in the verandah fell inside the verandah. Fortunately children were not there at the time of earthquake.

The seismic performance of random rubble stone masonry is discussed in Chapter 11. Schools fared no better than houses in an earthquake, whether in the Himalayan arc, the Deccan plateau, or the coastal region. Figure 13.6 shows the seismic performance of schools in diverse tectonic environments.

WHAT CAN BE DONE?

If lifelines and infrastructure are to continue to perform their intended function in a postearthquake scenario, then several aspects need to be considered at

(a) (b)

Fig. 13.6 Seismic performance of a: (a) stone masonry school in Killari due to the Latur earthquake of 1993, and (b) brick masonry boundary wall in the educational facility, Center for Ocean and Island Studies, in Port Blair due to the tsunami of Sumatra earthquake of 2004.

the planning and development stage, before any design or construction work is taken up. Some of the important ones are seismicity, seismotectonics and seismic hazards possible at the site, local geology and site conditions, seismic characteristics of the site, and seismic response of the site, and the design and construction of the foundation and the structure built on it.

CONCLUSION

Water, electricity, hospitals, and all medical facilities are lifelines of any community, and transport systems, communication facilities, industrial structures, and educational facilities are important infrastructure facilities, which ought to remain functional in the aftermath that follows an earthquake. This was not the case during several recent earthquakes. Failure of lifelines and infrastructure not only severely affects quality of life after the earthquake but also the planned economy of the country. Application of appropriate earthquake engineering interventions can go a long way in keeping these facilities intact and operational even in the postearthquake scenario.

REFERENCES

BIS: 1893–2002, *Indian Standard Criteria for Earthquake Resistant Design of Structures, Part I: General Provisions and Buildings* (Fifth Revision), Bureau of Indian Standards, New Delhi, 40 p.

Bose, P. R., A. Sinvhal, A. Bose and A. Verma, 2004, Impact of Kutch earthquake on non structural elements and appendages of buildings, in *Proceedings of the 13th World Conference in Earthquake Engineering*, Vancouver, Canada.

Oldham, R. D., 1899, Report on the Great Earthquake of 12th June 1897, *Memoirs Geological Survey of India*, Volume 29, 379 p.

Pandey, A. D., S. M. Pore and A. Sinvhal, 2006, Damage to the engineered constructions due to Kashmir Earthquake of 8 October 2005, in *100th Anniversary 1906 San Francisco Earthquake Conference* (Abstract Volume), 18–22 April 2006, San Francisco, USA, SSA 873.

Pore, S. M., A. D. Pandey and A. Sinvhal, 2005, Muzaffarabad (POK) earthquake and observations on water tanks, in *Proceedings of the Workshop on Liquid Retaining Structures*, Nov 10–11, 2005, IIT Roorkee, 6 p, CD.

Pore, S. M., A. D. Pandey and A. Sinvhal, 2006, Kashmir (Muzaffarabad) Earthquake of October 8, 2005: Performance of monumental buildings, in *Proceedings of All India seminar on Earthquake Resistant Design, Construction, Retrofitting and Rehabilitation of Buildings*, February 18–19, 2006, Banaras Hindu University, Varanasi, p 246–257.

Prakash, V., V. H. Joshi, and H. R. Wason, 2001a, Effects of the Kutch earthquake of 26th January, 2001 on municipal overhead water tanks, in *Proceedings of Workshop on Recent Earthquakes of Chamoli and Bhuj*, May 24–26, 2001, Roorkee, Indian Society of Earthquake Technology, p 433–441.

Prakash, V., V. H. Joshi and H. R. Wason, 2001b, Effects of the Kutch earthquake of 26th January, 2001 on Gujarat Electricity Board Substations, in *Proceedings of the Workshop on Recent Earthquakes of Chamoli and Bhuj*, May 24–26, 2001, Indian Society of Earthquake Technology, Roorkee, p 443–449.

Prakash, V., V. H. Joshi, H. R. Wason and A. Sinvhal, 2001c, Damage observed in temples, prayer halls and community centers due to the Kutch Earthquake of 26th January, 2001, in *Proceedings of Workshop on Recent Earthquakes of Chamoli and Bhuj*, May 24–26, 2001, Indian Society of Earthquake Technology, Roorkee, p 407–421.

Prakash, V., A. Sinvhal and P. R. Bose, 2001d, Effects of the Kutch earthquake of 26th January, 2001 on oil Industry, in *Proceedings of the Workshop on Recent Earthquakes of Chamoli and Bhuj*, May 24–26, 2001, Indian Society of Earthquake Technology, p 159–165, Roorkee.

Prakash, V. and A. Sinvhal, 2002, Seismic vulnerability of Taj Mahal, in *Proceedings of Strategy and Methodology for Conservation of Heritage Buildings*, April 4–5, 2002, Central Building Research Institute, Roorkee, CD and Abstract Volume, p 30.

Prakash, V., A. Sinvhal, P. R. Bose and A. Bose, 2004a, Effects of the Kutch earthquake on lifeline structures, in *Proceedings of International Conference on Seismic Hazards*, October 2001, IMD and DST, New Delhi, p 15–23.

Prakash, V., A. Sinvhal and P. R. Bose, 2004b, Oil industry and the Kutch Earthquake of 26th January, 2001, in *Proceedings of International Conference on Seismic Hazards*, Oct. 2001, IMD and DST, New Delhi, p 1–7.

Richter, C. F., 1958, *Elementary Seismology*, W. H. Freeman and Co., San Francisco, 768 p.

Sinvhal, A., P. R. Bose and R. N. Dubey, 1994, Damage report for the Latur Osmanabad earthquake of September 30, 1993, *Bull. Ind. Soc. Earthq. Tech.*, **31**(1), p 15–54.

Sinvhal, A. and P. R. Bose, 1996, Seismic performance of rural hospital At Killari, in *Proceedings of the VIIth All India Meeting of Women in Science (IWSA)—Role of Women in Science Society Interaction*, Roorkee, India, p 89–94.

Sinvhal, A., P. R. Bose, A. Bose and V. Prakash, 2001a, Damage observed to hospitals and medical facilities due to the Kutch earthquake of 26th January, 2001, in *Proceedings of the Workshop on Recent Earthquakes of Chamoli and Bhuj*, May 24–26, 2001, Indian Society of Earthquake Technology, Roorkee, 551 p.

Sinvhal, A., P. R. Bose, A. Bose and V. Prakash, 2001b, Damage observed to Surajbari Bridge due to the Kutch earthquake of 26th January 2001, in *Proceedings of the Workshop on Recent Earthquakes of Chamoli and Bhuj*, May 24–26, 2001, Indian Society of Earthquake Technology, Roorkee, p 423–431.

Sinvhal, A., P. R. Bose, A. Bose and V. Prakash, 2001c, Damage to railway bridge at Dholawa and railway station at Bhachau in Kutch earthquake of 26th January, 2001, in *Proceedings of Workshop on Recent Earthquakes of Chamoli and Bhuj*, May 24–26, 2001, Indian Society of Earthquake Technology, Roorkee, p 399–406.

Sinvhal, A., P. R. Bose, A. Bose and V. Prakash, 2001d, Damage observed to educational buildings due to the Kutch earthquake of 26th January 2001, in *Proceedings of Workshop on Recent Earthquakes of Chamoli and Bhuj*, May 24–26, 2001, Indian Society of Earthquake Technology, Roorkee.

Sinvhal, A., V. Prakash, H. Sinvhal and V. N. Singh, 2001e, Earthquake scenario and tourism in Uttaranchal, in *Proceedings of All India Seminar on Infrastructure Development in Uttaranchal (INDU)—Problems and Prospects*, Section VI, October 11–13, 2001, Institution of Engineers, Roorkee, India, p 25–40.

Sinvhal, A. and V. Prakash, 2002, The Kutch earthquake of January 2001 and heritage buildings and monuments, in *Proceedings of the Conference on Strategy and Methodology for Conservation of Heritage Buildings* (Abstract Volume), Central Building Research Institute, Roorkee, April 4–5, 2002.

Sinvhal, A., V. Prakash, H. Sinvhal and V. N. Singh, 2003, Impact of earthquakes on tourism in Uttaranchal, in *Souvenir, IGC 2003*, Geotechnical Engineering for Infrastructure Development, Roorkee, p 32–59.

Sinvhal, A., A. D. Pandey and S. M. Pore, 2006, Engineering aspects of the Kashmir earthquake of 8th October 2005 and the need for a blue print for the Himalayas, p 64–69, in *Proceedings of Seminar on Seismic Protection of Structures*, Chandigarh, Military Engineering Service, January 17, 2006, 103 p.

14 CHAPTER

Recording and Interpretation

INTRODUCTION

Even though earthquake shaking is of transient nature, it can be recorded on instruments and studied in considerable detail. Recorded earthquake data yield useful information not only about earthquake parameters such as location of the earthquake, i.e., its epicenter, depth of focus, time of origin, and magnitude of the event, but also about characteristics of materials through which seismic waves travel, such as velocity model and thickness of subsurface strata, attenuation of seismic waves, fault plane solutions, location of active lineaments, interior of the earth, etc. This chapter deals with some aspects of recording and interpretation of recorded data.

THE RECORDING INSTRUMENT

A seismometer senses the passage of seismic waves. Its basic assembly consists of a frame, a heavy mass, and a spring. The frame is rigid and in welded contact with the ground. The heavy mass is a magnet and is suspended from the frame through a spring. An electric coil, which is concentric to the magnet, is suspended from the same frame. When seismic waves arrive at such an assembly, the frame moves in accordance with passage of seismic waves and inertia of the heavy mass tries to resist this motion. A voltage develops in the coil because of its relative motion with respect to the magnet, and this is proportional to the ground movement. This is the principle of the seismometer.

Ground motion produced by an earthquake is usually very small. It is sensed by the seismometer, amplified, often more than a million times, and then written on a convenient device. The seismometer and the recorder together comprise a seismograph. The record written by a seismograph is the seismogram. A seismogram displays complex oscillations, which are a composite result of source characteristics of the earthquake, transmission path of the traveling seismic wave (through different strata and after reflection, refraction, dispersion, attenuation, etc.), and characteristics of the recording instrument.

Earthquakes can be distant or near, and they can have large or small magnitudes. So the amplitude of ground motion has a very large range, which has to be sensed and recorded. Amplitude of vibration is usually very large for large earthquakes near the epicenter. This requires instruments with special characteristics. On the other hand, small earthquakes, such as micro earthquakes, which have a magnitude equal to or less than 3.0, have smaller amplitudes even close to the epicenter. To meet all these varying needs, different kinds of sensing and recording instruments are currently in existence. When natural period of the instrument is very large compared to predominant period of ground motion, then the instrument records ground displacement or strain, and when natural period of the instrument is comparable to predominant period of ground motion, then the instrument records ground velocity. A third kind of instrument has a very short natural period compared to predominant period of ground motion, and this instrument records ground acceleration. To completely define the earthquake ground motion recording of three mutually perpendicular components is required; therefore, instruments are installed to record the vertical, and two horizontal components of ground motion.

Most earthquake instruments are designed to record continuously, 24 hours a day, because an earthquake can occur at any time. Recording can be either analogue or digital, or both. Analogue form of recording can either be on smoked paper, photographic paper, or on chart paper. Modern recording is now in digital form, on magnetic tape or on compact disk. Ground vibrations picked by seismometers at various sites can now be written very far from where these were sensed. Seismic signals can be conveniently transmitted to long distance via a cable, radio, or a satellite link and can be conveniently collected and recorded at a central station. This multichannel and multiplexed seismic data can also be interpreted online.

An array of several instruments is installed to monitor earthquakes. These may be distributed within a small geographical region (approximately 100 × 100 km or less), whence they are called local arrays. Results from these arrays are of special interest as they give a quick idea about current seismicity of the region. For this reason several micro earthquake arrays are deployed around sites of technoeconomic importance, where large civil structures are located, such as dams and nuclear power plants. Data collected by a micro earthquake network installed around the Tehri Dam site was subjected to several forms of conventional interpretation and, in addition, was also used to identify seismic micro zones, as given in Chapter 8. Recording instruments may be distributed in a limited area, or may be globally distributed, depending on the objective of recording.

Despite their many uses, deployment of seismometers and recording instruments in the field are fraught with many logistic and maintenance problems. Rugged and inhospitable terrain and extreme weather conditions are typical plate margin conditions around the Indian plate. Rain, snow, storm, and lightening disturb power lines. Power failure leads to loss of data for that duration. If located on a hilltop, the seismic signal is amplified due to topographical effects and becomes noisy due to wind conditions. Noise is added by grazing of animals and by cultural activity in the vicinity of the seismometer. To minimize noise effects, seismometers are located in remote environments, or are buried below the surface, and ideally are kept on deep rock exposures. All this puts additional constraints on seismic performance of the instrument, and subsequently on recording and interpretation of data.

When an earthquake ruptures subsurface rocks, body waves, i.e., primary and secondary waves, originate from the focus at the same time. The velocity with which these waves travel is different, hence they arrive at a point, e.g., at a seismometer, at different times. Since secondary waves are slower than primary waves, therefore the former arrive after the latter and this difference in travel times, particularly at large epicentral distances increases. This difference in travel time is of tremendous importance in determining epicenter of an earthquake. Some important results, which are obtained from seismograms immediately after an earthquake, are the parameters of an earthquake such as where and when the earthquake occurred and how big it was. These are usually given in terms of latitude and longitude of the epicenter, depth of focus, origin time, and magnitude. Their method of determination is given very briefly in this chapter. For magnitude see Chapter 7.

Epicentral distance is the distance between an epicenter and a recording station or a point of observation. For short distances, it is commonly given in kilometers, but for large distances it is given in terms of the angle subtended between the epicenter and the observation point at the center of the earth. Earthquakes recorded at an epicentral distance of less than 10° (i.e., less than approximately 1000 km) are termed as near events and events recorded at epicentral distances of greater than 10° are called tele-seismic events. Hypocentral distance is the distance between the focus and the point of observation.

Focus of an earthquake is the region inside the earth where an earthquake originates. It is used as a synonym for hypocenter or an earthquake source. The term focus has the same connotation as in optics as it is the center of disturbance and represents the position of initial rupture of rocks. The earthquake focus may be expressed in different ways depending on which aspect of the earthquake phenomenon is of most concern. In the simplest case, it is considered as a point source. In a more elaborate context, like in the case of a fault or a rupture, a line source or a plane surface may be considered. Yet again, to represent finiteness a source region may be considered. In that case, the focus can be considered as a volume of irregular size and shape. It is expressed in terms of latitude, longitude, and depth.

Epicenter is the point on the surface of the earth vertically above the focus, as shown in Figure 14.1. This roughly gives the location of an earthquake. It is given in terms of latitude and longitude of the epicenter. Since the focus is usually a volume of irregular size and shape, its projection on the surface may be a region of irregular shape also. Damaging effects of an earthquake are usually most severe at the epicenter.

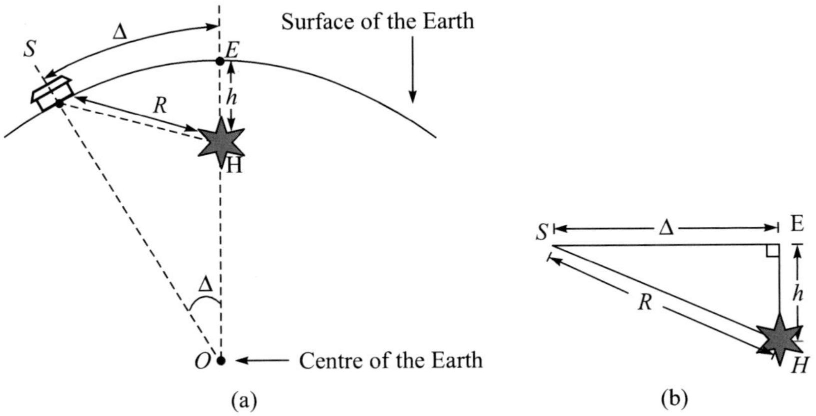

Fig. 14.1 (a) Concept of some commonly used terms are shown in this figure. Star and H denote focus of the earthquake, E is epicenter, *h* is focal depth, Δ is epicentral distance, S is point of observation, and R is hypocentral distance. (b) Simple geometry between focus H, epicenter E, and the observation point S is shown by the right angle triangle SEH.

Depth of focus, or focal depth, denotes vertical distance between focus and epicenter. Most earthquakes occur within a depth range of 0–70 km and these are called shallow focus earthquakes. Those originating between depths of 70 and 300 km are called intermediate focus, and those that occur between 300 and 700 km are termed as deep focus earthquakes. Earthquakes do not originate beyond a depth of 700 km, as at this depth pressure of overlying rocks does not permit rocks to break and release energy, which is a prerequisite for producing earthquakes.

Origin time is the instant at which the earthquake event (apart from foreshocks) starts at the focus. Origin times are usually given in terms of year, month, day, hour, minutes, and seconds. The last three are given in the form 08:46:39.3 or 08 h 46 min 39.3 s, which is equivalent to 08 h 46 min and 39.3 s.

This may be specified in local time like the Indian Standard Time (IST) or in Universal Coordinated time (UCT), or in Greenwich Mean Time (GMT), or in any other appropriate time frame. The Uttarkashi earthquake of October 20, 1991, originated at 02 h 53 min 16.4 s IST. On the Universal Coordinated Time this becomes 21 h 23 min 14.3 s on October 19, 1991, as IST = UCT + 5½ h.

DETERMINATION OF EPICENTER

If seismograms are available from at least three recording stations, A, B, and C, for the same earthquake, then the epicenter of this earthquake can be determined. Primary and secondary waves arrive at station A at times T_p and T_s, respectively. Arrival time of P- and S-waves is marked on the seismogram. This is laid parallel to the time axis and adjusted on the distance axis until the marks line up with the time distance curves, as shown in Figure 14.2. Epicentral distance, ΔA, is noted from the distance (x) axis. Thus, the difference in time of arrival of these two waves, i.e., (T_s – T_p), helps in determining the epicentral distance ΔA for station A.

Epicentral distance is determined for seismograms obtained from stations B and C, as ΔB and ΔC, in a similar way. The position of the three stations A, B, and C, is marked on a map and from each station a circle is drawn with a radius corresponding to its epicentral distance, ΔA, ΔB, and ΔC. This is shown in Figure 14.3. The point of intersection of these circles is the epicenter. This exercise is repeated for as many stations for which seismograms are available for the same earthquake. In practice, the number of stations for which

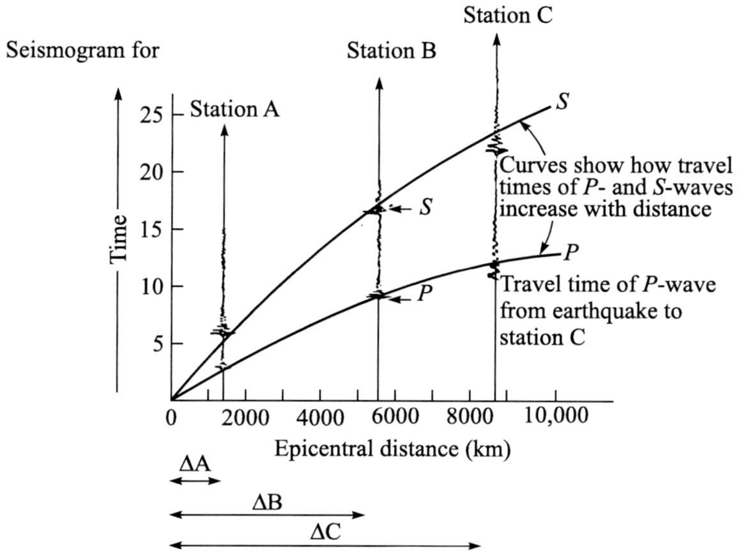

Fig. 14.2 Method of computing epicentral distance from seismogram recorded at three stations, A, B, and C.

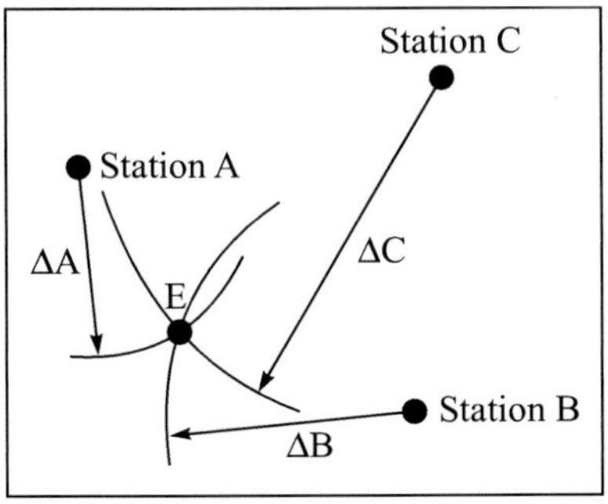

Fig. 14.3 Determination of epicenter, E, from data given in Figure 14.2.

observations are available is usually larger than three, and this increases the reliability of results. For a large epicentral distance, i.e., for a tele-seismic event, many different phases of *P*- and *S*-waves, some of which are shown in Chapter 3, arrive at different times. These are identified with the help of standard travel timetables and curves given by Jeffreys and Bullen, (1940, 1958). Velocity of media in which seismic waves travel is either known or assumed.

DETERMINATION OF DEPTH OF FOCUS

Determination of depth of focus is a problem, which is more difficult than determining epicenter, mainly because of the uncertainties involved while introducing a third dimension in the vertical direction. Focal depth of shallow focus earthquakes can be determined by the method given here. To simplify this complex problem, the Gutenberg-Hodgson's method (Macelwane and Sohon, 1932, 1936) makes a few assumptions to allow for a convenient representation of ray paths. The surface of the earth and a shallow discontinuity are horizontal and parallel planes. A constant velocity is considered for each of these two layers so that a straight-line ray path can be considered. The effect of the weathered layer on travel times is ignored.

As depicted in Figure 14.4, F is the focus, E is the epicenter of the earthquake, h is depth of focus, z is thickness of top layer, and V_0 and V_1 are

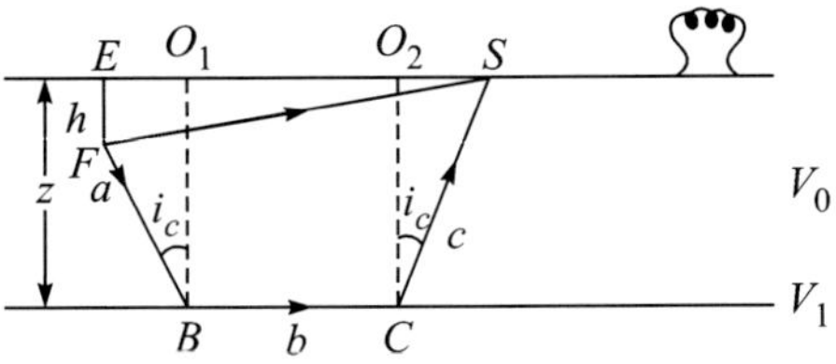

Fig. 14.4 Schematic representation of an earthquake that has its focus in the upper layer, and direct and refracted ray paths originating from the earthquake.

velocities in top and lower layers, respectively. S is the location of the seismometer station. Epicentral distance, ES, is denoted by Δ. The *P*-wave, which originates from the focus F, reaches the recording station, S, by a straight-line path (FS) with a travel time T_1.

Now, $$T_1 = \frac{\text{FS}}{V_0} = \sqrt{\frac{h^2 + \Delta^2}{V_0^2}}.$$

Another ray, which also starts from the focus, travels downward and is refracted by the discontinuity twice to reach the surface and is picked up at S. This refracted ray path is given by FBCS, or by $(a + b + c)$. Paths 'a' and 'c' are traversed with velocity V_0, while b is traversed with velocity V_1. Let travel time of path FBCS be denoted by T_2.

Then, $$T_2 = \frac{a}{V_0} + \frac{b}{V_1} + \frac{c}{V_0}.$$

According to Snell's law $(\sin i/\sin r) = V_0/V_1$. For critical reflection, $r = 90°$, whence $\sin r = 1$, therefore, $\sin i_C = V_0/V_1$. Also, $\cos^2 i_C + \sin^2 i_C = 1$. Converting trigonometric terms to velocities,

$$\sin i_C = \frac{V_0}{V_1}, \cos i_C = \sqrt{1 - \left(\frac{V_0}{V_1}\right)^2}$$

Now, $$a = \frac{z-h}{\cos i} = \frac{z-h}{\sqrt{1-\sin^2 i}} = \frac{z-h}{\sqrt{1-\left(\frac{V_0}{V_1}\right)^2}} = \frac{(z-h)V_1}{\sqrt{V_1^2 - V_0^2}}$$

And $$c = \frac{z}{\cos i} = \frac{z}{\sqrt{1-\left(\frac{V_0}{V_1}\right)^2}} = \frac{zV_1}{\sqrt{V_1^2 - V_0^2}}$$

And $$b = \Delta - \text{EO}_1 - \text{O}_2\text{S} = \Delta - a \sin i - c \sin i = \Delta - \sin i(a + c)$$

$$= \Delta - \frac{V_0}{V_1} = \left\{\frac{z-h}{\cos i} + \frac{z}{\cos i}\right\} = \Delta - \frac{V_0}{V_1} \cdot \frac{(2z-h)}{\cos i}$$

$$= \Delta - (a + c) \sin i = \Delta - \frac{V_0}{V_1} \frac{(2z-h)}{\sqrt{1-\left(\frac{V_0}{V_1}\right)^2}}$$

$$= \Delta - (2z - h) \frac{V_0}{\sqrt{V_1^2 - V_0^2}}$$

Substituting for a, b, and c in the expression for T_2

$$T_2 = \frac{(z-h)V_1}{V_0\sqrt{V_1^2 - V_o^2}} + V_0\frac{\{\Delta - (2z-h)\}}{V_1\sqrt{V_1^2 - V_0^2}} + \frac{zV_1}{V_0\sqrt{V_1^2 - V_0^2}}$$

$$= \frac{(2z-h)}{\sqrt{V_1^2 - V_0^2}}\left\{\frac{V_1}{V_0} - \frac{V_0}{V_1}\right\}$$

$$T_2 = \frac{\Delta}{V_1} + \frac{1}{\sqrt{1-\left(\frac{V_0}{V_1}\right)^2}}\left[\frac{2z-h}{V_0} - \frac{V_0(2z-h)}{V_1}\right] = \frac{\Delta}{V_1} + (2z-h)\sqrt{\frac{1}{V_0^2} - \frac{1}{V_1^2}}.$$

Therefore, $T_1 - T_2 = \sqrt{\frac{h^2 + \Delta^2}{V_0^2}} - \frac{\Delta}{V_1} - (2z-h)\sqrt{\frac{1}{V_0^2} - \frac{1}{V_1^2}}.$

Epicentral distance, Δ, is calculated as given earlier in this chapter and velocity V_1 of the lower layer is assumed. The expression for $(T_1 - T_2)$ contains three unknowns, h, z, and V_0. If seismograms are available from at least three stations, depth of focus, h, can be determined. If seismograms are available from several stations for the same earthquake, then the reliability of the determined depth of focus increases.

Since velocity of the media through which seismic waves travel is assumed, velocity model for the region depends on the interpreter's judgment of variations within the earth. However, when seismometers are near the epicenter, effects of transmission paths are minimized, as shown in Figure 14.5, and results are more reliable. Not only this, initial conditions at the source are more complicated than assumed, e.g., dislocation of rocks at the source is not instantaneous but is spread out in time and space. Also, heterogeneity exists within the earth and imperfections exist in elasticity. With all these variations in the earth and assumptions made in various formulations to justify these, slightly different source parameters emerge for the same earthquake, as shown in Tables 14.1 and 14.2. A seismogram for the Latur earthquake is shown in Figure 14.6.

Parameters of the Kashmir earthquake of 2005 and for the Kutch Earthquake of 2001 are given in Tables 14.1 and 14.2, respectively, as given by

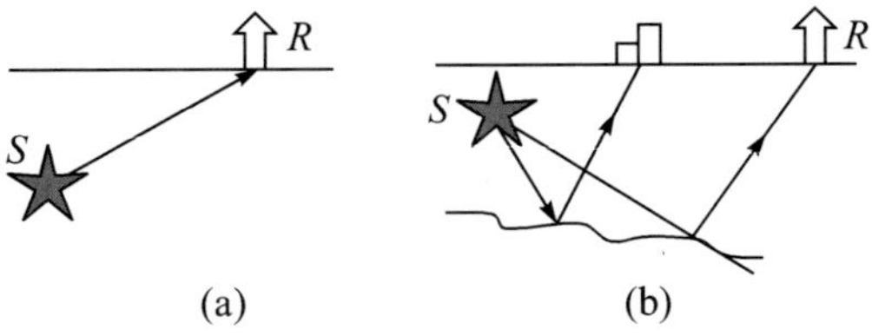

Fig. 14.5 Source and receiver geometry, earthquake source is shown by star and seismometer by R. The transmission path is shown for a (a) direct ray from source to receiver, and (b) rays reflected from a boundary.

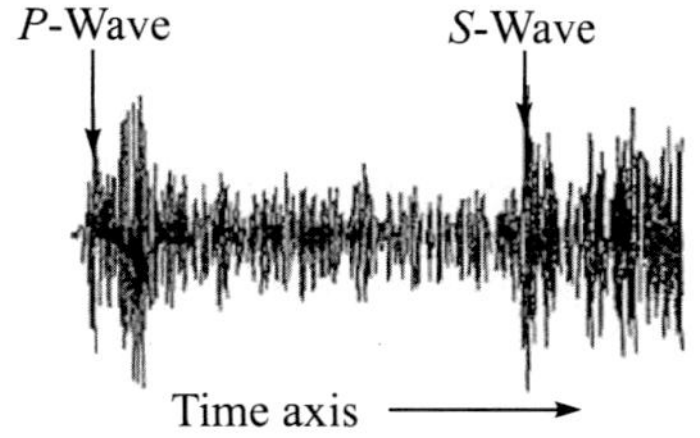

Fig. 14.6 First 3 min of the seismogram of the Latur earthquake. *P* and *S* denote the arrival of *P* (longitudinal) and *S* (shear) seismic waves. Arrival time of *P*-wave is 03 h 58 min 37.43 s, arrival time of *S*-wave is 04 h 00 min 50.75 s, origin time is 03 h 55 min 45.26 s (IST). The time difference of 2 min and 13.32 s between the arrival of the two types of seismic waves shows that the epicenter of this earthquake was about 1308 km away from the recording station. Analysis of the record reveals that the focus of the earthquake was in the upper crust, 6–8 km below the ground surface.

Table 14.1 Parameters for the Kashmir earthquake of Octobers 2005, as given by different agencies.

Agency	*Epicenter*		*Origin time UTC*	*Magnitude*	*Depth of focus*
	Latitude (North)	*Longitude (East)*			
IMD	34.6	73.0	03h 50 min 35.8 s	7.4	33
GSI	33.586	73.474	03 h 50 min 56.3 s	M_b 7.3 M_s 6.8	—
USGS	34.402	73.560	03 h 50 min 38 s 08 h 50 min 38 s LT	7.6	10
NEIC	34.432	73.537	03 h 50 min 38.63 s	M_w 7.3	20
Harvard	34.37	73.47	03 h 50 min 52.2 s	M_w 7.6	12

IMD—India Meteorological Department; GSI—Geological Survey of India; USGS—United States Geological Survey; NEIC—National Earthquake Informatics Centre; UTC—Universal Time Coordinated; LT—Local time in Pakistan.

Table 14.2 Parameters for the Kutch earthquake of January 26, 2001, as reported by different agencies. IMD subsequently revised the epicenter to 23.4°N 70.28°E (Bandhadi village).

Agency	*Latitude (North)*	*Longitude (East)*	*Origin time, IST (GMT + 5h 30min)*	*Magnitude*	*Depth of focus (km)*
IMD	23.6	69.8	08:46:39.3	M_L 6.9	15 k
USGS	23.36	70.34	08:46:41	M 7.7 M_w 7.6	22 k
NEIC	23.419	70.232	08:46:40.5	M_b 6.9 M_w 7.5	16
GSI	23.31	70.41	08:46:47.16	M_s 7.6	—

IMD—India Meteorology Department; USGS—United States Geological Survey; NEIC—National

different agencies. At times, there is some variation in determination of these parameters by various agencies due to (a) data set used, (b) method and software used, and (c) errors in reading time markings on seismograms.

DETERMINATION OF DEPTH OF BEDROCK

When an earthquake occurs, rocks are fractured and move relative to one another on opposite sides of the fracture. Such a rupture generates seismic waves that travel away from the fracture surface. These waves are picked up at various stations by seismometers. These data are used to determine source parameters, i.e., information about epicenter, origin time, depth of focus, magnitude, etc. after making many assumptions. Moreover, epicentral distance, hypocentral distance, and energy released at the focus are large.

In a related context, it is also known that seismic response of the built environment improves and is desirable when it is founded on bedrock, compared to the case when it is founded on soft sediments. Effects of this have been discussed in Chapters 3, 11, and 12. Foundations of structures must preferably rest on firm bedrock, especially for large and important civil structures such as dams, nuclear power plants, bridges, and industrial, coastal, and marine structures, and multistory buildings. Therefore, depth of bedrock should be determined for important structures. This can be estimated from seismic waves that are generated from simulated earthquakes, with a formulation that is similar to and simpler than that for determining depth of focus of an earthquake.

An artificial source of energy is used to generate seismic waves. This energy can be provided by several means, for example, by a hammer striking a steel plate on the ground, a hand-operated 'tamper', by a weight dropped on the ground, or by a small explosion. The source of energy is very small, simple, controllable, and movable, can be spread in a definite pattern, and above all its location (akin to epicenter and depth of focus of an earthquake) and origin time are known precisely. This reduces a lot of the complexity and assumptions involved in understanding source parameters of earthquakes and their seismic signals.

Instruments required for sensing and recording earthquakes and for simulated earthquakes are also similar, but differ only in detail. An artificial earthquake generates seismic waves that have higher frequency content, and therefore, high-frequency seismometers are required to detect the resulting ground motion. The number of recording channels is also increased, recording is at a higher speed, and the seismic signal is sampled at smaller intervals. On the other hand, for earthquakes frequency and attenuation of seismic waves are lower, fewer recording channels are needed, which work at lower recording speed, and have a larger sampling interval.

A continuous coverage of the subsurface is required to ascertain the continuity of bedrock and its depth. This requires installation of several seismometers in a planned pattern, the simplest of which is a straight line. Often several channels are used for recording, where each channel corresponds to a seismometer. From knowledge of travel times to the various seismometers and the velocity of waves, it is generally possible to reconstruct paths of direct, reflected, and refracted seismic waves and determine the depth of bedrock. Distance between source of energy and seismometers is relatively small (when compared to the case of epicentral distance in earthquakes) and is known accurately. Time taken by seismic waves to travel this distance, after reflection and refraction from the bedrock, is measurable. Depth of penetration is also small and can be determined more accurately than for an earthquake because of multiple coverage of the subsurface.

In the simplest case, bedrock is assumed to be horizontal, homogeneous, continuous, and a half space. It is buried below a single layer, which is also homogeneous and continuous. The two layers are separated by a horizontal boundary between them. Velocity of P-waves in the upper and lower layers is V_0 and V_1, respectively, with the constraint that V_1 is greater than V_0. A high-velocity contrast between bedrock and the overlying medium is desirable.

To determine depth of bedrock, seismic energy is generated on the surface or very close to the surface by artificial means. This seismic energy travels away from the source, A, as shown in Figure 14.7. D is the seismometer that picks the seismic waves. The seismogram is interpreted in terms of seismic wave paths, which fall into three main categories, the direct wave, and the waves that are refracted and reflected from the bedrock. Principles relevant to seismic waves are discussed in Chapter 3. The direct wave, AD, travels on the surface with velocity V_0. The refracted wave path is traced by the geometry shown by ABCD. It starts from the source A, travels downward and is incident on the boundary with bedrock at critical angle of reflection. A wave that is incident at critical angle gives total internal reflection or refraction. Beyond the critical angle of reflection, i_C, the waves refract and travel in the lower layer. At some point, it may again be critically reflected and

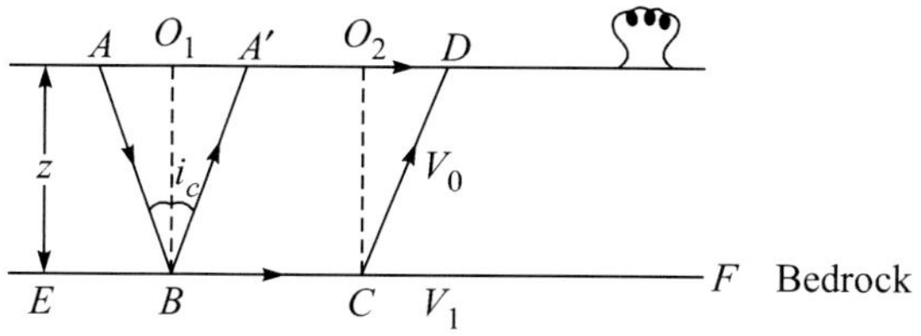

Fig. 14.7 Different wave paths are shown for horizontal bedrock buried beneath another layer. A is the source and D is the receiver. AD is direct wave; ABA′ is reflected wave, ABCD is wave refracted from bedrock, and *z* is the depth of bedrock.

emerge at the surface. The principal portion of the path ABCD is along bedrock and hence is approximately horizontal. On the other hand, the reflected wave initially travels downward and is then reflected back to the surface, the overall path being essentially vertical.

Since an almost continuous coverage of the subsurface boundary is required, an array of seismometers is installed in a pattern, the simplest of which is a straight line along AD. These seismometers detect the motion of the ground created by the seismic source. At D, the seismometer picks the seismic waves due to the direct wave AD and also due to the refracted wave ABCD. Time taken by seismic waves to travel from the source to the seismometer is recorded. Travel time depends on several factors such as distance between A and D, velocity of the two media V_0 and V_1, and depth of the bedrock. From the travel times of direct, refracted, and reflected waves, it is generally possible to estimate velocity of seismic waves in the two media and then depth of bedrock is known, on which large and important civil structures can be founded.

Time–Distance Relation

Let distance between source and seismometer, i.e., AD, along the surface of the earth be denoted by x. Let travel time of this direct ray be T_1.

Then, $$T_1 = x/V_0 \tag{14.1}$$

Let the refracted ray, which follows path ABCD, arrive at D at time T_2. Because BC is parallel to surface of the earth and is horizontal, therefore segment AB will be equal to segment CD. Paths AB and CD are traversed with velocity V_0. The segment BC, which is the refracted segment, travels with velocity V_1. Therefore,

$$T_2 = \frac{AB}{V_0} + \frac{BC}{V_1} + \frac{CD}{V_0} = \frac{2AB}{V_0} + \frac{BC}{V_1}.$$

Perpendiculars drawn from B and from C to the surface AD correspond to depth of bedrock, z, i.e., $BO_1 = CO_2 = z$. Using trigonometric relations, we get

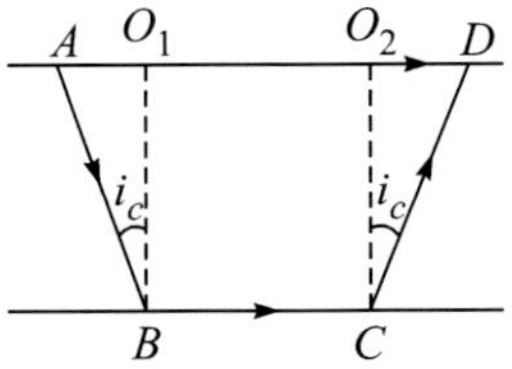

Fig. 14.8 ABCD is the refracted path.

$$\frac{BO_1}{AB} = \cos i_C; \quad \text{so,} \quad AB = \frac{z}{\cos i_C}.$$

Also, $$\begin{aligned} BC &= O_1O_2 = AD - AO_1 - O_2D \\ &= x - O_1B \tan i_C - O_2C \tan i_C \\ &= x - 2z \tan i_C \end{aligned}$$

Therefore,

$$T_2 = \frac{2AB}{V_0} + \frac{BC}{V_1}$$

$$T_2 = \frac{2z}{V_0 \cos i_C} + \frac{x - 2z \tan i_C}{V_1}$$

$$= \frac{x}{V_1} + 2z\left(\frac{1}{V_0 \cos i_C} - \frac{\tan i_C}{V_1}\right)$$

According to Snell's law ($\sin i/\sin r$) = V_0/V_1. For critical reflection $r = 90°$, and $\sin r = 1$, therefore, $\sin i_C = V_0/V_1$. See Figure 14.8. Also, $\cos^2 i_C + \sin^2 i_C = 1$. Converting trigonometric terms to velocities,

$$\sin i_C = \frac{V_0}{V_1}$$

$$\cos i_C = \sqrt{1 - \sin^2 i_C} = \sqrt{1 - \left(\frac{V_0}{V_1}\right)^2} = \sqrt{\frac{V_1^2 - V_0^2}{V_1^2}}$$

and $$\tan i_C = \frac{\sin i_C}{\cos i_C} = \frac{V_0}{\sqrt{(V_1^2 - V_0^2)}}$$

Therefore, $$T_2 = \frac{x}{V_1} + 2z\left(\frac{V_1}{V_0\sqrt{(V_1^2 - V_0^2)}} - \frac{V_0}{V_1\sqrt{(V_1^2 - V_0^2)}}\right)$$

$$= \frac{x}{V_1} + \frac{2z}{\sqrt{V_1^2 - V_0^2}}\left(\frac{V_1}{V_0} - \frac{V_0}{V_1}\right)$$

$$= \frac{x}{V_1} + \frac{2z}{V_0 V_1}\left(\frac{V_1^2 - V_0^2}{\sqrt{V_1^2 - V_0^2}}\right) = \frac{x}{V_1} + \frac{2z}{V_0 V_1}\sqrt{V_1^2 - V_0^2} \qquad (14.2)$$

A graph is plotted between distance x and arrival time T, as shown in Figure 14.9. This is the time–distance curve, and is referred to as the T–x curve. Depth of bedrock can be determined from this curve. The curve consists of two segments AP and PQ. AP corresponds to the direct wave and PQ corresponds to the refracted wave ABCD. The distance at which both the direct and the refracted ray arrive at the same time is called the cross-over distance, x_C. At this distance $T_1 = T_2$. After this distance, the first arrivals on a seismogram are refracted waves.

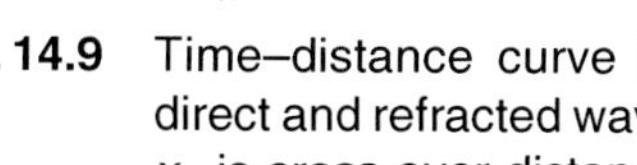

Fig. 14.9 Time–distance curve for direct and refracted wave. x_C is cross over distance.

In Equations (14.1) and (14.2), substituting x_C for x yields

$$\frac{x_C}{V_0} = \frac{x_C}{V_1} + \frac{2z\sqrt{V_1^2 - V_0^2}}{V_0 V_1}$$

$$x_C\left(\frac{1}{V_0} - \frac{1}{V_1}\right) = \frac{2z\sqrt{V_1^2 - V_0^2}}{V_0 V_1}$$

or,

$$x_C\left(\frac{V_1 - V_0}{V_0 V_1}\right) = \frac{2z\sqrt{V_1^2 - V_0^2}}{V_0 V_1}$$

$$z = \frac{x_C}{2}\frac{V_1 - V_0}{\sqrt{V_1^2 - V_0^2}}$$

or,

$$z = \frac{x_C}{2}\sqrt{\frac{V_1 - V_0}{V_1 + V_0}} \tag{14.3}$$

The equation for the first segment AP, a straight line, is $T = (x/V_0)$, with slope $1/V_0$. This yields the velocity in the top layer, i.e., V_0. Likewise, the second segment, PQ, also a straight line with slope $(1/V_1)$ gives velocity in bedrock, i.e., V_1. The cross-over distance, x_C, is determined from the graph from the point of intersection of the two slopes. Thus, the velocity of the two layers, V_0 and V_1, is determined from the graph. Therefore, z, i.e., depth of bedrock, can be estimated from Equation (14.3).

The ordinate of the point where PQ produced backward meets the time axis is the intercept time, T_i. For this point, the abscissa is zero, i.e., $x = 0$ and this indicates a vertical reflection from the bedrock. Therefore, $T_2 = T_i$ and Equation (14.2) becomes

$$T_i = \frac{2z}{V_0 V_1}(\sqrt{V_1^2 - V_0^2})$$

Therefore, depth of bedrock can be determined from the intercept time also, as given below.

$$z = \frac{T_i}{2}\frac{V_0 V_1}{\sqrt{V_1^2 - V_0^2}}$$

It is hence possible to determine depth of bedrock by this method. The time-distance graph and the corresponding subsurface geometry interpreted from this are shown in Figure 14.10.

The advantage of this seismic refraction method, which is based on refraction of seismic waves, is that depth of bedrock can be inferred from surface investigations, without actually digging or boring a well, with a small source of energy and simple instrumentation, and that too quite rapidly, with high accuracy and resolution. This method is also valuable for reconnaissance

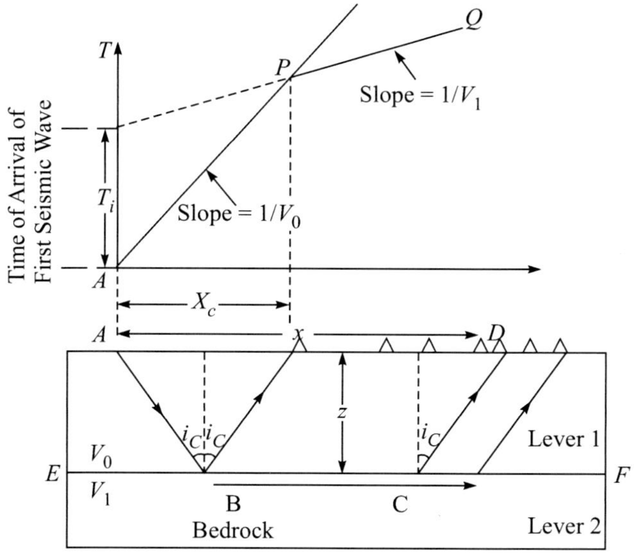

Fig. 14.10 Ray paths for direct, reflected, and refracted wave and corresponding travel time curve for a two-layer horizontal case. A = Shot point, i.e., source of energy; Δ, D = seismometers; x = distance between shot point and seismometer for direct ray; EF is boundary between upper layer and bedrock; BC is refracted path in bedrock; i_C = critical angle of refraction; T_2 = time taken to travel AD by refracted path ABCD; T_1 = time taken to travel AD by direct ray; T_i = intercept time; X_C = cross-over distance; V_0 = velocity in upper layer; V_1 = Velocity in lower layer; and z = depth of bedrock.

survey in areas where information about the subsurface strata is almost nonexistent. Horizontal and vertical extent of sedimentary basins and stratification of subsurface rocks is also revealed by reflection and refraction of seismic waves. When the objective is to find out more about depth, then reflection of seismic waves proves to be more useful, and the method is then known as seismic reflection method.

Depth of bedrock below the Karchham dam, across river Satluj in Kinnaur District of Himachal Pradesh, was investigated by the seismic refraction method given above (Sinvhal et al., 1995). The hammer and steel plate assembly, dropping of 35 kg of weight, and also explosives provided the source of energy. The turbulent flow of the river created noise in the seismic signal. Velocity in the top layer and basement was estimated as 430 m/s and 2200–2500 m/s, respectively. Because of this large velocity contrast, it was possible to record prominent reflections and refractions from the basement. The top layer was composed of unconsolidated river fill material, while the basement was composed of partially weathered, fractured, and fissured quartz, biotite, and gneiss. The basement rock was not horizontal but followed the profile of the river valley.

The seismic method outlined for determination of bedrock, with modifications, is of paramount importance in the oil industry. Almost all major oil companies rely on this method for finding oil and in selecting sites for exploratory oil wells. Despite the indirectness of the method, most seismic work results in mapping of geological structure rather than finding hydrocarbons directly, the likelihood of a successful venture is improved more than enough to pay for the seismic work. Likewise, engineering surveys, mapping of water resources, and other studies requiring accurate knowledge of subsurface structure derive valuable information from data collected by the seismic method.

Artificial earthquakes are increasingly used to determine the earth's structure down to and below the Mohorovicic discontinuity. Since depth of penetration is large, it is known as deep seismic sounding (DSS). Depth of mantle in the Pamir and Hindu Kush region in the western syntaxis was determined by this method. Seismic waves from nuclear explosions have been similarly used to study the interior of the earth. In addition, subsurface structure of regions where earthquakes are rare can be determined.

CONCLUSION

Instrumentally recorded seismic waves can be studied in considerable detail. Seismograms yield useful information about earthquake parameters such as location of the earthquake, i.e., its epicenter, depth of focus, time of origin, and magnitude of the event. Simple methods of their determination are given in this chapter. In addition, the usefulness of artificial earthquakes in determining depth of bedrock are also included. It is hoped that one day soon recorded data together with enhanced interpretation techniques may lead to earthquake prediction, and finally to mitigation of earthquake disasters. The next chapter deals with long- and short-term aspects that can be adopted in making a safer built environment.

REFERENCES

Jeffreys, H. and K. E. Bullen, 1940, 1958, *Seismological Tables*, British Association, Gray - Milne trust, 50 p.

Macelwane, J. B. and F. W. Sohon, 1932, 1936, *Theoretical Seismology*, Parts I and II, Wiley, New York.

Sinvhal, H., V.N. Singh, A. K. Jain, S. Singh and A. Sinvhal, 1995, Geophysical Investigations at Karchham dam Site, Department of Earth Sciences, University of Roorkee, Roorkee, Project and Report , 88 p (Unpublished).

15 CHAPTER

What Can Be Done

INTRODUCTION

In the aftermath of a tragic earthquake once rescue and relief operations are over, one question that is always asked is 'can earthquakes be predicted?' Perhaps if enough warning existed, most lives could have been saved. Several attempts have been made at prediction but the luxury of global success is limited. To predict earthquakes, one has to first understand their cause and mechanism, which itself is a complex problem, and then look for phenomena that could help in predicting them.

EARTHQUAKE PREDICTION

Earthquake prediction involves estimation of earthquake parameters, i.e., the place where an earthquake will occur, the time frame within which it will occur, and the magnitude range of the expected event. This information should be accompanied by a statement of odds that an earthquake of the predicted kind would occur by chance alone and without reference to any special evidence.

Earthquake prediction involves the precise measurement of variation in several physical parameters within seismically active areas. Five promising parameters are velocity of P-wave, uplift and tilting of ground, emission of radon gas from wells, electrical resistivity, and seismicity of the region. Precursory changes in velocity of P-waves are of particular interest as properties of rocks change before an earthquake, and lead to change in velocity of seismic waves. The second parameter that can be used in prediction is precursory change in ground level, such as ground tilt. The third parameter is release of radon, an inert gas, into the atmosphere along active fault zones, particularly from deep wells. The fourth parameter is electrical conductivity of rocks as electrical resistance of water-saturated rock changes drastically just before rocks fracture, as happens before an earthquake occurs. Variation in seismicity is the fifth parameter. A marked change in distribution of earthquakes in time and space is observed, usually an increase of small earthquakes.

Variations in these parameters take place in five stages, and are manifest in the strained rocks before, during, and just after a large earthquake. The first stage is a slow buildup of elastic strain due to the underlying tectonic forces. During this period, all seismic parameters have their normal values. In stage II, microcracks develop in rocks in fault zones, and volume of rock increases, or it dilates. As cracks open, velocity of P-waves through the dilatant volume decreases, the ground surface rises, radon gas escapes, electrical resistivity increases, and there may be a change in the incidence of micro-earthquake activity in the vicinity. In stage III, water diffuses from surrounding rocks into microcracks, leading to unstable conditions. As water fills the cracks, velocity of P-waves begins to increase again, uplift of ground ceases, emission of radon from fresh cracks tapers off, and electrical resistivity decreases. Stage IV is the onset of the earthquake. This is immediately followed by Stage V, during which numerous aftershocks occur in the area. The precursory period leading to Stage IV depends on the volume of rock involved in ultimate fault rupture of the main event. Rough estimates indicate that precursory events may continue for several months for an earthquake of magnitude 6, and for a much longer time, from 1 to 3 years for an earthquake of magnitude 8.

Other precursory phenomena include movements in the crust, detection of strain in the crust by geodetic surveys, occurrence of foreshocks, variation in ratio of compressional to shear wave velocity, change in water level, and identification of gaps in regular occurrence of earthquakes in both time and space. Although long-term changes in these parameters have been observed instrumentally, yet the number of measurements is limited and results have thus far conflicted. In some, unusual behavior of these parameters has been indicated before a local earthquake. In other regions, nothing significant was seen before an event, or variations occurred that were not associated with earthquakes (Press, 1975; Rikitake, 1976; Srivastava, 1983, 1992). The first correct prediction was for a small earthquake of magnitude between 2.5 to 3.0 in the state of New York, in August 1973 (Aggarwal et. al., 1973). It was based on an increase in travel time of P-waves.

The next significant earthquake predicted correctly was in China, on February 4, 1975. It was for a large earthquake of magnitude 7.3. This earthquake is known as the Haicheng earthquake and also as the Liaoning earthquake. Prediction was based on several factors. In late 1973 and 1974, variations in several physical parameters were reported. The ground surface rose at 20 times its normal rate near a fault; an increase in elevation of about

2.5 mm was observed in 9 months. Unusual fluctuations in the earth's magnetic field were reported, as were changes in elevation of shoreline on a nearby peninsula. Changes in ground water level were widespread. Tilt meters showed that direction of tilting changed in some places but not in others. Throughout the region, people recounted incidents of peculiar animal behavior. By the end of June 1974, these were considered symptomatic of a local earthquake of moderate size within the next 2 years. In early February of the following year, many small earthquakes were instrumentally recorded nearby. It was this increase in background seismicity that led to prediction. On February 4, 1975, it was sufficiently evident that a strong earthquake would probably occur within the next 24 hours. Thousands of people in the city of Haicheng and nearby towns and villages were officially urged to remain outdoors even though it was severe winter. Then, at 7.36 pm the predicted earthquake (magnitude 7.3) shook the Haicheng region (Bolt, 2004). Within the meizoseismal area, more than 90% houses collapsed; factories and machinery, dams, bridges, and irrigation works were damaged. Without this prediction, a large number of the 3,000,000 people in the densely populated province of Liaoning would have been killed inside collapsed buildings. The exact number of dead was not known, but may have reached a few hundred.

Several later predictions turned out to be false alarms, like a prediction in August 1976 in Kwangtung province near Kwangchow (Canton) in China. During the earthquake alert, many people slept outdoors in tents for nearly 2 months. No earthquake occurred. The most publicized lack of forewarning was the tragic earthquake of July 27, 1976, which almost razed Tang Shan, an industrial city of one million people in China. Unofficial reports estimated a death toll of about 650,000 in the meizoseismal area, and an additional 780,000 persons injured.

Historical world seismicity patterns make it possible to predict the probable place at which a damaging earthquake can be expected to occur. Pattern recognition techniques and probabilistic and deterministic methods are also being applied to the catalogue of known earthquakes for predicting earthquakes. Even in China, where between 500 and 1000 destructive earthquakes have occurred within the past 2700 years, statistical studies have not revealed any noticeable periodicities between great earthquakes, but indicate that long periods of quiescence can elapse between them. However, this record does not enable one to forecast the precise time of occurrence.

Lack of success in earthquake prediction in the United States of America, Union of Soviet Socialist Republic, Japan, and China, where numerous sophisticated instruments were operational, shows that the possibility of predicting time and place of an earthquake has limitations, and more investigations and research may give better results in future. However, after the damage caused by recent earthquakes, like at Uttarkashi, Latur, Kutch,

Sumatra, and Kashmir, at Los Angeles and San Francisco in U.S.A., and at Kobe in Japan, and many others, it became obvious that goals of earthquake prediction have yet to be achieved in practice. Since the luxury of correct earthquake prediction is still remote, it is imperative that those who are caught in the strong shaking can perhaps know some vital short-term safety measures that can be taken when caught in strong earthquake shaking for personal safety. The list given here is indicative of such possible measures.

WHAT TO DO WHEN CAUGHT IN AN EARTHQUAKE

There are certain actions, if taken during an earthquake, can be of immense benefit if one is caught in strong earthquake shaking. An open space away from the built environment is usually the safest place during an earthquake. If you are there stay there till the strong shaking stops.

If you are indoors when you feel the strong ground-shaking, try to leave the building as quickly as possible. If you cannot exit the building quickly enough, look for protection within the building. Take shelter under a strong and stable piece of furniture as soon as possible. This could be a study table, a bench, or a bed. Brace yourself against it, or lie down beside it on the floor if you cannot go underneath it, and stay there till the shaking stops. Do not worry about being embarrassed if you hide under a desk if at school or in a meeting. Others will follow your example soon. Those who wait to see whether or not such an action is necessary are the ones who are most likely to be hurt by falling debris. If you are in bed, roll out and lie next to or beneath it. Some of these measures are illustrated in 'Earthquake Problem Dos and Don'ts for Protection', authored by Chandra, Sinvhal, and Gupta, in 1994, in English and in Hindi.

Alternatively, stand under an interior doorframe, or in a corner of the room, or near a column to seek protection from falling wall and roof. This also offers a possible escape route later. Stay away from balconies, parapets, railings, or projections, as a sudden jolt, caused by an aftershock, can throw you off balance. Tall, heavy, and movable pieces of furniture like refrigerators, bookshelves, wardrobes, and machinery can topple over, move or slide against the floor and may cause injury. Toppled appliances such as a fridge, geyser, heater, gas stoves, etc. are liable to become a fire hazard or cause electrocution.

Face away from windows and mirrors so that breaking glass and splinter do not hurt you. Cover your head and face with a piece of cloth to protect yourself from breathing the thick dust that is thrown up if the building is damaged. As you go outdoors, put your arms over your head to protect yourself from falling debris and to reduce disorientation produced by seismic waves. Do not stay to collect your belongings or valuables. Most earthquakes will last for less than a minute or so. Your life is more precious than any of your belongings.

Strong ground shaking lasts for a very short time, and may be of the order of about 4–8 sec for an earthquake of magnitude 5.0, to about 43–86 sec for an earthquake of magnitude 8.5 (Kramer, 2004). It may extend for a very large event, such as the Sumatra earthquake of December 26, 2004, in which the strong shaking lasted for almost 3 min.

While cooking, or near a fire or a flame, extinguish these immediately, if you can, and move to a safer place. Shut off gas valves if there is any chance of a gas leak. Detect gas by smell, never by lighting matches or lighters. Do not operate electrical switches and appliances as these can create sparks that can ignite any leaking gas. Do not touch downed power lines, electrical wiring, or objects touched by these.

In tall buildings, upper floors shake more than lower floors because of swaying. Do not jump from upper floors in panic. Do not use the lift as it could be jammed or may not be working due to power failure. Use the staircase, if it is not already jammed with people or worse, is broken, as happened in many tall buildings in Ahmedabad, as shown in Figure 12.5. Wait till the shaking is over and then leave calmly. If in a crowded building, such as in a shopping complex or a cinema hall, evaluate exit and emergency routes carefully before you enter. Plan in advance how to leave the building in case of an emergency. Do not join a stampede; instead seek safety by ducking in between seats or in a corner.

When moving on the road, or if inside a moving vehicle, such as in a car, slow down to keep control. Move to the edge of the road, away from possible hazards such as tall buildings, high boundary walls, balconies, chimneys, overhead water tanks, slopes, power lines, electric poles, hoardings, bridges, fly overs, fallen debris, or any other structure that can be injurious. Avoid narrow streets that may get clogged with rubble falling from both sides. If possible, do not stop on or below a bridge or a flyover as these sway and are liable to be damaged during an earthquake. Stay inside the car until shaking is over. If you have to continue the journey after the shaking stops, be cautious and look out for damaged, weakened, or collapsed bridges, flyovers, and fissured roads. In hilly regions, also look out for landslides and rock falls.

WHAT TO DO AND NOT TO DO ONCE YOU ARE SURE THAT THE EARTHQUAKE IS OVER

Every one who lives in an earthquake-prone area should think deliberately and frequently about what to do before being confronted by the next earthquake disaster. Since one never knows when an earthquake may occur, it is better to be prepared in advance against such a calamity. Being prepared and knowing what to do can save lives and reduce injuries. Those who wish to have greater security should resort to these and other simple actions as soon as possible.

Panic is an additional hazard during and after an earthquake. By remaining calm you can take immediate and sensible actions to protect yourself and can thus increase your chances of being safe. Furthermore, other people near you may benefit from your calm attitude and follow your example. Behave responsibly and help and reassure young children and others who may suffer psychological trauma from the earthquake. Avoid upsetting other people by shouting or running around.

Check yourself and those around you for injuries. Assist and provide first aid if necessary. If at home during the earthquake, assist your family and neighbors in coping with the disaster. When you have done what you can, consider how you can help others. If you are at work when the earthquake originates, assist in every way you can then make your way home.

The first priority after an earthquake is to rescue people trapped and hurt within the debris of collapsed houses, schools, offices, and the built environment. Respond to rescue missions from neighbors, police, fire fighting, and civil defense organizations. Cover the injured with blankets to keep them warm in winter. Administer emergency first aid if necessary. Seek medical help for those who need it. Able-bodied persons in a community should organize themselves to look after the needs of the stricken community, as per need and ability.

Use great caution when entering or moving about in a damaged building as these can collapse without any warning. Inspect chimney, parapet, and balcony carefully for any damage. Protect your head by wearing a helmet, a turban, or at least a towel. Wear sturdy slippers or shoes when moving around a damaged area as these offer protection from sharp debris and broken glass. Expect aftershocks to follow the main event.

In case of power failure, first use food from the refrigerator that will spoil, then turn to other foods. Do not eat or drink anything from open containers, especially near shattered glass. Use the telephone sparingly at least for a few days after the earthquake, if the telephone lines are still functional. Keep it free for high-priority use such as to call for help, rescue, emergency, fire, medical services, etc. When the emergency is clearly over, inform relatives and friends about your safety.

Use water sparingly. It may be needed for fire fighting and for other emergency purposes. Shut off water mains if water pipelines are broken. Do not flush toilets until sewer lines are checked. In due course of time, report utility damage to the concerned authorities and follow their instructions. After this, damage can be assessed and remedial measures begun. Clean up and warn others of any spilled materials that are dangerous, such as chemicals, kerosene oil, diesel, petrol, and medicines.

Do not go sightseeing nor occupy streets unnecessarily in damaged areas unless your help is needed. Keep roads free for rescue and relief operations. Curfew is sometimes imposed after an earthquake to keep away the unscrupulous and looters. Do not go near beaches and other large water bodies. Tsunamis and seiches could visit even long after an earthquake. Do not believe or spread rumors, astrological predictions, and prophecies as these only add to the prevailing confusion.

Ensure that relief materials, such as food, water, and blankets are distributed equally and in abundance to everyone after an earthquake, and are bereft of any political, ethnic, religious, or any other bias. This will prevent the unscrupulous from snatching, hoarding, and black-marketing relief material. In addition, relief material will trickle down to the really needy and weak, who cannot reach distribution centers such as the old, the infirm, very young, and nursing mothers. It will also assuage the feelings of those who lost their dear ones and promote loyalty and goodwill toward the government and community. Schools can be used as temporary shelters. Disease and epidemics spread rapidly in temporary shelters due to inadequate sanitation and this must be prevented.

Someone who is awake at the time of the earthquake may be able to follow some of the dos and don'ts given here. However, this is not always possible. The Kutch earthquake of January 26, 2001, the Sumatra earthquake of December 26, 2004, and the Kashmir earthquake of October 8, 2005 originated in broad daylight, when most people were awake, and had a chance to scurry to safety, yet these three earthquakes together killed almost four lakh people. The dos and don'ts given here become even more redundant when an earthquake occurs at night when most people are sleeping. Moreover, when the electric supply fails and complete darkness engulfs the stricken, it only adds to the prevailing confusion. Then even an awake person may not be able to do much in the short span of approximately half a minute or so that the strong shaking lasts. The true extent of the calamity may become evident only after daybreak. This scenario was witnessed in the 6.4 magnitude Latur earthquake of 1993 in which more than 10,000 died and in the 6.9 magnitude Uttarkashi earthquake of 1991. No rules can make us completely safe from the fury of earthquakes; some rules will apply only in certain situations and must be altered, abandoned, or endured under other circumstances. However, some preparation can be made to meet the next earthquake.

HOW TO PREPARE FOR THE NEXT EARTHQUAKE

All passages, doorways, and exits should be useable and uncluttered at all times. During an emergency, these should be available for escape, for evacuation of the trapped and the injured, and for entry and exit of rescue workers.

Make an "emergency kit." It should have all the supplies that may be needed for a day or two after an earthquake. As two liters of water per person per day is adequate for drinking purposes, store accordingly for at least a day for your family. Include nonperishable food in the kit. Include a strong torch with spare batteries, candles, matches, blankets, and first aid supplies. If anyone in the family is on regular medication, have an extra supply in the emergency kit. This kit must be stored in a convenient and accessible place and everyone should know its storage place so that it can be carried away while fleeing the house.

Responsible members in the family must know the location and operation of main electric fuse box, and gas and water shut off valves of their home. These can be switched off after the first seismic vibrations. Keep a wrench of the proper size near the gas shut off valve. Overhead electric fixtures such as fans, bulbs, tube-light, etc. should be properly anchored in to the wall or ceiling. Use flexible gas and hot water connections wherever possible. Secure, fasten, bolt, or strap to wall or floor heavy appliances that use gas or electric power.

Keep beds away from large glass windows. Make sure that there are no heavy objects hanging above your bed or places where you spend a large amount of time as these can swing, hit a wall or a window, come off their hooks and fall during an earthquake, and can be injurious. These can be picture frames, mirrors, hanging plants, and light fixtures. Keep a torch beside your bed. Keep a battery-operated radio or a transistor set at a place where you spend a great amount of time. It is usually, the best way to get information and instructions after an earthquake emergency.

Fasten to walls any bookcases, or other heavy pieces of furniture that might topple and cause injury. Almirah usage should be so planned that large and heavy objects are kept in lower shelves, so that the almirah itself does not topple over due to strong shaking. Keep these latched. Shaken objects may fall outward and upon you when opened. Avoid stacking heavy unsupported objects on high open shelves or other high projections inside rooms. In case you must, make sure that the stack is stable and will not topple easily when shaken by an earthquake, as frequently happens with bags and suitcases, stored in the overhead storage space in a moving bus.

Chemicals, such as flammable liquids, kerosene oil, petrol, diesel, and pesticides, insecticides, and poisons should be stored in a secure place where they will not fall and break open. Flowerpots kept on edges of balconies and ledges are likely to fall during strong shaking and could be injurious. Move these to a safer place.

Discuss, plan, and develop evacuation procedures with members of your family, neighbours, coworkers, and community. If family members are usually at different places, decide in advance how the family will establish contact with each other. Identify a common person, away from your own locality, who can be contacted by telephone or cell phone (provided these work in the postearthquake scenario). Select a common meeting place locally that is outdoors, away from tall buildings, walls, and power lines, where the family will reunite after the earthquake. If family members know about bank account number, insurance policies, and business papers, then this will facilitate a smoother restart in the postdisaster scenario. For small children, it may usually be best if they stay at school until they can be collected. Make sure that all members of the household, as well as coworkers and others, know what to do during and after an earthquake.

Identify in advance facilities in your area that could be of help in the postdisaster scenario, such as medical centers, fire fighting stations, police posts, or any organized rescue and relief society. Make a list of important telephone numbers, and include ambulance, hospital, fire, and police services. Also learn first aid procedures. After a major disaster, hospitals may be overcrowded and medical personnel may be occupied with more serious cases.

Farmers should store seeds and grains in a safe place, such as in an open field with proper protection, so that these will be available after an earthquake and famine can be averted the following year. Keep immunization up to date for all family members. If your home is prone to damage have available some plywood and sheets of plastic to cover broken windows and other openings as a postearthquake protection from hostile weather.

Earthquakes damage the built environment and cause death, injury, and displace people. It is prudent to remember that most of the time it is not earthquakes that kill people, but a poorly built environment that does. To safeguard from such calamities, it is of utmost importance that the built environment is made in such a way that it remains safe and does not claim human lives in the event of an earthquake. This is possible only if we resort to making a built environment that can resist an earthquake. This entails several feasible long-term measures.

LONG-TERM MEASURES

Complete protection of all life and all property in all earthquakes is still a distant dream. However, efforts are on to make a built environment in which loss of life and property is minimized, and lifelines and infrastructure continue to function during and after the earthquake. Damage to the built environment at any location depends on several interrelated factors. These are magnitude of the earthquake, frequency of seismic waves amplitude and duration of ground shaking, distance from causative fault, fault pattern in the area, plate environment, epicentral distance, depth of focus, local geology, soil conditions, topography, seismic response of the structure and population density. For a safe built environment two considerations are necessary: choice of a suitable site, and an earthquake-resistant built environment. These are viable long-term solutions for mitigating earthquake disasters.

Site Considerations

Selection of a site where the built environment exists or is proposed is a very important mitigation aspect and alas, is overlooked in most cases. Damage to the built environment at any site depends on several factors, seismicity is one of them. Appropriate earthquake-resistant measures are required in regions of high seismicity. Meizoseismal areas of the great earthquakes, as discussed in Chapter 6, are the quintessential candidate areas for the origin of future earthquakes. Seismically vulnerable areas are identified in the seismic zoning map of India, BIS: 1893–2002. Seismic hazards that can afflict a site should be identified, like faulting, liquefaction, flash floods, landslides, tsunamis, etc.

Large-magnitude earthquakes cause more damage that is spread in a large area compared to small-magnitude earthquakes. This is borne out by the ground damage documented for the great earthquakes compared to other smaller earthquakes in the same area. In most cases, damage is maximum close to the epicenter and decreases away from this. But sometimes a large amount of damage takes place at a large epicentral distance. This depends on several factors; one important reason is the interplay between frequency content of seismic waves and their interaction with soft sediments and long and tall structures, and the other is the tsunami generated by some submarine earthquakes.

Local geology plays a very important part too. If the foundation at a site is on hard and competent rock then the seismic response is more desirable compared to a corresponding site on soft soil, all other parameters being same. Ground shaking is minimum in stable rock, therefore structures founded on such strata are less prone to earthquake damage. On the other hand soft soil, which may be in the form of alluvium, unconsolidated soil, filled ground, or geologically recent sediments, is prone to severe shaking and heavy damage, more so if these are thick and subsurface layers are saturated with water. Such strata absorb a significant amount of seismic energy and amplify seismic waves. Such a situation causes compaction of soft soil, subsidence, slumping, and liquefaction. Therefore, structures founded on this kind of soil are prone to heavy damage.

When different types of soil are in close contact with each other, damage may vary. This difference becomes more prominent when soft soil is in contact with a ridge of hard rock. The latter resists severe shaking and the damage is confined to regions of surrounding alluvium. Such effects became spectacular in the Bihar–Nepal earthquake of 1934 (Auden et al., 1939), and in the Kutch earthquake of 2001 (Sinvhal et al., 2001).

Casualties and injuries due to the primary effect of the earthquake alone, i.e., faulting are rare, but the ground and the built environment located in the fault zone or close to it are susceptible to damage (Figure 15.1(a)). Relative displacement of two sides of a fault involves forces that can be very destructive to the built environment. A fault can give rise to seismically induced ground damage in the form of liquefaction in soft soil, fissures, earthquake fountains, water falls, sand boils, offsets, land slides, and rock falls. If the earthquake has a marine origin and the causative fault has vertical displacement, it can cause a destructive tsunami in coastal areas. The most recent example of this was provided by the Sumatra earthquake of December 26, 2004. Therefore faults are of tremendous importance in the context of earthquake disasters. When the location of important structures and vital installations is under consideration, their proximity to known faults needs to be investigated thoroughly, especially for their potential of getting seismically activated in the near future. It is best to avoid faults altogether, but in practice this is not always possible.

Three mega faults in the Himalayas, which extend from Kashmir in the west to Arunachal Pradesh in the east, show current seismic activity. These are the Main Central Thrust (MCT), the Main Boundary Thrust (MBT), and the Frontal Foothill Thrust (FFT). These are associated with plate margin environments, neotectonics, surface deformation, and a tremendous amount of earthquake-induced damage. Three recent damaging earthquakes originated on these faults: the earthquakes of Uttarkashi in 1991, Chamoli in 1997, and Kashmir in 2005. It is pertinent to be aware that the rivers Indus, Ganga, and Brahmaputra and their many tributaries are tectonically controlled by these faults in their upper reaches and are tapped for their hydroelectric potential in the Himalayas. Design earthquake parameters and site investigations are carried out for all these large projects (Sinvhal and Prakash, 2004).

Since damage potential of faults is of such tremendous importance, their seismic response can be better understood if they are theoretically and computationally modeled. In the simplest case, a fault can be modeled as a plain rectangular surface, with a finite length, downward extension, dip, and strike. During an earthquake, rupture propagates along the fault plane and its response is studied at different locations. An example of this is given in Figure 4.5 for the Uttarkashi earthquake of 1991.

Topography of the site plays an important role in the seismic response of structures. In rugged mountainous and hilly terrains, landslides damage or bury houses (Figure 15.1(b)), obstruct roads and rivers, and disrupt transmission and distribution of electricity. The Kashmir earthquake of 2005 provides ample examples of this. On the other hand, the built environment located in coastal regions with a flat topography is prone to the disastrous effect of tsunamis

(a) (b) (c)

Fig. 15.1 Different kinds of damage caused by an earthquake: (a) Faulting at Moti Undo, near Mandvi due to Kutch earthquake of 2001, (b) Landslide at Sarai Bandi in Uri due to Kashmir Earthquake of October 8, 2006, (c) Partially submerged house in slush produced by transgression of sea at Car Nicobar after the tsunami generated by the Sumatra earthquake of December 26, 2004. (See color figure also.)

(Figure 15.1c). Widespread damage observed in the Sumatra earthquake of 2004 was mainly because of this factor, and is discussed in Chapter 10.

An Earthquake-resistant Built Environment

Experimental and computational setups to simulate earthquake forces are too expensive for validating building designs and, even in the best cases, cannot replicate all actual field situations. Therefore the most enduring lessons have to be learnt from the seismic response of ground and the built environment in the largest natural laboratory, i.e., the earth, from damage observations in all earthquakes, whether inter- or intra-plate. Complete protection of all life and the entire built environment in all earthquakes is still a distant dream. However, efforts are on to have a built environment in which loss of life is minimized, and lifelines and infrastructure continue to function during and after an earthquake disaster. Construction activities in seismically prone and hazardous areas that are vulnerable to different damaging effects of earthquakes are best avoided. Most of the time such situations are unavoidable; in that case appropriate strengthening measures are required. Structures should be preferably made on firm ground. For construction in soft soil, the ground should be strengthened, and the foundations should be sufficiently deep, wide, and strong. Subsequently, application of appropriate interventions regarding earthquake-resistant design of structures goes a long way in saving human lives.

Ideally, the built environment should be able to withstand strong ground shaking caused by seismic waves. Designing such a built environment is a challenging task. In seismically prone regions, small earthquakes may occur frequently near a structure during its lifetime; moderate earthquakes may occur once or twice; and a major earthquake may have a low probability of occurrence in its lifetime. In the first case, a well-designed ordinary building is expected to remain functional, together with all its nonstructural components. In the second case, the building is still expected to remain functional, but may suffer nonstructural damage, without claiming human lives. In the third case, which is the most severe of all the three cases, it is expected that the building may deform beyond economic repair and may have to be demolished later. To circumvent such a severe case a building can be made earthquake-resistant, if enough ductility is built into it, all elements of the building are securely tied together, and adequate bracing is provided against earthquake forces.

Building a seismically resistant house will keep one's family and possessions safe in an earthquake. It costs almost the same amount to build an earthquake-resistant house as it costs to build a non-earthquake-resistant house. However, for large important structures like dams, powerhouses, bridges, etc., earthquake-resistant design can be incorporated at an increased expense of the order of 5–10% of the project cost. Existing buildings should be strengthened and made earthquake-resistant. Old, weak, and unsafe buildings and structures should be removed, replaced, or strengthened.

As the earthquake force is proportional to the amount of ground shaking and to the mass of the building, therefore light, flexible, and strong materials are preferable in earthquake-prone regions of the world. This makes bamboo and timber ideal construction material. Small houses made of such materials just slide about in strong ground shaking without causing serious injury to their inhabitants, and even total collapse is not fatal. Consequently, indigenous architecture using locally available material with these desirable qualities developed in several vulnerable areas. Their commendable seismic response was amply demonstrated in several recent earthquakes, as shown in Figure 15.2. The seismic performance of traditional rural construction such as bhoongas, circular huts in Banni depression of Kutch, timber-framed

(a) (b) (c) (d)

Fig. 15.2 Diverse construction practices show a desirable seismic response: (a) A hut with a thatch roof supported on a timber frame survived the Latur earthquake of September 30, 1993 at Killari, while the stone masonry house next to it collapsed, (b) The timber frame of a triple-storey house with walls made of random rubble stone masonry (RRSM) at Kamalkote in Uri survived while the walls showed partial collapse due to the Kashmir earthquake of 2005, (c) A timber jetty continues to perform its function even after the all-pervasive devastation by the tsunami, after the Sumatra earthquake of 2004 in South Andaman Island, (d) Note the marked contrast in seismic response of an engineered and a nonengineered construction and desirability of engineering solutions. A stone masonry house with an RCC roof collapsed while an elevated municipal water tank in the background is intact.

construction in Kashmir and Latur, and Nicobarese huts provided very good examples. A traditional Nicobarese hut is supported on long timber stilts and located on high ground in the interior of the island. This makes choice of site, building material, and design suitable for the region. In strong ground shaking produced by the disastrous Sumatra earthquake of 2004, these huts showed no visible signs of structural stress, and life in these huts continued as usual after the earthquake and the tsunami. These desirable considerations were sometimes abandoned when no damaging earthquakes occurred in a period of rapid change and development. Unfamiliar and vulnerable designs, with no consideration of site and other factors, were unfortunately adopted in many regions, which claimed a heavy death toll later.

CONCLUSION

Since disasters caused by several earthquakes are known and well-documented, lessons learnt from these should be propagated and popularized. There is an urgent need for education and awareness on understanding earthquake disasters in the entire population (Bose et al., 1992, 2002; Jain and Sinvhal, 2002; Sinvhal et. al., 1992, 1993, 1994, 1996, 1997, 2006). In the long term, this will have a tremendous advantage in mitigating earthquake disasters. It will lead to a voluntary compliance and implementation of earthquake codes, which in turn will reduce casualties in future earthquakes. Tragic consequences of flouting these, with recent examples, should also be widely publicized. In the long term, this knowledge gained will have a tremendous advantage in mitigating earthquake disasters.

Earthquake codes and guidelines should be widely disseminated and be easily available to all. In addition, if these codes become comprehensive, detailed, are written in a reader friendly format, and include commentaries explaining the background, they stand a better chance of being understood and accepted by users. In view of the disaster caused by the tsunami, the same applies to environmental and coastal codes and regulations also.

A destructive earthquake retards the planned development and economy of the affected community by decades. Because of the urgency created by an earthquake disaster, scarce resources allocated to health, education, and other sectors are diverted for the emergency that develops after an earthquake, for rescue, relief, shelter, recovery, rehabilitation, and rebuilding purposes. Therefore, if earthquake-related knowledge and resource base of the people is strengthened then it will help them to make informed decisions, and the planned growth of a community and country can continue as envisaged. Knowledge, money, material, time, and labor will then be optimally utilized for mitigation of earthquake disasters, and human life will be safer in future earthquakes.

REFERENCES

Aggarwal, Y. P., L. R. Sykes, J. Armbruster and M. L. Sbar, 1973, Premonitory changes in seismic velocities and prediction of earthquakes, *Nature*, **241**, p 101–104.

Auden, J. B., J. A. Dunn, A. M. N. Ghosh, D. N. Wadia and S. C. Roy, 1939, The Bihar–Nepal Earthquake of 1934, *Memoirs of GSI*, Volume 73, 391 p.

BIS: 1893–2002, *Indian Standard Criteria for Earthquake Resistant Design of Structures, Part I: General Provisions and Buildings* (Fifth Revision), Bureau of Indian Standards, New Delhi, 40 p.

Bolt, B. A., 2004, *Earthquakes* (Fifth Edition), W. H. Freeman and Company, New York, 378 p.

Bose, P. R., A. Bose and A. Sinvhal, 1992, Manual of lecture notes for Short term Q.I.P Course on Earthquake Engineering for Architects and Planners, University of Roorkee, Roorkee, India, 350 p.

Bose, P. R., A. Verma, A. Bose, A. Sinha and A. Sinvhal, 2002, Needed modifications in civil engineering curriculum for earthquake disaster mitigation, in *Proceedings of the 12th Symposium on Earthquake Engineering*, Roorkee, India, p 686–692.

Chandra, B., A. Sinvhal and I. Gupta, 1994 a, Earthquake Problem Do's and Don'ts for Protection, Department of Earthquake Engineering, University of Roorkee, Roorkee, Sponsored by Rajiv Gandhi Foundation, New Delhi, India, 24 p.

Chandra, B., A. Sinvhal and I. Gupta, 1994 b, "Bhukamp Samasya Kya Karen Kya na Karen". Department of Earthquake Engineering, University of Roorkee, Roorkee, India, Sponsored by Rajiv Gandhi Foundation, New Delhi, India, 24 p.

Jain, R. K. and A. Sinvhal, 2002, Manual of Lecture Notes for Short term NPEEE Course on Earthquake Resistant Design for Built Environment, IITR, Roorkee, India, 119 p.

Kramer, S. L., 2004, *Geotechnical Earthquake Engineering*, Pearson Education, 653 p.

Press, F., 1975, *Earthquake Prediction*, Scientific American, 232 p.

Rikitake, T., 1976, *Earthquake Prediction*, Elsevier Scientific Publishing Company, Amsterdam.

Sinvhal, A. and L. S. Srivastava, 1992, Manual of lecture notes for Short term Continuing Education Course on Earthquake Hazards Evaluation, Design of Structures and Foundations, University of Roorkee, Roorkee, India, 282 p.

Sinvhal, A., P. R. Bose and B. V. K Lavania, 1993, Manual of lecture notes for Short term IGS Course on Earthquake Engineering for Geotechnical Engineers, University of Roorkee, Roorkee, India, 273 p.

Sinvhal, A., V. Prakash and M. K. Gupta, 1997, Manual of Lecture Notes for Short term QIP Course on Understanding Earthquake Hazards, University of Roorkee, Roorkee, India, 361 p.

Sinvhal, A., Bose, P. R., V. Prakash, A. Bose, A. K. Saraf and H. Sinvhal, 2001, Damage, seismo-tectonics and isoseismals for the Kutch earthquake of 26th January, 2001, in *Proceedings of Workshop on Recent*

Earthquakes of Chamoli and Bhuj, May 24–26, 2001, Indian Society of Earthquake Technology, Roorkee, p 61–70.

Sinvhal, A. and V. Prakash, 2004, Seismotectonics and design earthquake parameters for the Brahmaputra basin, in *Brahmaputra Basin Water Resources*, Eds. V. P. Singh, N. Sharma and C. S. P. Ojha, p 578–610, Kluwer Academic Publishers, The Netherlands..

Sinvhal, H., 1994, Manual of Lecture Notes for Short term Continuing Education Course on Managing Earthquake Disaster for Working Professionals, University of Roorkee, Roorkee, India, 147 p.

Sinvhal, H. and A. Sinvhal, 1996, Earthquake disaster mitigation in Himachal Pradesh; issues and scenarios, in *Proceedings of Interaction meet on Natural Hazards in H.P. and their Mitigation, Simla* (Extended Abstract), p 25–26.

Sinvhal, H. and A. Sinvhal, 2006, Understanding earthquakes, in *Proceedings of Seminar on Seismic Protection of Structures*, Chandigarh, Military Engineering Service, January 17, 2006, p 1–12.

Srivastava, H. N., 1983, *Earthquake Forecasting and Mitigation*, National Book Trust, New Delhi, 344 p.

Srivastava, H. N., 1992, *Earthquake Prediction Studies in Himalayas—Critical Evaluation: Himalayan Seismicity*, Geological Society of India, Volume 23.

Appendix I

MAGNITUDE ENERGY RELATION

Consider a point source that radiates seismic waves uniformly in all directions, as shown in Figure AI.1. Consider a wave that reaches the epicenter. At the epicenter, ground displacement x and velocity v at any time t are given by

$$x = a_0 \cos (2\pi t/T_0) \tag{AI.1}$$

$$v = dx/dt = -(2\pi a_0/T_0) \sin (2\pi/T_0) \tag{AI.2}$$

where a_0 is amplitude of wave at free surface and T_0 is period of wave. If e is density of kinetic energy of ground motion per unit volume, ρ is density of medium, v is instantaneous particle velocity, and m is mass, then kinetic energy of ground motion per unit volume is $e = ½ (mv^2)$, where v is instantaneous

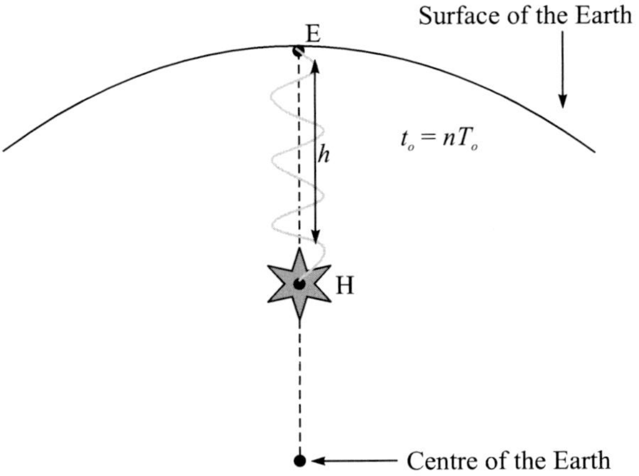

Fig. AI.1 Energy released in an earthquake. Schematic diagram of part of a wave train originating from a point source approaching a station at the epicenter. $t_0 = nT_0$, part of a wave train, source = point, h = focal depth, H = focus.

particle velocity. We know that mass = density × volume. If we consider a unit volume then mass = density. Therefore, $e = ½\,(\rho v^2)$ then kinetic energy of wave due to period T_0 is given by

$$e = (\rho/2T_0) \int_0^{T_0} v^2 dt \qquad \text{(AI.3)}$$

Substituting equation AI.2 in AI.3, implies

$$e = (\rho/2T_0)\,(2\pi\, a_0/T_0)^2 \int_0^{T_0} (\sin^2 2\pi t/T_0)\, dt$$

$$= (\rho/2T_0)\,(2\pi\, a_0/T_0)^2 \int_0^{T_0} [1 - \cos(4\pi t/T_0)]/2\, dt$$

$$= (\rho/2T_0)\,(2\pi\, a_0/T_0)^2\, \frac{1}{2} \left[t - \frac{\sin \dfrac{4\pi t}{T_0}}{\dfrac{4\pi}{T_0}} \right]_0^{T_0}$$

$$= (\rho/4)\,(2\pi\, a_0/T_0)^2$$

If t_o is duration of the wave train, n is number of wave periods in it ($t_0 = nT_0$), and c is velocity of propagation of wave within the medium, then energy flow per unit area at the station is ($ct_0\, e$), ignoring surface reflections. Energy flow per unit area (energy/area) is given by $ct_0(\rho/4)\,(2xa_0/T_0)^2$. If h is depth of focus of the earthquake, and $4\pi h^2$ is a spherical surface of radius h, then kinetic energy emanating from the source through a spherical surface of radius h is given by E_k = (area) × (velocity of propagation) × (wave train duration with period n) × (density of kinetic energy per unit volume), i.e.,

$$E_k = (4\pi h^2)\,(ct_0 e)$$
$$= 4\pi^3 h^2\, ct_0\, \rho (a_0/T_0)^2 \qquad \text{(AI.4)}$$

To calculate total seismic energy, E, released at the focus, following aspects are also taken into account. Since mean potential energy is equal to mean kinetic energy, therefore total energy E is a sum of potential energy and kinetic energy; $E = 2E_2$, in other words kinetic energy is doubled, therefore (2 × kinetic energy) = $2E_k$, = $8\pi^3 h^2 cT_0\, \rho(a_0/T_0)^2$. Also because of the free surface at the epicenter the amplitude doubles, i.e., $a_0 = 2a$, where a_0 is amplitude recorded at epicenter, i.e., at free surface, and a is amplitude at surface, i.e., $a = a_0/2$, therefore, $E_s = 8\pi^3 h^2 ct_0\, \rho(a_0/T_0)^2 = 2\pi^3 h^2 ct_0{}^2\, \rho(a_0/T_0)^2$.

Further, this calculation deals with waves of maximum energy, which at short distances are S-waves. The energy of P-waves is assumed to be half that of S-waves, and must be added (Gutenberg and Richter, 1956a). Therefore,

$$E_p = (2\pi^3 h^2 ct_0\, \rho(a_0/T_0)^2)/2 \qquad \text{(AI.5)}$$
$$E_p = \pi^3 h^2 ct_0\, \rho(a_0/T_0)^2 \qquad \text{(AI.6)}$$

So, total seismic energy is $E = E_s + E_p$. (AI.7)

$$E = 2\pi^3 h^2 c t_0 \rho (a_0/T_0)^2 + \pi^3 h^2 c t_0 \rho (a_0/T_0)^2$$

$$E = 3\pi^3 h^2 c t_0 \rho (a_0/T_0)^2 \quad \text{(AI.8)}$$

Therefore, E_k is given by

$$E = 3\pi^3 h^2 \, c t_0 \rho (a_0/T_0)^2 \quad \text{(AI.9)}$$

E is in ergs, h is in cm, t_0 and T_0 are in seconds, a_0 in cm, c is in cm/sec, and ρ is in g/cc. Taking $c = 3.4$ km/sec, $\rho = 2.7$ g/cc, $h = 16$ km as probable focal depth in south California

$$\log E_k = 12.34 + 2 \log (a_0/T_0) + \log t_0. \quad \text{(AI.10)}$$

For southern California, Gutenberg and Richter (1956) gave the following empirical relations.

$$\log t_0 = -1 + 0.4 \log (a_0/T_0) \quad \text{(AI.11)}$$

$$\log (a_0/T_0) = m_b - 2.3. \quad \text{(AI.12)}$$

Therefore,

$$\log E = 5.8 + 2.4\, m_b. \quad \text{(AI.13)}$$

This magnitude–energy relation gives the energy released in an earthquake of magnitude m_b. Substitute ($m_b = 2.5 + 0.63\, M$) in the above equation, by making appropriate substitution, the magnitude–energy relation becomes

$$\boxed{\log E = 11.8 + 1.5\, M} \quad \text{(AI.14)}$$

Thus, magnitude can be related to energy released during an earthquake (Bullen, 1965; Bullen and Bolt, 1985). Limitations of estimating energy by this method are due to several assumptions that are made in the derivation of kinetic energy of waves, e.g., a very simple waveform is considered, azimuthal effect of wave radiation is ignored, attenuation effects during propagation are neglected, wave behavior near the surface is neglected, effect of layering in the earth, which causes refraction, reflection, transmission, and attenuation of seismic waves is ignored. In addition, seismograms do not give precise local movement of the earth; therefore, only approximations can be made. Therefore, the estimate of energy released in an earthquake may have an uncertainty of the order of about 10. Therefore, this particular quantity cannot accurately represent all the energy processes at the seismic source.

REFERENCES

Bullen, K. E., 1965, *An Introduction to the Theory of Seismology* (Third Edition), Cambridge University Press, Cambridge, UK, 381 p.

Bullen, K. E. and B. A. Bolt, 1985, *An Introduction to the Theory of Seismology* (Fourth Edition), Cambridge University Press, Cambridge,

Appendix II

Some commonly used intensity scales are given here.

Table AII.1 The most commonly used form of Rossi-Forel Intensity scale.

Intensity	*Main earthquake effects*	*Implication to buildings*
I	Microseismic shock	Recorded by a single seismograph or by seismographs of the same model, but not by several seismographs of different kinds; the shock felt by an experienced observer.
II	Extremely feeble shock	Recorded by several seismographs of different kinds; felt by a small number of persons at rest.
III	Very feeble shock	Felt by several persons at rest; strong enough for the direction or duration to be appreciable.
IV	Feeble shock	Felt by persons in motion; disturbance of movable objects, doors, windows, cracking of ceilings.
V	Shock of moderate intensity	Felt generally by every one; disturbance of furniture, beds, etc., ringing of some bells.
VI	Fairly strong shock	General awakening of those asleep; general ringing of bells, oscillation of chandeliers, stopping of clocks, visible agitation of trees and shrubs, some startled persons leaving their dwellings.
VII	Strong shock	Overthrow of movable objects; fall of plaster; ringing of church bells, general panic, without damage to buildings.
VIII	Very strong shock	Fall of chimneys, cracks in walls of buildings.
IX	Extremely strong shock	Partial or total destruction of some buildings.
X	Shock of extreme intensity	Great disaster; ruins, disturbance of the strata, fissures in the ground, rock falls from mountains.

Table AII.2 The Oldham intensity scale (Oldham, 1899).

S.No.	*Isoseismal number*	*Description of effects*
1.	First	Includes all places where the destruction of brick and stone buildings was practically universal.
2.	Second	Those places where damage to masonry or brick buildings was universal, often serious, amounting in some cases to destruction.
3.	Third	Those places where the earthquake was violent enough to damage all or nearly all brick buildings.
4.	Fourth	Those places where the earthquake was universally felt, severe enough to disturb furniture and loose objects, but not severe enough to cause damage, except in a few instances, to brick buildings.
5.	Fifth	Those places where the earthquake was smart enough to be generally noticed but not severe enough to cause any damage.
6.	Sixth	All those places where the earthquake was only noticed by a small proportion of people who happened to be sensitive, and being seated or lying down were favorably situated observing it.

Table AII.3 The Mercalli intensity scale (as given in Auden et al., 1939).

I	Instrumental shock, that is, noted by seismic instruments only.
II	Very slight, felt only by a few persons in conditions of perfect quiet, especially on the upper floors of houses, or by many sensitive and nervous persons
III	Slight, felt by several persons, but by inhabitants in a given place: said by them to have been hardly felt, without causing any alarm, and in general without their recognising it was an earthquake until it was known that others had felt it.
IV	Sensible to moderate, not felt generally, but felt by many persons indoors, though by few on the ground floor, without causing any alarm, but with shaking of fastenings, crystals, cracking of floors, and slight oscillation of suspended objects.
V	Rather strong, felt generally indoors, but by few outside, with waking of those asleep, with alarm of some persons, rattling of doors, ringing of bells, rather large oscillations of suspended objects, stopping of clocks.
VI	Strong, felt by everyone indoor, and by many with alarm and flight into the open air; fall of objects in houses, fall of plaster with some cracks in badly built houses.
VII	Very strong, felt with general alarm and flight from houses, sensible also out of doors; ringing of church bells, fall of chimney, pots, and tiles; cracks in numerous buildings, but generally slight.
VIII	Ruinous, felt with great alarm, partial ruin of some houses, and frequent and considerable cracks in others; without loss of life, or only with a few isolated cases of personal injury.
IX	Disastrous, with complete or nearly complete ruin of some houses and serious cracks in many others, so as to render them uninhabitable; a few lives lost in different parts of populous places.
X	Very disastrous, with ruin of many buildings and great loss of life, cracks in the ground, landslips from mountains, etc.

Table AII.4 Comparison of different Intensity Scales.

A	*B*		*C*		*D*	
Oldham Scale	*RF Scale*	*MMI Scale*	*RF Scale*	*MMI Scale*	*MSK*	*Richter Magnitude*
—	I	I	—	I	I	
6	II	II	HI	II	II	
	III	III	III	III	III	3.0
5	IV	IV	IV–V	IV	IV	3.5
	V	V	V–VI	V	V	4.0
4	VI	VI	VI–VII	VI	VI	5.0
	VII	VII	VIII	VII	VII	6.0
		VIII	VIII+–	VIII	VIII	
3	VIII		IX			6.5
2	IX	IX	IX+	IX	IX	7.0
1	X	X–XII	X	X	X	
				XI	XI	7.5–8.5
				XII	XII	

(A) Oldham scale is related to RF scale (Oldham, 1899). (B) Comparison of MMI and RF scales as given by Richter (1958) p. 651. (C) Comparison of MMI and Medvedev–Sponhover–Karnik (MSK) scale. (D) Magnitude relates to comparison with MMI in epicentral area of an earthquake.

REFERENCES

Auden, J. B., J. A. Dunn, A. M. N. Ghosh, D. N. Wadia and S. C. Roy, 1939, The Bihar–Nepal Earthquake of 1934, in *Memoirs of GSI*, Volume 73, 391 p.

Oldham, R. D., 1899, Report on the Great Earthquake of 12th June 1897, in *Memoirs Geological Survey of India*, Volume 29, 379 p.

Richter, C. F., 1958, *Elementary Seismology*, W. H. Freeman and Co., San Francisco, 768 p.

Bibliography

Agarwal, P. and M. Shrikhande, 2006, *Earthquake Resistant Design of Structures*, Prentice Hall of India, new Delhi, 634 p.

BIS: 13920-1993, *Indian Standard Code of Practice for Ductile Detailing of Reinforced Concrete Structures Subjected to Seismic Forces*, Bureau of Indian Standards, New Delhi, India.

Bolt, B. A., 2004, *Earthquakes* (Fifth Edition), W. H. Freeman and Company, New York, 378 p.

Bullen, K. E., 1965, *An Introduction to the Theory of Seismology* (Third Edition), Cambridge University Press, Cambridge, UK, 381 p.

Bullen, K. E. and Bolt B. A., 1985, *An Introduction to the Theory of Seismology* (Fourth Edition), Cambridge University Press, Cambridge, 499 p.

Brown, G., C. Hawkesworth and C. Wilson, 1992, *Understanding the Earth*, Cambridge University Press, Cambridge.

Chummar, A. V., 1989, A Manual of Earthquake Resistant Non-Engineered Construction, Indian Society of Earthquake Technology, University of Roorkee, Roorkee, India, 158 p.

Condie, K. C., 1976, *Plate Tectonics and Crustal Evolution*, Pergamon Press, Elmsford, NY.

Dewey, J. F., 1972, Plate tectonics, *Scientific American*, **226**, p 56–57.

Dewey, J. F. and J.M. Bird, 1973, Mountain belts and the new global tectonics, *J. Geophys. Res.*, **75**, p 2625–2647.

Dimri, V. P., 1992, *Deconvolution and Inverse Theory*, Elsevier, Amsterdam.

EQ 85-4, 1985, A report on micro earthquake studies for Chamera hydro electric project, H. P., Earthquake Engineering Studies, Department of Earthquake Engineering, University of Roorkee, Roorkee, India (Unpublished).

EQ 86-02, 1986, Report on collection, analysis and interpretation of data (April 1980–March 1983) from seismological laboratories in the Ganga valley region of Himalayas, Volume VI, Earthquake Engineering Studies, Department of Earthquake Engineering, University of Roorkee, Roorkee, India.

EQ 87–16, 1987, Report on collection, analysis and interpretation of data (April 1985–March 1987) from seismological laboratories in the Ganga Valley region of Himalayas, Volume IX, Earthquake Engineering Studies, Department of Earthquake Engineering, University of Roorkee, Roorkee, India.

Fowler, C. M. R., 1990, *The Solid Earth: An Introduction to Global Geophysics*, Cambridge University Press, Cambridge, UK, 472 p.

Gansser, A., 1964, *Geology of the Himalayas*, Interscience Pub., London, 269 p.

Gupta, H. K. and B. K. Rastogi, 1976, *Dams and Earthquakes*, Elsevier Scientific Publishing Company, Amsterdam, 229 p.

Gutenberg, B. and C. F. Richter, 1954, *Seismicity of the Earth and Associated Phenomena*, Princeton University Press, New Jersey.

ISI: 4326-1993, *Indian Standard Code of Practice for Earthquake Resistant Design and Construction of Buildings*, Bureau of Indian Standards, New Delhi, India.

ISI: 13935–1993, *Repair and Seismic Strengthening of Buildings—Guidelines*, Bureau of Indian Standards, New Delhi, India, 22p.

Jain, A. K., 1987, Kinematics of transverse regional tectonics and Holocene stress field in the Garhwal Himalayas, *Journal of Geological Society of India*, **30**, p 160–186.

Jain, A. K. and S. Singh, 2009, *Geology and Tectonics of the Southeastern Ladakh and Karakoram*, Geological Society of India, Bangalore, 181 p.

Kaila, K. L. and H. Narain, 1971, A new approach for preparation of quantitative seismicity maps as applied to Alpide Belt-Sunda Arc and adjoining areas, *BSSA*, **61**, p 1275–1291.

Kramer, S. L., 2004, *Geotechnical Earthquake Engineering*, Pearson Education, 653 p.

Lawson, A. C., 1908, Landslides in the California Earthquake of April 18, 1906, Report of the State Earthquake Investigation Commission, Volume 1, Part 2, p 384–401, Carnegie Institution of Washington, Washington, D.C.

Lillie, R. J., 1999, *Whole Earth Geophysics*, Prentice Hall Inc, New Jersey, 361 p.

Lowrie, W., 1997, *Fundamentals of Geophysics*, Cambridge University press, Cambridge, UK, 354 p.

McGeary, G. and C. C. Plummer, 1994, *Physical Geology: Earth Revealed* (Second Edition), Wm. C. Brown Communication, Inc., Dubuque, Iowa, 539 p.

Mittal, R. S. and L. S. Srivastava, 1959, Geotectonic position and earthquakes of Ganga--Brahmaputra region, in *Proceedings of the First Symposium on Earthquake Engineering*, University of Roorkee, Roorkee, India.

Molnar, P., 1990, A review of the seismicity and the rates of active underthrusting and deformation at the Himalaya, *Journal of Himalayan Geology*, **1**(2), p 131–154.

Monroe, J. S. and R. Wicander, 2001, *The Changing Earth: Exploring Geology and Evolution* (Third Edition), Thomson Learning Academic Resource Center, USA, 733 p.

Mussett, A. E. and M. F. Khan, 2000, *Looking into the Earth: An Introduction to Geological Geophysics*, Cambridge University Press, Cambridge, UK, 470 p.

Narula, P. L., S. K. Acharyya and J. Banerjee, 2000, *Seismotectonic Atlas of India and Its Environs*, Geological Survey of India.

NEIC: National Earthquake Information Centre (USA).

Oxburgh, E. R., 1971, Plate Tectonics, in *Understanding the Earth*, Eds. I. G. Gass, P. S. Smith and R. G. L. Wilson, p 263–286, Open University Set Book, Science Foundation Course, Artemis Press Limited, Sussex, 383 p.

Pande, P. and I. A. Parvez, 2008, *Seismic Microzonation: The Indian Scene*, CMMACS, Bangalore, IUGS.

Plummer, C. C., D. McGeary and D. H. Carlson, *Physical Geology* (Ninth Edition), McGraw-Hill Higher Education, New York, 574 p.

Press, F. and R. Siever, 1986, *Earth* (Fourth Edition), W. H. Freeman and Co., San Francisco, 649 p.

Runcorn, S. K. (Ed.), 1967, *International Dictionary of Geophysics*, Pergamon Press, Oxford, 1728 p.

Shultz, C. H., 1990, *The Mechanics of Earthquakes and Faulting*, Cambridge University Press, Cambridge.

USGS: United States Geological Survey.

USGS fact sheet, 2004–3072

Varnes, D. J., 1978, Slope movement types and processes, in *Landslides, Analysis and Control*, Eds. R. L. Schuster and R. J. Krizek, NRC, Washington, DC, Transportation Research Board, Special Report 176, p 11–33.

Vita-Finzi, C., 1986, *Recent Earth Movements: An Introduction to Neo Tectonics*, Academic Press, London, 226 p.

Wadia, D. N., 1966, *Geology of India* (Third edition), Macmillan and Co. Ltd., London.

Yeats, R. S., K. Sieh and C. R. Allen, 1997, *Geology of Earthquakes*, Oxford University Press, New York, 568p.

Websites:
http:// neic.usgs.gov/neis/epic/epic-global.html.
http://landslides.usgs.gov
http://pbs.usgs.gov/fs/2004/3072
http://en.wikipedia.org/wiki/
http://en.wikipedia.org/wiki/landslide
http://www.ce.washington.edu/html/what/what1.html

Glossary

The glossary of terms given here aims to define the terms used in this book and in other common references. Most of the terminology is in accordance with Bolt (2004), Press and Siever (1986), and Runcorn (1967).

Abyssal plain: This is the deepest part of the ocean, maybe as deep as 5000 m.

Accelerogram: The record from an accelerograph showing acceleration as a function of time.

Accelerometer: A seismograph for measuring ground acceleration as a function of time.

Accelerograph: A strong motion earthquake instrument recording ground acceleration.

Active fault: A fault along which slip has occurred in historical (or Holocene) time or on which earthquake foci are located.

Aftershocks: Smaller earthquakes that follow the largest earthquake (main shock) of a series, concentrated in a restricted volume within the crust, within a span of several months.

Alluvium: Loose materials like clay, silt, sand, gravel, and larger rocks, washed down from hills and mountains and deposited in low areas.

Amplitude (wave): The maximum height of a wave crest or depth of a trough.

Amplification: An increase in earthquake motion as a result of resonance of the natural period of vibration with that of the forcing vibration.

Andesite: Volcanic rock (name derived from the Andes mountains)

Anisotropy: Any material in which physical properties (for example, light transmission or seismic wave velocity) vary quantitatively with the direction in which they are measured.

Aseismic region: One that is relatively free of earthquakes. Actually, all areas show some seismicity over a sufficiently long interval.

Aseismic ridge: A submarine ridge that is actually a fragment of continental crust; distinguished from a mid-oceanic ridge, which is seismically active.

Asthenosphere: The soft and weak layer below the lithosphere, characterized by low seismic wave velocities and high seismic attenuation. It is probably partially molten where it may also be the site of convection. The layer or shell of the earth below the lithosphere, which is weak, in which isostatic adjustment takes place and magmas may be generated.

Attenuation: Reduction of amplitude or change in wave due to energy dissipation over distance with time.

Axial load/force: Force coincident with primary axis of a member.

Azimuth: The arc of the horizon between the meridian of a place and a vertical circle passing through any celestial body.

Bar: An international unit of pressure equal to 10^6 dynes/cm^2, approximately one atmosphere ($\equiv$ 0.982 atmospheres). 1 kilobar = 1000 bars.

Basalt: A fine-grained, dark, mafic igneous rock composed largely of plagioclase feldspar and pyroxene. Oceanic crust is mostly basalt. In rift valleys, found mostly in oceanic crust, basic, volcanic rock, (Deccan Traps).

Base shear: Total shear force acting at the base of a structure.

Basement rock: The oldest rocks recognized in a given area, a complex of metamorphic and igneous rocks that underlies all the sedimentary formations. It is usually of Precambrian or Palaeozoic age.

Basic rock: Any igneous rock containing mafic minerals rich in iron and magnesium, but containing no quartz and little sodium-rich plagioclase feldspar (preferred term *mafic rock*).

Basin: In tectonics, a circular, and syncline like depression of strata. In sedimentology, it is the site of accumulation of a large thickness of sediments.

Benioff zone: It is a characteristic of destructive plate margins and is marked by an oceanic plate sinking into the mantle at a trench. This zone has high seismicity and is usually the locus of intermediate and deep focus earthquakes. It is tens of kilometers thick and extends to the depths of up to 700 km below the surface of the earth.

Block fault: A structure formed when the crust is divided into blocks of different elevation by a set of normal faults.

Braced frame: One that is dependent upon diagonal braces for stability and capacity to resist lateral forces.

Brittle failure: Failure in material that generally has a very limited plastic range; material subject to sudden failure without warning.

Capable fault: A fault along which it is mechanically feasible for sudden slip to occur.

Causative fault: A fault that causes an earthquake.

Centre of mass: The point through which the resultant of the mass of a system acts. This corresponds to the center of gravity of the system.

Centre of stiffness: The point through which the resultant of the restoring forces of a system acts.

Compression: To press together, to force into a narrower space, to condense or concentrate.

Continental crust: It consists largely of granite and granodiorite (upper continental crust, lower continental crust).

Continental drift: The horizontal displacement or rotation of continents relative to one another.

Continental shelf: The gently sloping submerged edge of a continent, extending commonly to a depth of about 200 m or the edge of the continental slope. It may have as much as 200 m of seawater above it.

Continental slope: The region of steep slopes between the continental shelf and continental rise. It may have as much as 1200 m of water above it.

Convection: A mechanism of heat transfer through a liquid in which hot material from the bottom rises because of its lesser density, while cool surface material sinks.

Convection cell: A single closed flow circuit of rising warm material and sinking cold material.

Convergence zone: (*Destructive boundary, sink*) a band along which moving plates collide and land area is lost either by shortening and thickening of the crust or by subduction and destruction of crust. This is also the site of volcanism, earthquakes, trenches, and mountain building.

Core: Innermost shell of the earth. It is at a depth of about 2900 km from the surface of the earth. It is thought to be composed of iron, nickel, and silicates and to be molten on the outside with a central solid inner core.

Creep (along a fault): Very slow periodic or episodic movement along a fault trace unaccompanied by earthquakes.

Creep (slow fault slip): Slow slip occurring along a fault, without producing earthquakes.

Critical damping: The damping beyond which the motion will not be oscillatory. The minimum damping that will allow a displaced system to return to its initial position without oscillation.

Crust: Outer most thin shell of the earth. The outermost layer of the lithosphere, consisting of relatively light materials. The continental crust consists largely of granite and granodiorite; the oceanic crust is mostly basalt. General composition is silicon–iron–aluminium.

Dam: An embankment to restrain water; to keep back water by a bank.

Damping: A rate at which natural vibration decays as a result of absorption of energy. The effect of internal friction, imperfect elasticity of material, slipping, sliding, etc. in reducing the amplitude of vibration and is expressed as a percentage of critical damping.

Deflection: Displacement of a member due to application of external force.

Density: (ρ) The mass per unit volume of a substance, commonly expressed in g/cm^3.

Depositional remnant magnetization: A weak magnetization created in sedimentary rocks by the rotation of magnetic crystals into line with the ambient field during settling.

Depth of focus: See *focal depth.*

Design acceleration spectrum: An average smoothened plot, of maximum acceleration, as a function of frequency or time period, of vibration for a specified damping ratio for earthquake excitation at the base of a single degree of freedom system.

Design horizontal acceleration coefficient (A_h): It is a horizontal acceleration coefficient that is used for design of a structure.

Diaphragm: Generally a horizontal girder composed of a web (such as a floor or roof slab) with adequate flanges, which distributes lateral forces to the vertical resisting elements.

Dilatancy: (Of rocks) increase in volume of rocks mainly due to pervasive micro cracking.

Dip: The angle that the fault plane makes with the horizontal. The angle by which a stratum or other planar feature or fault plane deviates from the horizontal. The angle is measured in a plane perpendicular to the strike.

Dip slip fault: A fault in which the relative displacement is along the direction of dip of the fault plane; the offset is either normal or reverse.

Drift: In buildings, the horizontal displacement of basic building elements due to lateral earthquake forces.

Dispersion: (Of wave) the spreading out of a wave train due to each wave length traveling with its own velocity.

Ditch: A trench dug in the ground, any long narrow receptacle for water.

Divergence zone: A region along which tectonic plates move apart and new crust is created. It is the site of mid oceanic ridges, earthquakes, and volcanism.

Ductility: Ability to withstand inelastic strain without fracturing. Ductility of a member or structure is the capacity to undergo large inelastic deformations without significant loss of strength or stiffness.

Duration (of strong shaking): It is the time interval between the first and last peaks of strong ground motion above specified amplitude.

Dynamic: Having to do with bodies in motion.

Earthquake (event, shock): It is a sudden transient motion of the ground, which originates within a limited subsurface region and spreads in all directions.

Earthquake parameters: See *parameters*.

Eclogite: An extremely high-pressure metamorphic rock containing garnet and pyroxene.

Elasticity: The ability of a material to return to its original form or condition after a displacing force is removed.

Elastic limit: The maximum stress that can be applied to a body without resulting in permanent strain.

Elasto plastic: Total range of stress, including expansion beyond elastic limit into the plastic range.

Elastic rebound theory: The theory of earthquake generation proposing that faults remain locked while strain energy slowly accumulates in the surrounding rock, and then suddenly slip, releasing this energy.

Energy absorption: Energy is absorbed as a structure (or ground) distorts inelastically.

Energy dissipation: Reduction in intensity of earthquake shock waves with time and distance, or by transmission through discontinuous materials with different absorption capabilities.

Eon: The largest division of geologic time, embracing several eras, (for example, the Phanerozoic, 600 mya to present; Proterozoic and Archaean); also any span of one billion years.

Epicenter: The point on the surface of the earth vertically above the focus. This also roughly gives the location of an earthquake. Latitude and longitude are needed to locate it.

Epicentral distance: The distance between an epicenter and a recording station or point of observation. For short distances (i.e., less than approximately 1000 km), it is commonly given in kilometers. For large distances (>10°), it is given in terms of the angle subtended between the epicenter and the observation point at the center of the earth.

Epoch: One Subdivision of a geologic period, often chosen to correspond to a stratigraphic series. Also used for a division of time corresponding to a paleomagnetic interval.

Era: A time period including several periods, but smaller than an eon. Commonly recognized eras are Precambrian, Paleozoic, Mesozoic and Cenozoic.

Event: See *earthquake*.

Failure mode: The manner in which a structure fails (column buckling, overturning of structure, etc.).

Fault: It is a fracture along which observable displacement of blocks in the crust occurs parallel to the plane of break. A fracture or zone of fractures in rock along which two sides are displaced relative to each other parallel to the fracture. The total fault offset may range from a few centimeters to kilometers. A planar or gently curved fracture in the earth's crust across

Fault block mountain: A mountain or range formed as a horst when it was elevated (or as the surrounding region sank) between parallel normal faults.

Fault line: This is the surface trace of a fault.

Fault plane: The plane that best approximates the fracture/rupture surface of a fault.

Fault surface: The breakage of ground along the surface trace of a fault caused by the intersection of the fault surface ruptured in an earthquake with the earth's surface.

Fault zone: Faults are rarely single planar units. Normally they occur as parallel sets of planes along which movement has taken place to a greater or lesser extent. Such sets are called fault zones or fracture zones. Instead of being a single clear fracture, the zone is hundreds or thousands of meters wide the fault zone consists of numerous interlacing smaller faults. The zone of disturbed rocks between fault blocks.

Fault mega: A fault with linear dimensions of several thousand kilometers.

Fault minor: A fault with linear dimensions of a few kilometers.

Fault major: A fault with dimensions between a thousand kilometer and a few kilometers.

Feldspar: General term for a group of alumino-silicate minerals containing sodium, calcium, or potassium and having a framework structure. Feldspars are the most common minerals in the earth's crust.

Fissure: A narrow opening or chasm, a cleft, slit, or furrow.

Faulting: The movement that produces relative displacement of adjacent rock masses along a fracture.

Felsic: An adjective used to describe a light colored igneous rock that is poor in iron and magnetism and contains abundant feldspars and quartz.

Felt area: Total extent of area where an earthquake is felt.

First motion: On a seismogram, the direction of motion at the beginning of the arrival of a *P*-wave. Conventionally, upward motion indicates a compression of the ground; downward motion a dilation.

Flexible system: A system that will sustain relatively large displacements without failure.

Focal depth: It is the distance between focus and epicenter. Shallow focus earthquakes have depths less than 70 km; intermediate earthquakes have depths between 70 and 300 km; and deep focus earthquakes have depths between 300 and 700 km.

Focal mechanism: The direction and sense of slip on a fault plane at the hypocenter of an earthquake. This is inferred from the first seismic waves that are recorded at various stations. These are drawn on maps with 'beach ball' symbols. Black areas denote compression, and white areas denote dilation. The fault plane that moved is parallel to one of the two planes dividing the sphere in half.

Focus (hypocenter, source): It is the region inside the earth where an earthquake originates, i.e., where initial rupture of rocks takes place. It is the point from which seismic waves originate. It is expressed in terms of latitude, longitude, and depth.

Foot wall: It is that face of the rock, which lies below the fault plane. It is called the footwall because where inactive faults have been 'filled in' with mineral deposits and then mined, this is the side on which miners walk.

Force: Any cause that changes the direction or speed of the motion of a portion of matter.

Foreshocks: Smaller earthquakes that precede the main shock of a series concentrated in a restricted volume of the crust, within a short time span, say of several months.

Fracture zone: See strike–slip fault.

Free oscillations: Natural vibrations of the whole world induced by very large earthquakes.

Frequency: Referring to vibrations; the number of wave peaks which pass through a point in a unit of time, usually measured in cycles per second.

Friction breccias: Breccias formed in a fault zone or volcanic pipe by the relative motion of two rock bodies.

Fundamental period: The largest period (duration in time of one full cycle of oscillatory motion) for which a structure or soil column shows a response peak, commonly the period of maximum response.

Gabbros: A black, coarse grained, intrusive igneous rock, composed of calcic feldspars and pyroxene. Its intrusive equivalent is basalt.

Gable: The triangular part of an exterior wall of a building between the top of the side walls and the slopes on the roof.

Geodimeter: A surveying instrument to measure the distance between two points on the earth's surface.

Gneiss: A coarse-grained regional metamorphic rock that shows compositional banding and parallel alignment of minerals.

Gouge: Crushed, sheared, and powdered rock altered to clay.

Graben (rift valley): A long and narrow block in the crust that has dropped down along normal faults relative to the adjacent rocks. Tensional crustal forces cause these down-dropped fault blocks.

Granite: A coarse-grained, intrusive igneous rock composed of quartz, orthoclase feldspar, sodic plagioclase feldspar, and micas. Acidic rock, plutonic rock, found mostly in continental crust.

Granitization: The formation of metamorphic granite from other rocks by re-crystallization with or without complete melting.

Granodiorite: Found mostly in continental crust

Great earthquake: An earthquake that has a magnitude greater than or equal to 8.

Ground acceleration: Acceleration of the ground due to earthquake forces.

Ground displacement: The distance that ground moves from its original position during an earthquake.

Ground failure: A situation in which the ground does not hold together such as land sliding, mudflows, and liquefaction.

Ground movement: A general term; includes all aspects of motion (acceleration, particle velocity, displacement).

Ground velocity: Velocity of the ground during an earthquake.

Guyot (sea mounts): Submerged mountain or seamount found in the ocean.

Hade: It is the angle between the fault plane and the vertical plane. Hade is complement of the dip of the fault plane.

Hanging wall: It is that face of the rock that lies above the fault plane. It is called the hanging wall because where inactive faults have been 'filled in' with mineral deposits and then mined, this is the side on which miners can hang their lanterns.

Hazard (seismic): Dangerous physical effects of earthquakes caused by ground shaking such as ground damage, landslides, liquefaction, tsunamis, the built environment, human casualties and injuries, etc.

Heat flow: The rate at which heat escapes at the earth's surface, related to the nature of the surface rocks and the rate at which heat is supplied to the crust from below.

Hertz: The unit of frequency equal to one cycle per second or 2π radians per second.

Higher modes of vibration: Structures and elements have a number of natural modes of vibration.

Himalayas: Rugged part of the Alpine Himalayan mountain chain, geologically the youngest, seismically most active mountain chain in the world.

Hooke's law: The principle that the stress within a solid is proportional to the strain. It holds only for strains of a few percent or less.

Holocene: The most recent geologic era; from about 10,000 years ago to the present. The Holocene is the latest epoch of the Quaternary period.

Horst: An elongated, elevated block of crust forming a ridge or plateau, typically bounded by parallel, outward-dipping normal faults.

Hot spot: The surface expression of a mantle plume.

Hypocenter: See *focus*.

Hypocentral distance: Distance between the focus and the point of observation.

Igneous rock: A rock formed by congealing rapidly or slowly from a molten state.

Importance factor (I): A factor used to obtain the design seismic force depending on the functional use of the structure, characterized by hazardous consequences of its failure, its postearthquake functional need, historic value, or economic importance.

Index of refraction: The ratio of the speed of light in vacuum to the speed in a material; this ratio determines the amount of light that is refracted as it passes into a crystal.

Inelastic behavior: Behavior of an element beyond its elastic limit.

Inertia: Inertness, the inherent property of matter by which it tends to remain forever at rest when still, and in motion when moving.

Infrastructure: Telecommunications, transport systems: roads, highways, railway, bridges, airports, seaports, industry, and schools.

Intensity: It is an estimate of quality and quantity of damage, based on macro seismic effects, caused by an earthquake. It is an estimate of ground shaking that occurred at a place. It is a space-dependent rating assigned by an experienced observer, using a descriptive scale of damage to ground, the built environment and effects on people caused by an earthquake. A measure of ground shaking obtained from the damage done to structures built by man, changes in the earth's surface and felt reports.

Intensity scales: The damage caused by an earthquake is assigned in terms of a descriptive scale, with grades indicated by Roman numerals, usually from I to XII, or I to X.

Interplate earthquake: An earthquake with its focus on a plate boundary.

Intraplate earthquake: An earthquake with its focus within a plate.

Inundation: Horizontal extent of water penetration.

Island arc: A linear or arc-like chain of volcanic islands formed at a convergent plate boundary. The island arc is formed in the overriding plate from rising melt derived from the subducted plate and from the asthenosphere above that plate. Chain of islands above a subduction zone

Isoseismal lines: Lines joining points of equal earthquake intensity. Contour lines drawn on a map to separate one level of intensity from another.

Isoseismal map: This map represents the type and extent of damage and severity with which the earthquake was felt in an area.

Isostacy: The mechanism whereby areas of the crust rise or subside until the mass of their topography is buoyantly supported or compensated by the thickness of crust below, which 'floats' on the denser mantle. The theory is that continents and mountains are supported by low-density crustal 'roots'.

Landslide: A portion of land that falls down, generally from the side of a hill, usually due to undermining effect of water. The rapid down slope movement of soil and rock material, often lubricated by groundwater, over a basal shear zone. It is the down slope mass movement of earth resulting from any cause.

Lateral: Force coefficients. Factors applied to the weight of a structure or its parts to determine lateral force for a seismic structural design.

Lava: Magma or molten rock that has reached the surface.

Left lateral fault: A strike slip fault on which the displacement of the far block is to the left when viewed from either side.

Lifeline: Network of essential facilities and services like water supply, electricity, communication facilities, and public health facilities

Lintel: The piece of timber or stone over a doorway, the headpiece of a door or casement.

Liquefaction: Process of soil and sand behaving like a dense fluid rather than a wet solid mass during an earthquake. Transformation of a granular material (soil) from a solid state into a liquefied state as a consequence of increased pore water pressure induced by vibrations.

Lithosphere: The outer, rigid shell of the earth situated above the asthenosphere. It contains the crust and the upper mantle. Tectonic plates are composed of lithosphere. The outer most layer of the earth, distinguished from the subjacent asthenosphere by its greater rigidity and strength; commonly includes the crust and part of the upper mantle, which are distinguished by differences in seismic wave speeds rather than rheological properties.

Load: Burden to weigh down, to weight by something specially added.

Local magnitude: See M_L.

Love waves: Transverse vibration of seismic surface wave. Seismic surface waves with only horizontal shear motion transverse to the direction of propagation.

Low velocity zone (LVZ): A region in the earth, especially a planar layer that has lower seismic wave velocities than the region immediately above it.

Lumped mass: For analysis purposes, assumed grouping of mass at specific locations.

Lurching of ground: Disruption of soil by lateral spreading under gravity.

m_b : Body wave magnitude.

M_L (local magnitude): A measure of the strain energy released by an earthquake within 100 km of its epicenter. It is defined by Richter as the base 10 logarithm of the amplitude, in microns, of the largest trace deflection that would be observed on a standard torsion seismograph at a distance of 100 km from the epicenter.

M_S : Surface wave magnitude. A magnitude determined at tele seismic distances using the logarithm of the amplitude of 20 sec period surface waves generated by an earthquake.

M_W : Moment magnitude. The seismic moment of an earthquake, converted to a magnitude scale that roughly parallels the original Richter magnitude scale. However, since it is not based on the same measurements as Richter (local or surface wave) magnitudes, the different magnitudes do not always agree, particularly for very large earthquakes. Because it relates directly to the energy released by an earthquake, it has become the standard in modern seismology.

Macro seismic effects: Those earthquake effects that can be observed on a large scale in the field, without the aid of any instrument.

Macro zones: Large zones of earthquake activity (such as zones designated by Seismic Zoning Map of India IS 1893–2002).

Mafic mineral: A dark-colored mineral rich in magnesium and iron, especially a pyroxene, amphibole, or olivine.

Magma: Molten rock material that forms igneous rocks upon cooling.

Magnetic anomaly: The value of the local magnetic field remaining after the subtraction of the dipole portion of the earth's field.

Magnetic epoch: A geologically long period during which the earth's magnetic field was of predominantly one polarity; the epochs immediately before and after a given epoch would by definition be characterized by a field of opposite polarity.

Magnetic events: Geologically short periods within magnetic epochs during which the field had a reversed polarity.

Magnetic north pole: The point where the earth's surface intersects the axis of the dipole that best approximates the earth's field. The point where the earth's magnetic field dips vertically downward.

Magnetic reversal: A change of the earth's magnetic field to the opposite polarity.

Magnification factor: An increase in lateral forces at a specific site for a specific factor.

Magnitude (of an earthquake): It is a numerical scale for quantifying an earthquake. It indicates the amount of energy released at the source and is determined from seismograms. It is the logarithm (base 10) of the largest trace amplitude measured on a seismogram written by a standard instrument, placed at a distance of 100 km from the epicenter. Determined by taking the common logarithm (base 10) of the largest ground motion recorded during the arrival of a seismic wave and applying a standard correction of distance to the epicenter. Some common types of magnitude are Richter or local M_L; body wave magnitude m_b; surface wave magnitude M_S, and moment magnitude M_W.

Magnitude scales: Magnitude computed by different formulae, surface wave magnitude, body wave magnitude, local magnitude, moment magnitude, etc.

Main event, main shock: This is the event with the largest magnitude in a series of events. It should be at least half a magnitude unit larger than the next largest quake in the series. Otherwise, the series of quakes may be termed as a swarm.

Mantle: Middle shell of the earth between the crust and core. It comprises of the main bulk of the earth, varying in depth from about 40–2900 km.

Margin, active: This is a continental margin characterized by volcanic activity and earthquakes (i.e., location of transform fault or subduction zone).

Margin, passive: Continental margin formed during initial rifting apart of continents to form an ocean; frequently has thick deposits.

Meizo-seismal area: This is the area within the isoseismal of highest intensity. The area of strong shaking and significant damage in an earthquake.

Metamorphic rock: A rock whose original mineralogy, texture, or composition has been changed due to the effects of pressure, temperature, or the gain or loss of chemical components.

Metamorphism: The changes of mineralogy and texture imposed on a rock by pressure and temperature in the earth's interior.

Micro earthquake: It is an earthquake that has magnitude less than or equal to 3, and is not felt by people nearby.

Microseism: A weak vibration of the ground that can be detected by seismographs and which is caused by waves, wind, or human activity, but not by an earthquake. Weak, almost continuous background seismic waves or 'earth' noise that can be detected only by seismographs; often caused by surf, ocean waves, wind, or human activity. (Have nothing to do with micro seismic effects.) They are not small earthquakes. They are continuous disturbances in the ground recorded by instruments. They may be connected with weather. They are very puzzling and provoking phenomena. Microseisms are studied for the purpose of improving signal-to-noise ratio for detection of earthquake events.

Micro seismic: Effects are small-scale, observable only with instruments.

Micro zones: Breaking up of macro zones into much smaller zones of specific earthquake activity.

Mid oceanic ridge: A major linear elevated landform submerged in an ocean, which resembles a mountain range, with a central rift valley. It is many hundreds of kilometers (200–20,000 km) in length. A ridge crest rises 2–4 km above the level of the ocean floor, and near the axis slopes away from the crest, almost symmetrically. It consists of many small, slightly offset segments. It is a characteristic of a plate boundary occurring in a divergence zone, i.e., it is a site where two plates are pulled apart and new oceanic crust is created. Sometimes these ridges give off lava. If they are high enough to be exposed above the water level, they become islands.

Mineral: A naturally occurring, solid, inorganic element or compound, with a definite composition or range of compositions, usually possessing a regular internal crystalline structure.

Modal analysis: Determination of design earthquake forces based upon the theoretical response of a structure in its several modes of vibration to excitation.

Mode: The shape of the vibration curve.

Modified Mercalli Intensity Scale: An earthquake intensity scale that came up in 1931. It divides the macro seismic effects of an earthquake into 12 categories, from I (not felt by people) to XII (total damage).

Mohorovičić discontinuity (Moho, M discontinuity): The boundary between crust and mantle, marked by a rapid increase in seismic wave velocity to more than 8 km per sec. Depth: 5 (under oceans) to 45 (under mountains) km. Abbreviated "Moho" or 'M-discontinuity'.

Moment (of earthquakes): See *seismic moment.* The rigidity of the rocks times the area of faulting times the amount of slip. It is a measure of earthquake size.

Moment frame: One which is capable of resisting bending movements in the joints, enabling it to resist lateral forces or unsymmetrical vertical loads through overall bending action of the frame. Stability is achieved through bending action rather than bracing.

Moment magnitude: See M_W.

Mortar: A cement of lime, sand, and water, used to bind together stones or bricks in building.

Mountain: A steep sided topographic elevation larger than a hill, also a single prominence forming part of a ridge or mountain range.

Mountain belts: The mechanism whereby areas of the crust rise or subside until the mass of their topography is buoyantly supported or compensated by the thickness of crust below, which "floats" on the denser mantle. Continents and mountains are supported by low density crustal "roots".

Mud flow (or earth flows): This happens where there is plenty of ground water. The earthquake is accompanied or followed by a sudden burst of water from a locality where it normally appears as springs. This water carries sand and mud with it in a flow that may be destructive to buildings in its path.

Mud volcano: Mass movement of material finer than sand, lubricated with large amounts of water.

Natural frequency: The constant frequency of a vibrating system in the state of natural oscillation.

Near earthquakes: Or local earthquakes. An earthquake that has an epicentral distance less than 10°.

Net slip: It is the resultant of strike slip and dip slip. Strike slip is the slip component parallel to the strike of the fault and dip slip is the slip component parallel to dip of fault.

Nonstructural components: Those building components that are not intended primarily for the structural support and bracing of building (partitions, nonbearing masonry walls, claddings, staircases, water tanks, etc.).

Normal fault: A dip slip fault in which the block above the fault plane has moved downward relative to the block below. A fault under tension where the overlying block moves down the dip or slope of the fault plane. A fault that shows vertical displacement. This kind of a fault is a sign of tectonic extension.

Obduction: A process occurring during plate collision, whereby a piece of the subducted plate is broken off and pushed up onto the overriding plate; this mechanism probably explains why we find blocks of ophiolite on continents.

Oblique slip fault: A fault that combines some strike-slip motion with some dip slip motion. A combination of normal and slip or thrust and slip faults whose movement is diagonal along the dip of the fault plane.

Oceanic crust: Consists largely of basalt.

Oceanic ridge: A long, continuous mountain chain submerged in ocean. See *mid-oceanic ridge.*

Oceanic trench: Deep depression in the ocean floor. See *trench.*

Origin time: This is the instant at which the earthquake event (apart from foreshocks) starts at the focus. Origin times are usually given in terms of year, month, day, h, min, and sec. The last three are given as 08: 46: 39.3, or 08 h, 46 min, 39.3 s, which is equivalent to 08 h 46 min 39.3 s.

Orogenic belt: A linear region, often a former geosyncline that has been subjected to folding, and other deformations in a mountain building episode.

Orogeny: Mountain making, particularly by folding and thrusting of rock layers. In the framework of plate tectonics, orogeny occurs primarily at boundaries of two colliding plates, where intervening material is crumpled and volcanoes are initiated.

Out of phase: The state where a structure in motion is not at the same frequency as the ground motion; or where equipment in a building is at a different frequency from the structure.

P-wave: See *primary wave.*

Paleo magnetism: The science of the reconstruction of the earth's ancient magnetic field and the positions of the continents from the evidence of remnant magnetization in ancient rocks.

Pangea: All lands.

Panthalassa: All seas.

Parameters: Of an earthquake are latitude and longitude of the epicenter; depth of focus; origin time; and magnitude.

Peridotite: Ultrabasic plutonic rock.

Period: See natural frequency.

Period (wave): The time interval between successive crests in a sinusoidal wave train; the period is the inverse of the frequency of a cyclic event.

Period (geologic): The most commonly used unit of geologic time, representing one subdivision of an era.

Plate: A thin rigid body with a large horizontal dimension. It is composed of the lithosphere. At some depth (40–150 km) the plate (i.e., the lithosphere) is decoupled from the underlying material. It moves in relation to other plates over a deeper interior. Plates meet in convergence zones and separate in divergence zones. A relatively rigid segment of the earth's lithosphere, with a large horizontal dimension.

Plate boundary: This is the surface trace of the zone of motion between two plates. Oceanic ridges and trenches are considered to be diagnostic of plate boundaries. This is the surface trace of the zone of motion between two plates. Oceanic ridges and trenches are considered to be diagnostic of plate boundaries.

Plate margin: It is the marginal part of a particular plate. Two plate margins meet at a common plate boundary. Most tectonic activity is localized at plate margins. These are, therefore, regions of intense seismic activity. Differential motion may exist between adjacent plates.

Plate tectonics: A geological model in which the earth's crust and upper mantle (i.e., the lithosphere) are divided into a number of more or less rigid segments called plates. It deals with the theory and study of plate formation, movement, interaction, and destruction. It attempts to explain seismicity, volcanism, mountain building, and paleo-magnetic evidence in terms of large horizontal surface motions. A global theory of tectonics in which an outermost sphere (the lithosphere) is divided into a number of relatively rigid plates that collide with, separate from, and translate past one another at their boundaries.

Plinth: The square at the bottom of the base of a column, the projecting band at the bottom of a wall.

Plume: Hypothetical rising jet of hot, partially molten mantle material, supposed to be responsible for intraplate volcanism.

Pole of spreading: An imaginary point on the earth's surface that represents the emergence of an imaginary axis passing through the earth's center and about which one plate moves relative to another; thus, for each pair of plates, there is a unique pole.

Precursor: A change in the geological conditions that is a forerunner to earthquake generation on a fault.

Prediction (of earthquakes): For forecasting in time, place, and magnitude of an earthquake; the forecasting of strong ground motions.

Primary wave (P, longitudinal, compressional, irrotational, push): It is the fastest of all seismic waves and, therefore, the first to arrive at any location after the earthquake. It causes compressions and dilatations of the material. The particle movement is parallel to the direction of propagation of the wave. Its velocity is 5.5–7.2 km/sec in the crust and 7.8–8.5 km/sec in the upper mantle. Longitudinal waves are compressional waves with volume change.

Rayleigh waves: Seismic surface waves with ground motion only in a vertical plane containing the direction of propagation of the waves.

Reflection method: See seismic reflection method.

Refraction (wave): The departure of a transmitted wave from its original direction of travel at the interface with a material of different index of refraction (light) or seismic wave velocity.

Refraction method: See seismic refraction method.

Regional metamorphism: Metamorphism occurring over a wide area and caused by deep burial or strong tectonic forces of the earth.

Reid's theory: This is a theory of fault movement and earthquake generation that holds that faults remain locked while strain energy accumulates in the rock, and then suddenly slip and release this energy.

Reinforce: To enforce again, to strengthen with new force or support.

Reinforcement: Additional force or assistance.

Rending: To tear asunder with a force, to split, to tear away.

Resonance: Induced oscillations of maximum amplitude produced in a physical spectrum when an applied oscillatory motion and the natural oscillatory frequency of the system are the same.

Response: Effect produced on a structure by earthquake ground motion.

Response reduction factor: It is the factor by which the actual base shear force, which would have generated if the structure were to remain elastic during its response to Design Basis Earthquake shaking, shall be reduced to obtain the design lateral forces.

Return period of earthquakes: The time period (years) in which the probability is 63% that an earthquake of a certain magnitude will recur.

Reverse fault: A dip slip fault in which the upper block, above the fault plane, moves up and over the lower block so that older strata are placed over younger ones.

Richter magnitude: See *magnitude*.

Rift: A region where the crust has split apart and is usually marked by a rift valley.

Rift valley: A fault trough formed in a divergence zone or in other area of tension.

Right lateral fault: A strike slip fault on which the displacement of the far block is to the right when viewed from either side.

Rigidity (stiffness, μ): The ratio of the shearing stress to the amount of angular rotation it produces in a rock sample. (Reciprocal = flexibility). Relative stiffness of a structure or element. In numerical terms, equal to the reciprocal of displacement caused by a unit force.

Risk (seismic): The relative risk is the comparative earthquake hazard from one site to another. The probabilistic risk is the odds of earthquake occurrence within a given time interval and region

Rockslide: Land slide involving mainly large blocks of detached bedrock with little or no soil or sand.

Run up: Maximum vertical elevation of water on land.

Sag (fault): A narrow geological depression found in strike slip fault zones.

Sag pond: A pond occupying a depression along a fault.

Scarp: A cliff-like steep slope. Scarps are often produced by faulting, especially when dip–slip is significant.

Sea floor spreading: The mechanism by which new sea floor crust is created at ridges in divergence zones and adjacent plates are moved apart. It is the mechanism by which adjacent plates move apart at and new crust is created. This happens at constructive plate margins, i.e., at oceanic ridges. The rate of spreading is approximately 0.5–10 cm per year, and may continue through many geological periods.

Sea mounts (see also *Guyot*): An isolated tall mountain on the sea floor that may extend more than 1 km from base to peak.

Secondary wave (S, shear, transverse, rotational, standing, shake): It consists of elastic vibrations transverse to the direction of wave propagation. It travels more slowly than the P-wave. It cannot propagate in a liquid.

Seiches: Oscillations (standing waves) of water in a bay or lake.

Seismic: Pertaining to earthquake activities.

Seismic belt: A narrow, well-defined, semicontinuous geographical area along which earthquakes are confined.

Seismic discontinuity: A surface within the earth across which *P*-wave or S-wave velocities change rapidly, usually by more than ±0.2 km/s.

Seismic event: See *earthquake*.

Seismic Gap: A segment of an active fault zone that has not experienced a major earthquake during a time interval when most other segments of the zone have. Seismic gaps are supposed to have a high future earthquake potential.

Seismic moment, M_0: A measure of the strength of earthquake, equal to the product of the force and the moment arm of the double couple system of forces that produces ground displacements equivalent to that produced by the actual earthquake slip. It is also equal to the product of the rigidity modulus of the earth material, the fault surface area, and the average slip along the fault. Therefore, both seismological and geological observations can produce the same result.

Seismic reflection method: A mode of seismic prospecting in which the seismic profile is examined for waves that have been reflected from near horizontal strata below the surface.

Seismic refraction method: A mode of seismic prospecting in which the seismic profile is examined for waves that have been refracted upward from seismic discontinuities below the profile. Greater depths may be reached than through seismic reflection.

Seismic retrofitting: It corresponds to upgradation of deficient structures that are in operation.

Seismic strengthening: This is the process of enhancing the capability of a structure for improved performance against specified earthquake hazard level.

Seismic upgradation: This is associated with the rehabilitation of the structure damaged by earthquakes.

Seismic wave: An elastic wave in the earth usually generated by an earthquake source or explosion.

Seismic zone: A region on the surface of the earth associated with active seismicity.

Seismicity: A general term for the number of earthquakes in a unit of time and space. The worldwide or local distribution of earthquakes in space and time; a general term for the number of earthquakes in a unit of time, or for relative earthquake activity.

Seismogram: Record produced by a seismograph.

Seismograph: An instrument that records motions of the earth's surface, as a function of time, that are caused by seismic waves.

Seismology: Science of study of earthquakes, seismic sources, and wave propagation through the earth.

Seismometer: The sensor part of the seismograph, usually a suspended pendulum, sensor, velocity transducer.

Seismoscope: A device that indicates the occurrence of an earthquake but does not write or tape a record. A simple seismograph recording on a plate without time marks.

Sg waves: Shear waves reflected from granite layers.

Shadow zone (seismic): Region on the far side of the earth's surface from an earthquake, not reached by *P*-waves from that earthquake because they have been deflected at the surface of the outer core. *S*-waves have a large shadow zone because they cannot travel through the liquid outer core.

Shale: Sedimentary rock, mainly clay.

Shear: A strain where compression is answered by elongation at right angles.

Shear distribution: Distribution of lateral forces along the height or width of a building.

Shear strength: The stress at which a material fails in shear.

Shear wall: A wall designed to resist lateral forces parallel to the wall. A shear wall is normally vertical, although not necessarily so.

Shield: A large region of stable, ancient basement rocks within a continent.

Shift: Shift denotes the relative displacement of point far enough removed from the fault to be unaffected by local disturbance in the fault zone. Dip shift and strike or lateral shift denotes components of shift parallel to strike or dip of the fault and the resultant of the two is called net shift.

Simple harmonic motion: Oscillatory motion of a wave, single frequency. Essentially a vibratory displacement such as that described by a weight, which is attached to one end of a spring and allowed to vibrate freely.

Slickensides: These are parallel grooves, ramps, and scratches on one or both of the inside faces of a fault, showing the direction of slip.

Slip: The motion of one face of a fault relative to the other. It is the relative displacement of formerly adjacent points, measured along the fault plane. Most faults slip only during earthquakes; in between earthquakes the two sides are locked.

Slumps: To fall or sink suddenly into water or mud; to fail or fall through helplessly; to fall suddenly or heavily.

Soil–structure interaction: The effects of the properties of both soil and structure upon response of the structure.

Source: See *focus*.

Spectra: A plot indicating maximum earthquake response with respect to natural period or frequency of the structure or element. Response can show acceleration, velocity, displacement, shear or other properties of response.

Stability: Resistance to displacement or overturning.

Stiffness: Rigidity, or the reciprocal of flexibility.

Stoneley wave: These are surface waves of Rayleigh type for the case of a finite layer overlying an infinite substratum.

Strain: A quantity describing the exact deformation of each point in a body. The geometrical deformation or change in shape of a body. Roughly, it is the change in a dimension in an angle, length, area, or volume divided by the original value.

Strain release: Movement along a fault plane; can be gradual or abrupt.

Strain seismograph: An instrument that measures changes of strain in surface rocks to detect seismic waves.

Stratum: A single sedimentary rock unit with a distinct set of physical or mineralogical characteristics or fossils such that it may be readily distinguished from beds above and below. Plural: strata.

Stress: A quantity describing the forces acting on each part of a body in units of force per unit area.

Stress drop: The sudden reduction of stress across the fault plane during rupture.

Strike: The angle between true north and the horizontal line contained in any planar feature (inclined bed, dike, fault, fault plane, etc.); also the geographic direction of this horizontal line. The direction that is perpendicular to the dip direction.

Strike slip fault (a trans current fault, transform fault, fracture zone lateral slip): A fault whose relative displacement is parallel to the strike of the fault. A fault whose relative displacement is purely horizontal.

Strong ground motion: The shaking of the ground near an earthquake source made up of large amplitude seismic waves of various types.

Subduction: The sinking of a plate under an overriding plate in a convergence zone.

Subduction zone: See *Benioff Zone.*

Subsidence: Settling or sinking.

Surface trace (fault surface): The intersection of a fault plane with the surface of the earth. Sometimes it is accompanied by geomorphic evidence such as ridges, valleys, saddles, etc.

Surface wave: A seismic wave that follows the earth's surface only, with a speed less than that of shear waves. The two types of surface waves are Rayleigh waves (forward and vertical vibrations) and Love waves (transverse vibrations).

Surface wave magnitude: See M_s.

Swarm (of earthquakes): A series of earthquakes in the same locality, no one earthquake being of outstanding size.

Tectonics: It is the large-scale deformation of the outer part of the earth resulting from forces inside the earth. It involves the study of movements and deformation of the crust on a large scale, including metamorphosis, folding, faulting, and plate tectonics.

Tectonic earthquakes: Earthquakes resulting from sudden release of energy stored by major deformation of the earth. Earthquakes associated with faulting or other structural processes.

Tele seismic: It is an earthquake recorded by a seismograph at a great distance. By international convention this distance is required to be over 1000 km from the epicenter. Earthquakes originating nearer the recording station are 'near earthquakes' or 'local earthquakes'.

Tele-seismic event: An earthquake that has an epicentral distance greater than 10°.

Tension: Act of stretching, strain in the direction of the length, or the degree of it, strain.

Time development response analysis: Study of the behavior of a structure as it responds to a specific ground motion.

Throw and heave: They are apparent displacements as seen in a cross-section normal to the fault plane. Throw is the vertical distance separating the faulted parts of a bed and heave is the horizontal distance.

Thrust, thrust fault: A reverse fault in which the dip of the fault plane is less than 45°. This kind of a fault indicates tectonic compression.

Torsion (rotation): Twisting around an axis.

Trans current fault: See *strike–slip fault.*

Transform fault: See strike–slip fault or fracture zone. A strike–slip fault connecting the ends of an offset in a mid-oceanic ridge or an island arc. Some pairs of plates slide past each other along transform faults.

Transition zone (seismic): A seismic discontinuity, found in all parts of the earth, at which the velocity increases rapidly with depth; especially the one at 400–700 km.

Travel time (or transit time): It is the time that elapses between the origin time and the arrival of a given seismic wave at a specified point, usually a seismograph station.

Travel time curve: A curve on a graph of travel time versus distance for the arrival of seismic waves from distant events. Each type of seismic wave has its own distinct curve.

Trench: It is a deep, long, narrow, and arcuate depression in the ocean floor. Its length may be several thousand kilometers and width may be 8–10 km. Along this trough, a plate bends down into a subduction zone and descends into the mantle. It is diagnostic of a destructive plate boundary.

Triple junction: A point that is common to three plates and which must also be the meeting place of three boundary features, such as divergence zones, convergence zones, or transform faults. A point where three plates meet.

Tsunami: A long ocean wave usually caused by sea floor displacement in an earthquake or landslide. A sea wave produced by large displacement of the ocean bottom, usually the result of earthquakes or volcanic activity. A standing wave on the surface of the water in an enclosed or semienclosed basin (lake, bay, or harbor).

Ultra-mafic rock: An igneous rock consisting dominantly of mafic minerals, containing less than 10% feldspar. Includes dunite, peridotite, amphibole, and pyroxene.

Unconformity: A surface that separates two strata. It represents an interval of time in which deposition stopped, erosion removed some sediments and rock, and then deposition resumed.

Vibration: A periodic motion that repeats itself after a definite interval of time.

Viscosity: A measure of resistance to flow in a liquid.

Volcanic earthquakes: Earthquakes associated with volcanic activity.

Volcanic tremor: The more-or-less continuous vibration of the ground near an active volcano.

Volcanism: Geological process that involved the eruption of molten rock.

Volcano: An opening in the crust that has allowed magma to reach the surface.

Wadati-Benioff zone: See *Benioff zone.*

Warp: To turn, to twist out of shape, to turn from the right course, to pervert, to bend.

Waterfall: A fall or a perpendicular descent of a body of water.

Wavelength: The distance between two successive crests or troughs of a wave.

Zone: See *seismic zone.*

Zone factor: It is a factor to obtain the design spectrum depending on the perceived maximum seismic risk characterized by maximum considered earthquake in the zone in which the structure is located.

Subject Index

Author's Profile

AMITA SINVHAL (*nee* Amita Agarwal) is Professor at the Department of Earthquake Engineering, Indian Institute of Technology, Roorkee. Her fields of specialization are Engineering Seismology, Seismotectonics, Engineering Geophysics, and Seismic Exploration.

Born and brought up in Lucknow, she studied at La Martiniere Girls' School, Loreto Convent and Lucknow University. She obtained her M.Tech and Ph.D. degrees from the University of Roorkee (now Indian Institute of Technology, Roorkee). Prior to joining IIT Roorkee as a faculty member in 1982, she worked at the Department of Geodesy and Geophysics of Cambridge University, U.K. and at Medical Research Council (MRC) Cambridge, U.K.

She has published more than 100 research papers in refereed international and national journals, as chapters in books, and in conference proceedings. She has co-authored a book "*Pattern Recognition in Oil Exploration*" published by Kluwer Academic Press, The Netherlands. The Rajiv Gandhi Foundation sponsored *Earthquake Problems Do's and Dont's* co-authored by her, published both in English and in Hindi. Dr. Sinvhal's research publications have fetched her four prestigious awards. These include the Khosla Research gold and silver medals in 1984 and 1985, and ISET award in 1998 and in 2000.

She has worked on more than 80 earthquake-related consultancy projects dealing with site-related problems for dams, bridges, nuclear power plants, refineries, historical monuments, etc., and conservation and management of lakes and rivers, and on gender issues.

She is a member of several international and national professional bodies, which include Indian Society of Earthquake Technology (ISET), Indian Geotechnical Society, Indian Women Scientists Association, Indian Society for Continuing Engineering Education, Indian Society for Wind Engineering, and the Third World Organization of Women in Science.

Dr. Sinvhal is regularly invited by several organizations to deliver talks on earthquakes- and oil- exploration-related topics. She organizes short-term courses on Earthquake Disaster Mitigation and Engineering Aspects of Seismology. Dr. Sinvhal has widely traveled and visited several countries of the continents of Europe, North America, Africa, and Asia.

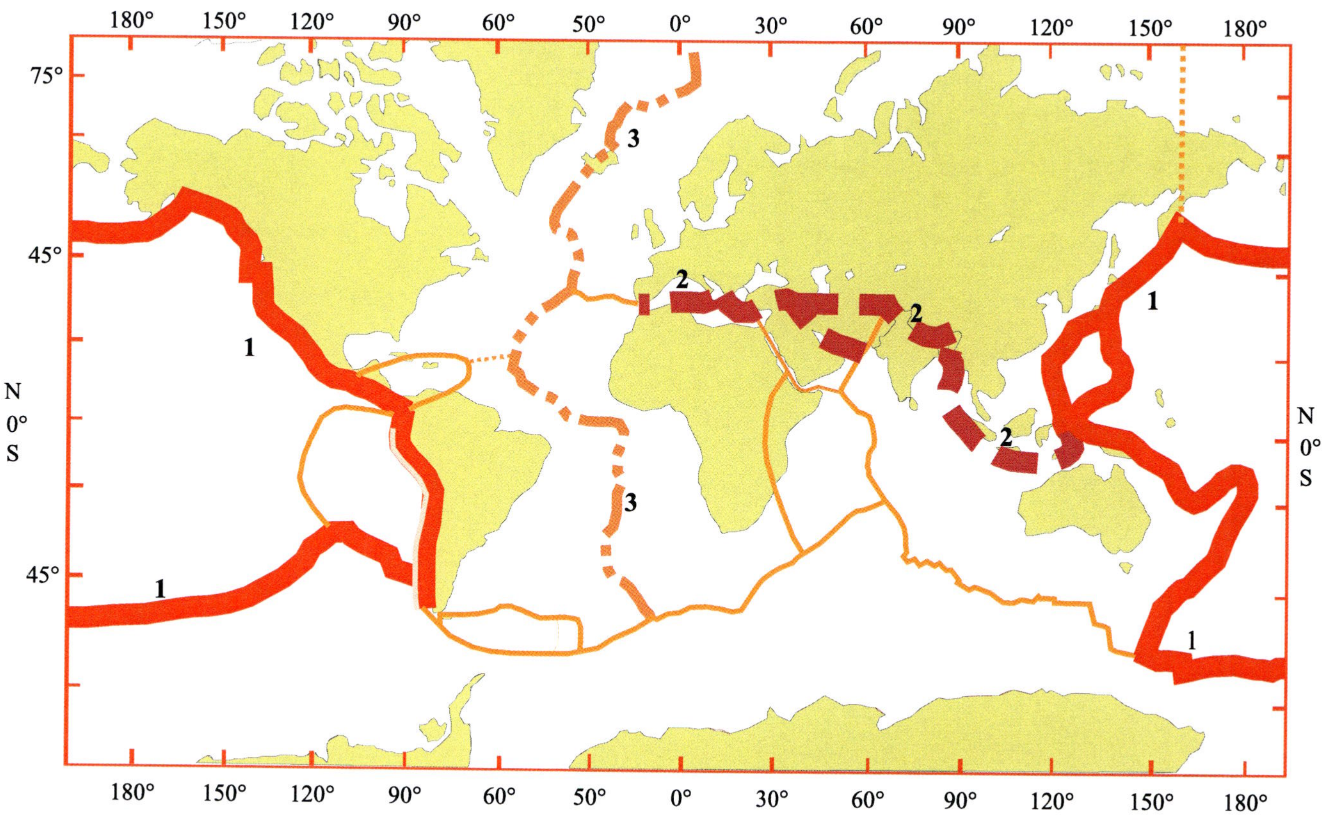

Fig. 1.1 The two main seismic belts are: (1) the Circum Pacific belt, and (2) the Alpine-Himalayan belt. (3) The Mid-Atlantic Ridge, forms a third, less active belt.

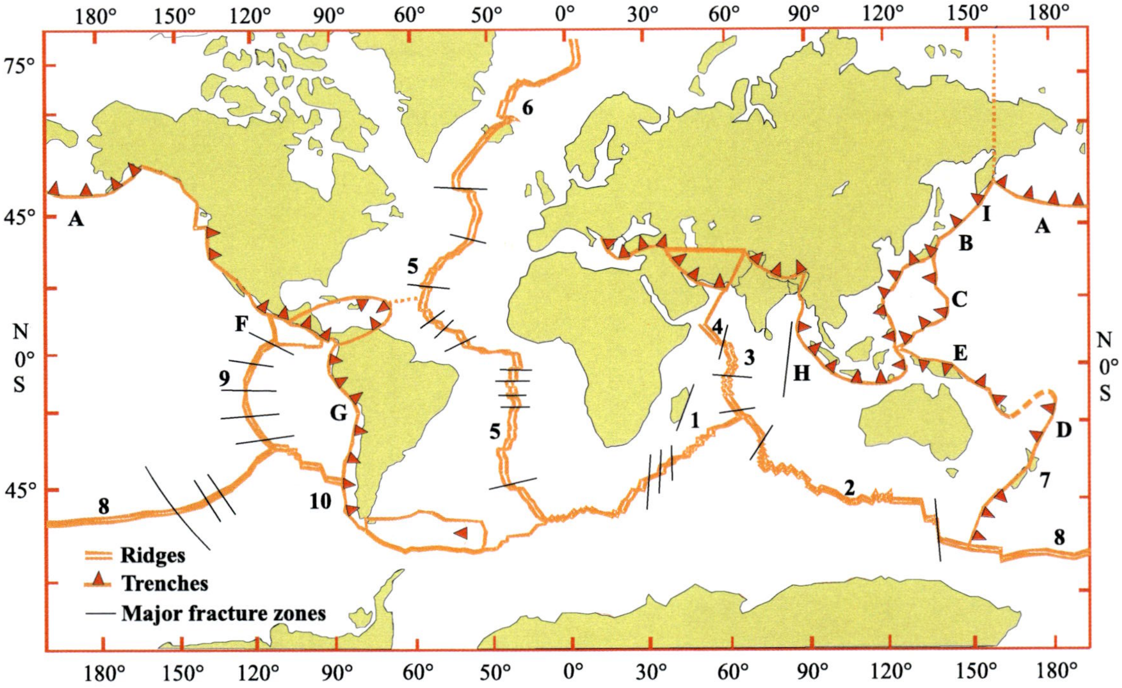

Fig. 1.4 Map showing the position of mid-oceanic ridges, trenches and major fracture zones. Parallel thick lines indicate crest of the mid-oceanic ridge system. The ridges shown on this map are: (1) South West Indian Ocean Ridge, (2) South East Indian Ocean Ridge, (3) Central Indian Ocean Ridge, (4) Carlsberg Ridge, (5) Mid-Atlantic Ridge, (6) Reykjanes Ridge, (7) Macquarie Ridge, (8) Pacific-Antarctic Ridge, (9) East Pacific Rise, and (10) Chile Rise. Thick lines with teeth indicate deep-sea trenches. These are: (A) Aleutian Trench, (B) Japan Trench, (C) Mariana Trench, (D) Kermadec-Tonga Trench, (E) New Hebrides Trench, (F) Middle America Trench, also known as Mexico Trench, (G) Peru-Chile Trench, (H) Andaman–Sumatra–Java–Sunda Trench, and (I) Kurile Trench. Thin solid lines indicate major fracture zones or transform faults.

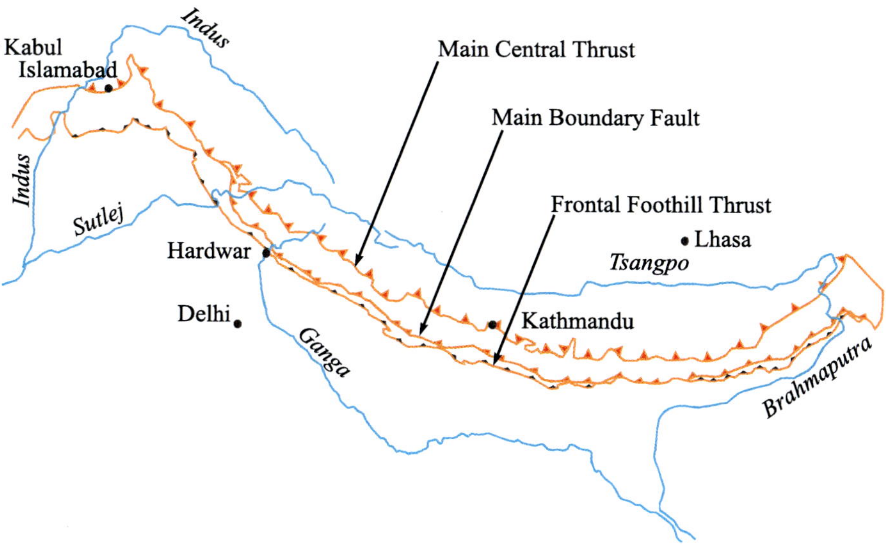

Fig. 4.4 Three mega thrusts along the Himalayas extend from Kashmir in the west to Arunachal Pradesh in the east. These are the Main Central Thrust Main Boundary Thrust, and Frontal Foothill Thrust.

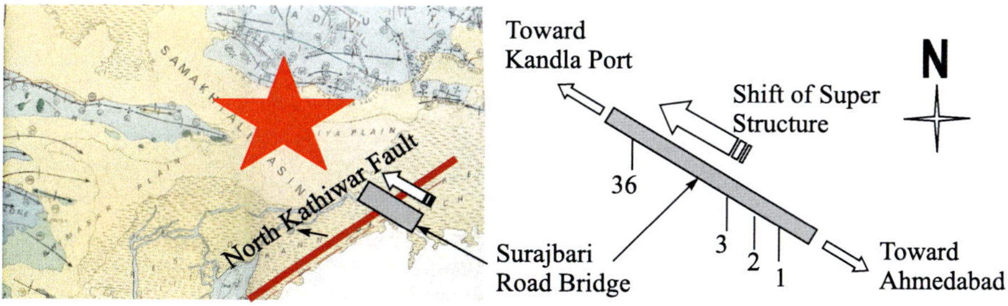

Fig. 4.6 The superstructure of Surajbari road bridge, located on the North Kathiawar fault, near Maliya Miyana in Kutch, was displaced by about 50 cm toward the north. Star depicts epicenter of the Kutch earthquake of 2001.

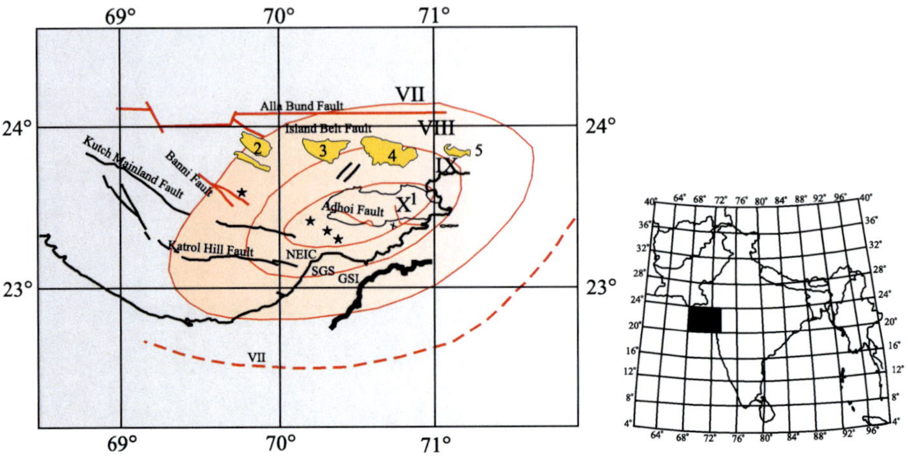

Fig. 7.8 Isoseismal map for the Kutch earthquake of January 26, 2001, shows intensity between X and VII. Map also shows major faults (in red) and ridges (in yellow). Higher isoseismals are elongated in an almost east–west direction. (1) Wagad ridge, (2) Pachcham uplift, (3) Khadir ridge, (4) Bela ridge, and (5) Charor uplift. Epicenters as provided by different agencies (as given in Table 14.2) are shown by star.

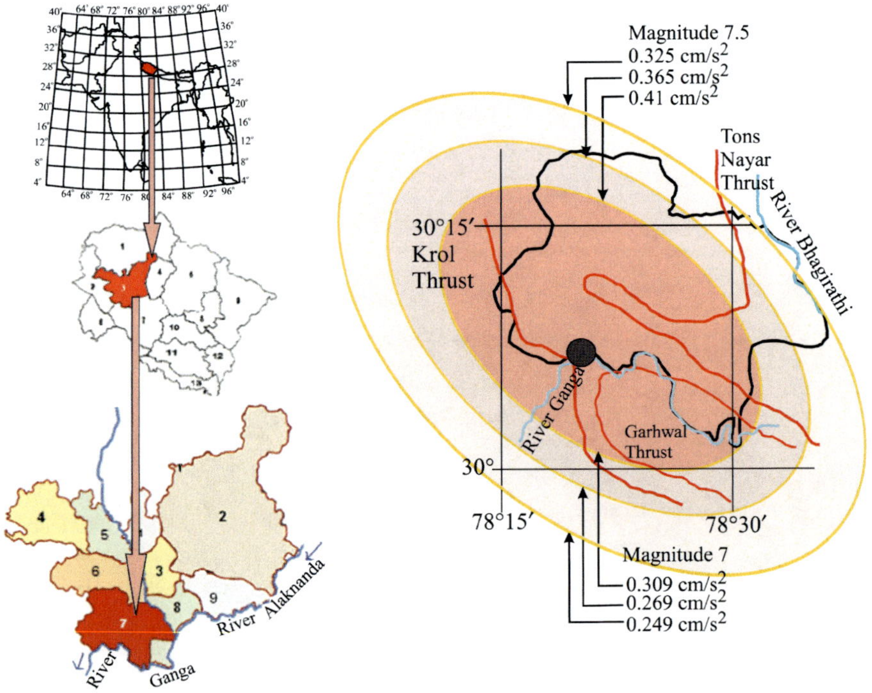

Fig. 8.5 Acceleration contours with epicenter at Tapowan (30°08′10″N and 78° 20′30″E) for different hypocentral distances elongated parallel to the trend of Main Boundary Thrust. 3 is Tehri Garhwal district, and 7 is Narengra Nagar block.

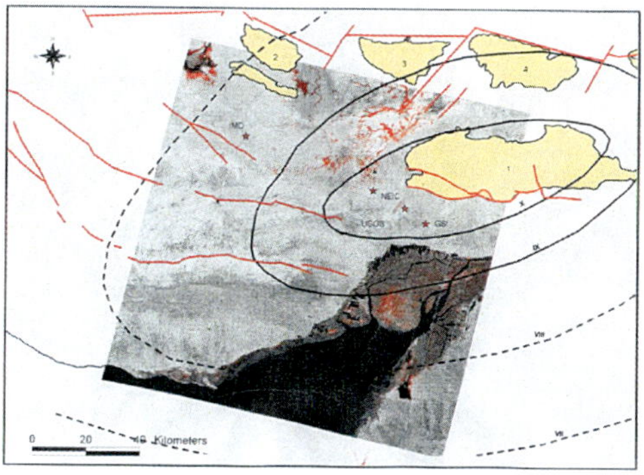

(a)

(b)

(c)

Fig. 9.3 (a) Pools of water that emerged after the earthquake, as revealed by satellite imageries, are shown by red dots. Isoseismal map for the Kutch earthquake of January 26, 2001 and major faults are also shown. (b) Evidence of pools of water that emerged as earthquake fountains from the fault along the Rukmavati River, near Mandvi in the Kutch earthquake of January 26, 2001. Faulted zone along the dry reservoir of the Wangdi dam, from which water gushed out. (c) Pool of water from fissure at Moti Undo, near Mandvi.

Fig. 9.11 Slope protection walls kept the winding roads functional in the Kashmir earthquake of 2005. The portion shown is in the region of the Main Boundary Thrust, near Kamalkote in Uri.

Fig. 10.6 Part of the berthing jetty at Junglighat, Port Blair, sheared off its piers, separated from the approach jetty and fell in the water. A barge capsized at this jetty. Piers on which jetty rested earlier are protruding from the sea.

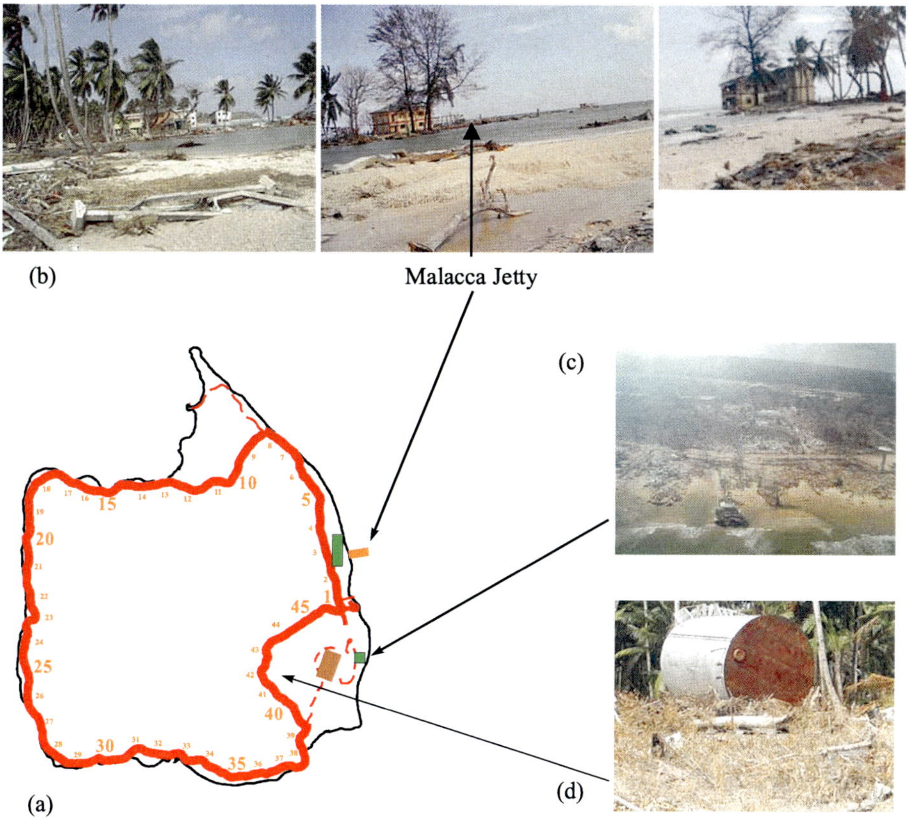

Fig. 10.4 (a) Map of Car Nicobar Island with the main road along the coast shown by kilometer markings. (b) The sea front at the former densely populated Malacca, collapsed and washed out Malacca jetty and a few marooned double-story houses that survived the deluge. (c) Aerial view of sea-facing Air force residential Colony at Malacca. Damage to sea front row of houses was extensive. This decreased on the landward side due to the shielding effect of the front row houses. Airstrip is seen in the background. (d) Oil tanks.

Fig. 11.3 Stonewalls and earthquake disasters are almost synonymous with human tragedy, whether in the Himalayan Arc, or in peninsular India. All houses shown here were made of random rubble stone masonry. (a) In Tangdhar, after the Kashmir earthquake of October 8, 2005, (b) In Bhuj, after the Kutch earthquake of January 26, 2001, (c) A collapsed house in Bhatwari after the Uttarkashi earthquake October 20, 1991. The heavy concrete roof was supported on walls made of a mixture of random rubble stone masonry and concrete blocks. (d) The devastated village of Killari after the Latur earthquake of September 30, 1993.

Fig. 12.5 Collapse of one tower and seismic performance of staircase in a tall building in: (a) Ahmedabad and (b) Gandhidham.

(a) (b) (c)

Fig. 13.2 (a) Longitudinal fissures on national Highway NH 8A, between Gandhidham and Bhachau. The Kohinoor salt factory, seen in top left corner, collapsed partially, (b) Transverse fissures across Andaman Trunk Road in Baratung due to the Sumatra earthquake of 2004, (c) Clogged streets in Gubal, due to the Latur earthquake of 1993 (left).

(a)

(b)

(c)

Fig. 15.1 Different kinds of damage caused by an earthquake: (a) Faulting at Moti Undo, near Mandvi due to Kutch earthquake of 2001, (b) Landslide at Sarai Bandi in Uri due to Kashmir Earthquake of October 8, 2006, (c) Partially submerged house in slush produced by transgression of sea at Car Nicobar after the tsunami generated by the Sumatra earthquake of December 26, 2004.

Made in United States
Orlando, FL
04 November 2026

90199926R00193